T0332494

Genomes, Browsers, and Databases

The recent explosive growth of biological data has led to a rapid increase in the number of molecular biology databases. These databases are located in many different locations and often use varying interfaces and non-standard data formats. Consequently, integrating and comparing data from them can be difficult and time-consuming. This book provides an overview of the key tools currently available for large-scale comparisons of gene sequences and annotations, focusing on the databases and tools from the University of California, Santa Cruz (UCSC), Ensembl, and the National Center for Biotechnology Information (NCBI). Written specifically for biology and bioinformatics students and researchers, it aims to give an appreciation for the methods by which the browsers and their databases are constructed, enabling readers to determine which tool is the most appropriate for their requirements. Each chapter contains a summary and exercises to aid understanding and promote effective use of these important tools.

PETER SCHATTNER is a research associate in computational biology at the University of California, Santa Cruz. His principal research interests are in the genome-wide identification and characterization of non-protein-coding RNA genes and cis-regulatory mRNA motifs. Dr. Schattner has taught bioinformatics courses at the University of California and California State University and has worked in the research and development of medical ultrasound and magnetic resonance instrumentation at SRI International (Stanford Research Institute) and Diasonics, Inc. He has been a Woodrow Wilson Fellow and was leader of the team that received the 1990 Matzuk Award for technical innovation of the American Institute of Ultrasound in Medicine.

Genomes, Browsers, and Databases

PETER SCHATTNER
University of California, Santa Cruz

*Data-Mining Tools for
Integrated Genomic Databases*

CAMBRIDGE
UNIVERSITY PRESS

CAMBRIDGE
UNIVERSITY PRESS

University Printing House, Cambridge CB2 8BS, United Kingdom

One Liberty Plaza, 20th Floor, New York, NY 10006, USA

477 Williamstown Road, Port Melbourne, VIC 3207, Australia

314-321, 3rd Floor, Plot 3, Splendor Forum, Jasola District Centre, New Delhi - 110025, India

79 Anson Road, #06-04/06, Singapore 079906

Cambridge University Press is part of the University of Cambridge.

It furthers the University's mission by disseminating knowledge in the pursuit of education, learning and research at the highest international levels of excellence.

www.cambridge.org
Information on this title: www.cambridge.org/9780521711326

© Peter Schattner 2008

First published 2008

A catalogue record for this publication is available from the British Library

Library of Congress Cataloging in Publication data
Schattner, Peter, 1948–
Genomes, browsers, and databases : data-mining tools for integrated genomic databases / Peter Schattner.
  p. ; cm.
Includes bibliographical references and index.
ISBN 978-0-521-88443-3 (hardback) – ISBN 978-0-521-71132-6 (pbk.)
1. Gene libraries. 2. Genomics – Data processing. 3. Databases. 4. Browsers (Computer programs) I. Title.
[DNLM: 1. Databases, Genetic. 2. Genomics. 3. Gene Library. 4. Molecular Biology.
5. Software. 6. User-Computer Interface. QU 470 S312g 2008]
QH442.4.S33 2008
572.8´602856312 – dc22   2008003901

ISBN 978-0-521-88443-3 Hardback
ISBN 978-0-521-71132-6 Paperback

To my wife, Sue

Contents

Preface

The idea behind this book developed in late 2004–early 2005 while I was working on two unrelated projects in computational genomics. The first project involved the computational detection of small nucleolar RNAs (snoRNAs) in genome sequences. In the course of this work, I noticed – as others had, as well – that, in mammals, snoRNA genes are located within introns of protein-coding genes (so-called snoRNA host genes), which are often genes that code for ribosomal proteins. This observation led to speculation as to whether there were additional common features of the introns and genes that contain snoRNAs. For example, are the host genes of homologous mammalian snoRNAs themselves homologous? Do those host genes have other shared functions beyond the fact that several of them code for ribosomal proteins? Are the introns that contain the snoRNAs consistently longer (or shorter) than the average introns found in these genes? Are the snoRNAs found at any characteristic distance from the nearest exon-intron junctions in their host gene? To answer these questions would require accessing sequence and annotation data for both the human and mouse genomes and performing some simple calculations and statistics on that data. Moreover, because there were some 200 human snoRNAs already known (and a similar number of mouse snoRNAs), performing this data acquisition and manipulation would require computer processing.

The second project involved regions of the mammalian genome exhibiting "extreme codon conservation," that is, extended regions (typically 150 nt or longer) in which homologous protein-coding genes from several species not only have identical amino acid sequences (which is not unusual), but also use identical codons. Although such extreme codon conservation is unusual, many such regions do exist in mammalian genes. One hypothesis for the existence of these regions is that they not only code for proteins but also contain motifs for post-transcriptional processing, such as RNA binding sites or secondary structures. To assess this hypothesis, we needed to determine whether the conserved regions overlapped known conserved alternative splice sites, whether they were enriched for known exonic splicing enhancers or RNA editing sites, and so forth. Answering these questions again required accessing sequence and annotation data from numerous genomic regions and performing statistical analyses on these data.

Of course, others were also addressing biological questions that required these kinds of bioinformatic analyses. For example, E. Levanon and colleagues (Levanon et al., 2004) and others discovered RNA editing sites by screening for genomic locations where DNA adenine bases align with mRNA/EST guanines. Similarly, S. Brenner and colleagues detected instances of "nonsense mediated decay" (NMD) via genome-wide screens for mRNAs with "premature stop codons" (Green et al., 2003). What was apparent was that although the biological phenomena in these examples were unrelated – snoRNA host genes, codon conservation in mRNAs, RNA editing sites, and NMD – the types of bioinformatics tasks required for addressing them were strikingly similar: identify a set of genomic regions, obtain the sequence and a set of annotations corresponding to each region, and perform some comparisons or manipulations on the resulting data to enable some inference to be made about the region.

It was also clear that integrated genome databases, such as those used by the Ensembl, MapViewer, and the UCSC Genome Browser, were excellent data sources for obtaining the required sequences and annotations. Although the genome browser interfaces were principally designed to examine one genomic region at a time, their underlying databases could access data from multiple genomic regions simultaneously – that is, in "batch" mode – as required by these types of bioinformatics analyses. Moreover, much of the software required to perform these analyses already existed in the genome database computer code, as this code was needed to create the genome browser displays.

However, I found very few papers in which this approach to genomic analysis was actually adopted. My impression was that this was in part because of limited documentation describing how one could use the genome browser databases for general genomic data mining. In addition, because at this time tools such as Ensembl's BioMart, the UCSC Table Browser, and Galaxy had only recently become available, this sort of strategy was still largely restricted to the biologist with programming skills.

As a result, the molecular biology and bioinformatics research communities generally took advantage of only a fraction of the capabilities provided by genome browsers and databases. Admittedly, there is a considerable learning curve to using these tools. My goal for this book is to ease that learning curve so that more researchers – including those with limited computer skills – are able to use these remarkable tools to address a much wider range of biological questions than they have previously.

This book is intended for graduate or advanced undergraduate students in bioinformatics or biology or for self-study by researchers or students who want to more fully exploit the power of the genome databases. I envision two distinct audiences: biologists with little or no programming experience and bioinformaticists and biologists with programming backgrounds. The first five chapters of the book, as well as Chapter 12, should be accessible to both groups. There are no formal programming prerequisites for these chapters – although, in some places, a general sense of how computer databases and data files are set up will be helpful. Chapters 6 through 11 do assume some programming background. Nevertheless, non-programmer biologists are encouraged to read these chapters as well. Even if they are unable to follow the programming details, they should get a sense of the types of biological questions that their more computationally oriented collaborators will be able to address with these tools. The biological background assumed is that of an introductory molecular biology course. With this background, the descriptions of the biological examples used should be reasonably self-contained. In addition, the book contains a glossary including both biology and computer science terms that may be unfamiliar to the reader.

I have focused the book on the techniques that I have found most useful in addressing practical biological queries. In particular, the book is primarily focused on tools that integrate data from multiple primary biological data repositories in a standardized and unified manner. As such, the book is not intended to be an "egalitarian" treatment of the three genome browser databases, in the sense of giving equal time to each one. Rather it is one researcher's perspective on useful tools for various bioinformatics tasks. That said, I realize that sometimes my emphasis may be biased by my having more experience with certain tools than with others. In this regard, I should note that (notwithstanding my affiliations with UCSC and the BioPerl project) I have not been involved in the design or development of any of the genome databases or browsers. Similarly, the opinions I express here are solely mine and not those of any of the browser teams.

It is also worth emphasizing that the genome databases and browsers are still rapidly evolving. Consequently, the book is primarily focused on the basic architecture and concepts of the genome databases and tools and where to obtain information on them, rather than on creating a comprehensive catalog of browser features. Such features are continually being added and enhanced, and their online documentation is generally quite good. In addition, the reader should not be surprised if some interface or display found on a browser has changed from the way it is described in the book. Moreover, some features described here will undoubtedly disappear, whereas in other cases statements like "such-and-such database system does not currently support such-and-such capability" will no longer be true. For brevity, I have not always included the word "currently" when describing features that a given browser or database does or does not have, but the reader should realize that I do imply this in every such statement of genome database features.

I am thankful for the help of many people, without whose assistance this book would never have been written. I am particularly indebted to Lincoln Stein for his thorough and insightful review of the entire book. I am also grateful to Deanna Church, Hiram Clawson, Sean Eddy, Xose Fernandez, Jim Kent, and Anton Nekrutenko for reading and commenting on parts of the manuscript. Without their input, there would certainly be far more errors in this book. Of course, any mistakes that remain are solely mine.

I would like to thank Ewan Birney, Mathieu Blanchette, Eli Hatchwell, Fan Hsu, Arek Kasprzyk, Bob Kuhn, Heikki Lehvaslaiho, Todd Lowe, Jason Stajich, Daryl Thomas, Heather Trumbower, David Wheeler, Ann Zweig, and especially Mark Diekhans for numerous discussions on genomes, browsers, and databases, and Katrina Halliday, Alison Evans, Katie Greczylo, and the entire team at Cambridge University Press for their assistance in the preparation of the manuscript. Last, but not least, I want to express my heartfelt gratitude to my wife for her support during this project and for her tolerating my sometimes extreme mood swings and work habits during the course of preparing the manuscript.

1
_____

The Molecular Biology Data Explosion

The explosion of genome sequence data in the last decade has been so widely noted as to have almost become a cliché. The first microbial genome was only sequenced in 1995. However, by late 2007, web sites that track genome sequencing projects, such as NCBI's Entrez Genome Project site (http://www.ncbi.nlm.nih.gov/genomes/static/gpstat.html) and the Genomes OnLine Database (GOLD) project (http://www.genomesonline.org) had cataloged approximately 1,000 complete microbial genome sequences. Similarly, the first complete genome of a multicellular organism (*C. elegans*) became available in 1998. Nine years later, there are complete or draft genome sequences available for more than 60 multicellular species, with low-coverage data or sequencing projects in progress for dozens of others. Figure 1.1 shows summary statistics for genomes that have been sequenced as of November 2007. Moreover, the rate at which genomes for new species and species variants are being sequenced continues to accelerate as novel sequencing technologies lower the cost of obtaining sequence data. For example, Figure 1.2 shows some of the 25 mammalian species whose genome sequencing is currently in progress. Meanwhile, along with increasing amounts of DNA sequence data, there has been a remarkable increase in the quantity of data describing how the information in the genome sequence is used to implement the functions of the organism.

With this explosion of data has come the opportunity – for those with the skill and ability to identify patterns and correlations among the data – to develop an ever more profound understanding of the way organisms function at their most fundamental levels. Genes are being identified that are involved in such basic human experiences as thinking, feeling, and communicating with spoken language. Genomes of hundreds of new microbial species are being sequenced, some of which may hold keys to humanity's most vexing problems in energy generation and environmental preservation. And genes and gene-regulatory mechanisms are rapidly being identified that underlie many of the most dread diseases, including cancer, heart disease, and the degenerative neurological diseases of aging.

Indeed, these are heady times in molecular biology. However, the same data explosion that is enabling all these advances is threatening to drown us in its very enormity.

ENTREZ Genome Project

Nucleotide | Protein | Genome | Structure | PopSet | Taxonomy | OMIM

Go | Clear

Genome sequencing projects statistics

Organism	Complete	Draft assembly	In progress	total
Prokaryotes	597	397	492	1486
Archaea	47	3	30	80
Bacteria	550	394	462	1406
Eukaryotes	23	131	184	338
Animals	4	54	89	147
Mammals	2	22	25	49
Birds		1	2	3
Fishes		3	6	9
Insects	1	19	20	40
Flatworms		1	3	4
Roundworms	1	3	13	17
Amphibians			2	2
Reptiles			2	2
Other animals		6	19	25
Plants	3	3	34	40
Land plants	2	2	27	31
Green Algae	1	1	7	9
Fungi	10	53	30	93
Ascomycetes	8	46	21	75
Basidiomycetes	1	5	5	11
Other fungi	1	2	4	7
Protists	6	19	27	52
Apicomplexans	1	10	6	17
Kinetoplasts	1	2	6	9
Other protists	4	7	14	25
total:	**620**	**528**	**676**	**1824**

Revised: Nov 07, 2007

Figure 1.1 Screenshot from NCBI's Entrez Genome Project (at http://www.ncbi.nlm.nih.gov/genomes/static/gpstat.html) showing the number of species whose genomes have been sequenced or whose genomes are in the process of being sequenced as of November 2007.

As the quantity of data increases, the task of discerning the critical interrelationships among this data has become increasingly difficult. Organizing biological information into dedicated databases of related data has been helpful. However, as the number of biological databases reaches into the thousands (the annual database review of the journal *Nucleic Acids Research* now regularly includes almost 1,000 new or significantly enhanced molecular biology databases each year), intelligently "mining" these data sources becomes ever more challenging.

To a large extent, the current difficulties of analyzing molecular biological data arise simply from the need to characterize such a large quantity of highly interrelated information. However, the biological research community has also brought some challenges of biological data integration and analysis onto itself by the way such data have historically been stored, transferred, and manipulated. Biology databases are located in many different locations. Many of these databases are only downloadable as flat files, as a result of which database searching may be awkward and slow, or else local relational databases may need to be set up. Varying data formats are used requiring the use of multiple data parsers for automated data analyses. As a result, integrating and comparing data from multiple biological databases is difficult and tedious.

Organism	Taxon ID	Genome Size (MB)	Number of Chromosomes	Method	Sequencing Center/Consortium
Lemur catta	9447			WGS	Broad Institute
Loxodonta africana	9785	3000	28	WGS	Broad Institute
Loxodonta africana	9785	3000	28	Clone-based	NISC - NIH Intramural Sequencing Center
Macaca fascicularis	9541				Washington University (Wash U)
Macropus eugenii	9315	3800	8	WGS & Clone-based	Baylor College of Medicine (more)
Manis pentadactyla	143292		20	WGS	Washington University (Wash U)
Nomascus leucogenys	61853				Washington University (Wash U)
Nomascus leucogenys	61853				Baylor College of Medicine
Ornithorhynchus anatinus	9258			WGS	Washington University (Wash U)
Ornithorhynchus anatinus	9258			Clone-based	NISC - NIH Intramural Sequencing Center
Oryctolagus cuniculus	9986	3500	22	Clone-based	NISC - NIH Intramural Sequencing Center
Otolemur garnettii	30611		31	Clone-based	NISC - NIH Intramural Sequencing Center
Pan troglodytes	9598	3100	24		Celera Genomics
Papio anubis	9555			Clone-based	NISC - NIH Intramural Sequencing Center
Papio anubis	9555			Clone-based	University of Oklahoma
Papio hamadryas	9557				Baylor College of Medicine
Pongo pygmaeus	9600			WGS	Baylor College of Medicine
Procavia capensis	9813		19		Baylor College of Medicine
Pteropus vampyrus	132908		19	WGS	Baylor College of Medicine
Pteropus vampyrus	132908		29	Clone-based	NISC - NIH Intramural Sequencing Center
Rhinolophus ferrumequinum	59479	1929		Clone-based	NISC - NIH Intramural Sequencing Center
Sorex araneus	42254	3000		Clone-based	NIH Intramural Sequencing Center (more)

Figure 1.2 Subset of mammalian genome sequencing projects in progress as of November 2007. Data taken from http://www.ncbi.nlm.nih.gov/genomes/leuks.cgi.

Genome databases offer solutions to these problems. By aggregating data from scores of primary databases and integrating data in a uniform and standardized manner, they enable researchers to formulate complex biological queries involving data that were originally located in multiple databases. Learning how to effectively query such interrelated biological data is the primary focus of this book. However, before we can begin this task, we need to spend a little time describing what a genome database is, what the main types of data that it includes are, and how such a database is designed and constructed.

1.1 What is a genome database?

By a genome database, we will mean a data repository (generally implemented via multiple relational databases) that includes all or most of the genomic DNA sequence data of one or more organisms. Generally, a genome database will also include additional data (usually referred to as "annotations") that either describe features of the DNA sequence itself or other biological properties of the species. A genome database typically also includes a web-based user interface – referred to as a "genome browser" – that offers the ability to visualize disparate annotations of genes and other genomic locations in ways that were not possible previously.

Early genome databases and browsers focused on integrating data from a single species, generally one of the biological research community's "model organisms." There was WormBase, for the nematode worm, *Caenorhabditis elegans*; FlyBase, for the fruit fly, *Drosophila melanogaster*; the Saccharomyces Genome Database (SGD) for budding yeast; the Mouse Genome Database (MGD); and so on. Since the completion of the sequencing of the human genome, three additional databases have been developed – EBI's Ensembl Database, the NCBI MapViewer Database, and the UCSC Genome Database – that contain not only integrated human genomic data but also data from many other species as well. This latter feature is important as it becomes increasingly apparent that to interpret the genome of a single species, we need to compare it with its evolutionary relatives. As we will see in detail later in this book, the NCBI, Ensembl, and UCSC Genome Database projects each provide somewhat different, largely complementary resources. Collectively, these projects provide tools and data for genomic analysis that have become indispensable for modern biological research, as evidenced by the fact that UCSC, Ensembl, and MapViewer papers have been referenced more than 3,000 times to date in the scientific literature.

1.2 What classes of annotations are found in the genome databases?

Annotations in the genome databases can be roughly separated into two different classes. The first class includes what might be called "local chromosomal" annotations, as they are associated with a specific region along a chromosome. Examples of such localized annotations include (definitions of unfamiliar terms can be found in the glossary):

- Locations of genes
- Gene-structure annotations indicating a gene's exon-intron boundaries
- Locations of known and putative gene regulatory regions such as promoters, transcriptional enhancers, CpG islands, splicing enhancers and silencers, DNase hypersensitive sites, nucleosome sites, and so on
- Transcript alignments indicating the genomic sources of observed proteins, mRNAs/cDNAs, and expressed sequence tags (ESTs)
- Alignments of protein, mRNA, and EST sequences from related species

- General chromosomal features such as repetitive sequences, recombination "hotspots," and variations in local CG%
- Alignments of genomic DNA from other species, which can provide clues regarding sequence conservation and chromosomal evolution
- Annotations of regions that vary within a population of individuals, including single nucleotide polymorphisms (SNPs), short indels, large structural or copy-number variations, and correlations among sequence variations, such as those that have been identified by the haplotype mapping projects (e.g., HapMap)
- Genome-wide RNA expression data from tiling-array and related projects
- Sequence features that are used in the process of assembling the genome, such as sequence tagged sites (STSs) from genetic and radiation hybrid maps

The other class of annotations includes those that are not directly associated with a genomic region, such as:

- Protein structure data
- Evolutionary data, including evolutionary relationships among individual genes as well as among chromosomal regions and entire genomes
- Annotations describing phenotype variations
- Metabolic- and signaling-pathway data
- Protein-interaction data, such as data from yeast two-hybrid system experiments and data derived from protein-chip expression analysis

To be sure, this distinction between annotations associated with a genomic region and other data is not rigid. However, it can be useful to consider to what extent any given annotation describes a local feature because of the powerful ability provided by the genome databases to address queries involving multiple annotations associated with the same region.

1.3 Building and maintaining a genome database

Building a genome database is a complex multiphase task. Although some of these tasks vary with the specific annotations included within the particular database and with the way the database is designed, certain basic tasks are necessary for the construction of essentially any genome database. These fundamental tasks include:

- Sequencing the genomic DNA
- Assembling the fragments of DNA sequence data into continuous pieces spanning all or most of the length of the organism's chromosomes
- Aligning transcript data to the genomic sequence
- Identifying the locations of the genes within the genome sequence
- Designing and implementing the data-storage architecture to house the data
- Maintaining and updating the database as additional data become available

In many cases, responsibility for the completion of each of these tasks belongs to a different project team. For example, genome sequencing is generally the responsibility of one of the major sequencing centers such as the Broad Institute, the Wellcome-Trust Sanger Center, Washington University, Baylor University, or the Joint Genome Institute. In contrast, sequence assembly is performed by other groups; for example, the human genome assembly was carried out initially by UCSC, and independently by Celera Genomics, and is now performed by the NCBI. Sequence annotation, particularly transcript alignment and gene prediction, have been carried out by yet other groups, for example, Ensembl and NCBI for the human genome. Finally, construction of the genome databases themselves is the responsibility of the groups that will actually provide the genome browser interfaces and maintain the databases, for example, Ensembl, NCBI, and UCSC.

In the following sections, we will introduce each of these tasks briefly. We will return to some of them in more detail later in the context of examining how they impact the information that is available from the genome databases. In addition, these topics are quite broad, and entire books could be (and in some cases have been) written about them. References to the literature are included for those readers who would like to learn more about these important topics.

1.3.1 Sequencing and assembly

To date, nearly all genomic sequencing has been carried out using the conventional Sanger sequencing protocols. With Sanger sequencing, the genome is first randomly cut (e.g., using mechanical shearing) into pieces of between 10 kilobases and 1 megabase, depending on the specific protocol. These pieces are then amplified and subsequently sequenced through a multistep process that involves fluorescent labeling, sequence priming, sequence extension using chain-terminator nucleotides, and electrophoresis (e.g., see chapter 7 of Primrose and Twyman, 2006, for a detailed description). It is worth noting that novel technologies are emerging that show promise for supplanting conventional Sanger sequencing, at least for some applications. These new methods significantly lower costs and increase sequencing output compared to conventional methods. We will describe the potential impact of these emerging technologies in the final chapter.

Because of the random nature with which the original genome is cut, sequencing protocols require that far more bases be sequenced than the number of bases in the entire genome. This is necessary to increase the likelihood that each base will occur and be sequenced from at least one clone. The average number of times any base in the genome has been sequenced in a sequencing project is referred to as the *coverage* of the genome (e.g., a sequencing project of a 100-megabase chromosome with five-fold (5x) coverage involves the sequencing of 500 megabases). For example, so-called draft sequences have 4x to 5x coverage, whereas a finished sequence typically has 8x to 9x coverage. In some cases, for economic reasons, only "low coverage" – that is, 1x to 2x coverage – sequencing is performed (for an interesting discussion of the trade-offs involved in low coverage sequencing, see Green, 2007).

Once a genome has been sequenced, or even partly sequenced, the sequence data needs to go through a process called *sequence assembly*. This is because current Sanger sequencing technology is limited to sequencing no more than approximately 1,000 base pairs in a single data acquisition or "read." (Note: The newer sequencing technologies, though faster and cheaper, have even shorter reads.) In contrast, chromosomes may be 100 million base pairs or more in length. Consequently, chromosome sequence assembly is a complex process in which a large number of overlapping reads are stitched together into longer contiguous regions called "contigs." Subsequently, contigs separated by distances of approximately known length are linked together into "scaffolds." Depending on the sequencing and assembly technology, an additional intermediate step may be necessary to determine the sequence of the individual clones, that is, the pieces of DNA into which the chromosomes were sheared in the initial phase of the sequencing process.

Although this assembly process is straightforward in principle, problems arise in regions where the sequence is highly repetitive or in regions where there are gaps between individual reads. To address these problems, two general strategies for genome sequence assembly have been developed – clone mapping and whole genome shotgun assembly (WGSA). In clone mapping, one first builds a genomic "map" of each chromosome, which includes a list of genetic features or landmarks (e.g., sequence tagged sites) with their relative positions along the chromosome. Using these landmarks, clone and contig sequences can be "anchored" to regions of the chromosome, making it possible to distinguish sequences that are duplicated in other parts of the genome.

In contrast, with WGSA the initial step of building a genomic map is skipped. Instead, the WGSA process includes the cloning of longer (20–50 KB) sequence fragments. One KB of both ends of these clones are then sequenced in individual sequence reads. Using these "paired-end reads," it is then possible to build a scaffold assembly that jumps over ambiguous, duplicated genomic regions without requiring a map of genetic landmarks. Initially, it was unclear whether WGSA would be capable of assembling large genome sequences. However, the effectiveness of WGSA was demonstrated in the assembly of the fly and human genomes, and WGSA has become the primary method of genome-sequence assembly.

Because of the ambiguities in determining the precise location of sequence fragments during genome assembly – no matter which assembly strategy is used – a feature, such as a gene, may be located precisely within a clone or a contig but its location within the entire chromosome might be much less well established. It is for this reason that feature locations are sometimes given in contig or clone coordinates as well as, or instead of, in chromosomal coordinates. Even so-called finished assemblies, such as the current assemblies of the human and mouse genomes, still have gaps. These sequence gaps – for example, those in the centromeric regions – can be quite large. In fact, "finished" assemblies are not really complete at all. Rather, they are simply assemblies that are considered to be as complete as possible within the limits of current technology.

For low coverage (1x–2x) sequences, assembly is much more difficult. In fact, low coverage sequences can generally only be assembled if the genome assembly of a closely related species is available to use as a reference scaffold for ordering the sequence fragments. In addition, with low coverage sequencing, identified genes are often missing exons or are otherwise incomplete. On the other hand, because the costs of sequencing a genome are roughly proportional to coverage, one can sequence approximately four times as many genomes at 2x coverage than one could sequence at 8x. Moreover, since for many comparative genomics applications it is more important to have data from many related species than to have complete-gene sequence data, low coverage sequencing is used in many sequencing projects (see Pontius et al., 2007, for an example of how low coverage sequencing data can be used in the analysis of mammalian genomes). For more details on sequencing and assembly methods, the reader is referred to any modern molecular biology or genomics text, such as Primrose and Twyman (2006).

1.3.2 Transcript alignment and gene prediction

Once the genome has been at least partially assembled, the next step is to locate important biological features – and particularly genes and their exon-intron boundaries – on the assembled sequence. This is not an easy task, and several different strategies, each with its own advantages and disadvantages, have been developed for this purpose. In general, these strategies can be divided into those that are based on transcript alignments, those generated by purely *ab initio* computational predictions, and those that include a combination of alignment and computational approaches. Transcript-based alignments include alignments of proteins, cDNA/mRNAs, and ESTs, both from the genome of the species being sequenced as well as from homologs from related species. In addition, the transcript alignments may be performed completely automatically by computer or may involve manual curation of computer-generated alignments.

In general, gene annotation methods involving manual curation yield fewer false positives – that is, pseudogenes that are annotated as functional genes – than purely computational approaches. However, manually curated approaches are much more labor intensive and tend to generate more false negatives, that is, missed genes. Consequently, depending on the requirements of the specific application (e.g., whether it is more important that one has high confidence that all annotations are correct than that no true genes are being missed), one approach may be preferred over the other.

1.3.2.1 Manually curated gene annotation

The two main projects for manual curation of transcript-based mammalian gene annotations are the Reference Sequence (RefSeq) Project of the NCBI and the Vertebrate Genome Annotation (VEGA) Project of the Sanger Institute. Although the specifics of the RefSeq and VEGA annotation algorithms vary considerably (see Ashurst et al., 2005, and Pruitt et al., 2007, for details), they both are based on manually curated alignments of transcripts to the genome. Consequently, the RefSeq and VEGA datasets often agree, particularly in gene detection and in distinguishing functional genes from pseudogenes. However, RefSeq and VEGA annotations do not always agree, especially in terms of their predicted exon-intron boundary locations.

To address the fact that RefSeq and VEGA annotations sometimes differ, yet another manual curation project, the Consensus Coding Sequence (CCDS) Project, has been started (http://www.ncbi.nlm.nih.gov/CCDS). The goal of CCDS is to identify highly reliable gene annotations, namely those for which there is 100% agreement between the RefSeq and VEGA annotations and that meet other quality tests developed by the CCDS Project, for example, tests to confirm that the predicted gene is neither a processed pseudogene nor produces a transcript that would be subject to nonsense mediated decay. Currently, the CCDS dataset is restricted to human gene annotations; however, expansion to other mammalian species (e.g., mouse) is planned.

1.3.2.2 Automated gene annotation

Compared to the curated RefSeq, VEGA, and CCDS datasets, the fully automated gene prediction systems provide larger numbers of gene annotations, and producing them is much less labor intensive. Such automated gene-prediction algorithms include both systems, such as the Ensembl Pipeline (Curwen et al., 2004; Potter et al., 2004), which are based largely on transcript alignments, and the *ab initio* computational gene-prediction programs. Furthermore, the *ab initio* programs can be partitioned into two major subclasses: single species gene-prediction programs, such as GENSCAN (Burge and Karlin, 1997), and newer programs that use multiple-species sequence alignments (Gross and Brent, 2006). Gene finders based on multiple sequence alignments rely on the fact that genes and gene structures are typically conserved in related species. Consequently, if a predicted gene splice junction has a consensus splice-site sequence that is conserved in other species, then the site is more likely to be genuine than if the splice-site sequence is not conserved. By using the additional information contained in sequence alignments, multispecies gene prediction programs usually have considerably lower false positive rates than single species programs (Brent, 2007).

Whether they are based on transcript alignment or on *ab initio* predictions, datasets produced by the automated pipelines generally have higher false positive rates or incorrect intron-exon boundaries than the manually curated datasets. However, despite their higher false positive rates, the non-curated datasets can be very useful. For example, many genuine genes are expressed only at low levels or in specific tissues or developmental stages. Consequently, transcripts of these genes may not have been experimentally detected to date, and such genes will generally not be included in the curated datasets. And of course, for non-model species for which there is little transcript data, non-transcript-based computational approaches are the only available tool. Nevertheless, it is important to remember that the higher false positive rates for non-curated datasets means that these datasets need to be viewed with more caution.

1.3.2.3 Accuracy of gene prediction methods

With so many different ways of predicting genes, it is important to be able to assess the relative accuracies of the different approaches. Addressing this question for human gene annotation is one of the goals of the "Encyclopedia of DNA Elements," or ENCODE Project (Birney et al., 2007). The ENCODE Project ultimately seeks to annotate all functional DNA motifs in the human genome. In its pilot phase, the ENCODE Consortium has generated a large number of annotations for a small (around 30 MB) subset of the human genome.

One of ENCODE's initial objectives has been to generate a complete list of functional protein-coding transcripts in the 30-MB ENCODE regions, the so-called GEN-CODE gene set. This list of transcripts was generated by gene predictions from multiple computational and curated gene annotation methods, followed by experimental PCR-based validation, in over 20 different types of human tissue. The results (Harrow et al., 2006) indicate that current methods of gene annotation are quite good – but far from perfect. In addition, they confirm that current manual curation methods are generally more specific, but less sensitive, than fully automated approaches. For example, ENCODE determined that Ensembl's automated gene prediction pipeline detected 84.0% of the validated gene exons, whereas RefSeq's manually curated algorithm detected 80.0%. On the other hand, 98.3% of RefSeq's exon predictions could be experimentally verified, as compared to 91.5% for Ensembl.

Because there are so many different approaches for genomic gene identification, with different strengths and limitations, one often finds multiple different "gene" annotations in the genome browsers. We will consider this topic in more detail in later chapters. For now, suffice it to say that, for example, in the hg18 build of the UCSC Human Genome Database, there are more than a dozen different sets of protein-gene annotations. So if one is searching for data about a specific human gene, which annotation set should one use? Although there are no hard-and-fast rules, a useful guide would be to first check whether the gene is annotated in the CCDS dataset.[1] If not, one could check if it is included in VEGA or RefSeq. If the gene is not found in any of the manually curated sets, one could check an automated gene annotation dataset such as UCSC genes or Ensembl genes, or a modern *ab initio* gene prediction program such as N-SCAN. Finally, we note that this discussion has been limited largely to annotations of vertebrate genes; other curated gene annotation datasets are available for the non-vertebrate model species, such as the SGD Gene Set for yeast genes and the FlyBase Gene Set for *D. melanogaster* fly genes.

---

[1] If the region of interest is within the ENCODE regions, using the GENCODE gene set would also be a good choice.

1.3.2.4 Non-protein-coding genes

The various gene datasets described in the previous paragraphs all refer to protein-coding genes. However, many biologically important genes are never translated into proteins, and instead perform their functions as RNA. Examples of such non-coding RNAs (ncRNAs) include the transfer RNA (tRNA), ribosomal RNA (rRNA), small nuclear RNA (snRNA), small nucleoloar RNA (snoRNA), and microRNA (miRNA) families.

For some species, such as human and mouse, for which there is extensive ncRNA experimental (transcript) data, genome browsers often provide sets of experimentally supported, ncRNA-gene annotations. Typically, these annotations are for a specific class of ncRNAs, such as the snoRNAs or miRNAs, and are directly extracted from a primary database that archives experimentally verified ncRNAs, such as the human snoRNA database at http://www-snorna.biotoul.fr (Lestrade and Weber, 2006). For species with limited transcript data, one is restricted to computationally generated annotations. Unfortunately, apart from tRNAs and rRNAs, ncRNAs are difficult to detect in newly sequenced genomes through purely computational techniques. One project that attempts to computationally detect members of known ncRNA families in newly sequenced genomes is the Rfam project (http://www.sanger.ac.uk/Software/Rfam) (Griffiths-Jones et al., 2005), and some genome browsers, such as Ensembl, include Rfam annotations. However, other than for tRNAs and rRNAs, computationally generated ncRNA datasets include numerous pseudogenes and, hence, should be viewed with caution.

The computational detection of novel classes of ncRNA genes in the genome (i.e., ones that do not belong to any of the well-known ncRNA families) is still extremely difficult. Algorithms to identify potential novel ncRNAs, generally on the basis of conserved, putative secondary structures, do exist, and their predictions can be found in some of the genome browsers – for example, the EvoFold annotations (Pedersen et al., 2006) in the UCSC Genome Browser. The predictions in these datasets undoubtedly contain some genuine ncRNAs and can be very useful in guiding research. However, they also include many false positives. In short, any purely computational annotation of a novel ncRNA should probably be considered mainly as an interesting prediction.

Attempts to identify novel ncRNAs in a genome-wide manner are also increasingly made experimentally. The approach is to measure RNA transcription across the entire genome. The hypothesis is that if a section of the genome is being transcribed, and especially if it is being transcribed at different levels in different tissues or developmental stages, the transcript must be doing *something*. In fact, widespread transcription has been observed across mammalian and other genomes, including in many chromosomal regions far away from any known genes and including no identifiable open reading frames. Some of these transcribed RNAs have been shown to be functional. The functions of others will undoubtedly be experimentally determined in the future. However, for the majority of these transcripts there is still no evidence that they are functional at all, as opposed to simply being some form of transcriptional "noise" (see, e.g., Hüttenhofer et al., 2005, for more details). Consequently, one should not automatically assume that an RNA is functional simply because it is transcribed.

1.3.3 Genome database design

One of the primary goals of a genome database is to aggregate data from multiple "primary" databases into a single database with a uniform interface, thereby enabling the user to make integrated queries involving data from the entire database. To this end, three main data integration approaches have been developed in the field of bioinformatics database design (see Stein, 2003, for a more detailed introduction to these topics). These approaches are generally referred to as link integration, data warehouses, and federated databases (also called view integration). In a link integrator, a user presents a query or a keyword (e.g., a gene name) and the system returns an annotated list of web hyperlinks to information pertinent to the query. In a link integration system, data from the primary databases are not stored locally by the link integrator – only links (i.e., web addresses) are stored in the integrating database.[2] Link integration has the advantage of being easy to implement and always having up-to-date data, as long as the link addresses remain current. The GeneCards web site (http://www.genecards.org; Figure 1.3) is one example of a link integration system.

Unfortunately, link integrators suffer from a significant limitation, namely, that it is generally not possible to make integrated queries involving data from different primary databases. To address this limitation, data warehousing takes an opposite approach to database integration, namely, making a local copy of all of the data from each underlying, primary database. The data are then usually reformatted into standardized data structures and loaded into the data "warehouse" – that is, one or more databases that can be queried in a unified manner by the warehouse software.

Because of its capabilities for integrating data from multiple data sources, the data warehouse approach has become important in the storage and access of many different types of molecular biology data. For example, the Atlas data warehouse (Shah et al., 2005) integrates taxonomy, disease, and protein interaction data. Biozon (Birkland and Yona, 2006) contains protein structural and alignment data, whereas BioWarehouse (Lee et al., 2006) focuses on integrating data from metabolic and signaling pathways. In each of these systems, complex sets of related data can be accessed

---

[2] Some authors (e.g., Stein, 2003) also define "link integration systems" as ones that make local copies or "caches" of primary database data but do not reformat that data. These authors reserve the term "data warehouse" to systems that both make local data copies and reformat that data. Pure data-caching systems are important in some applications (e.g., Google is largely a data-caching system) but they are less common in bioinformatics data integration. Consequently, we will find it useful to describe any data integration system that makes local copies of primary-database data as a "data warehouse," reserving the term "link integration" for systems that only store pointers (e.g., web links) to primary data.

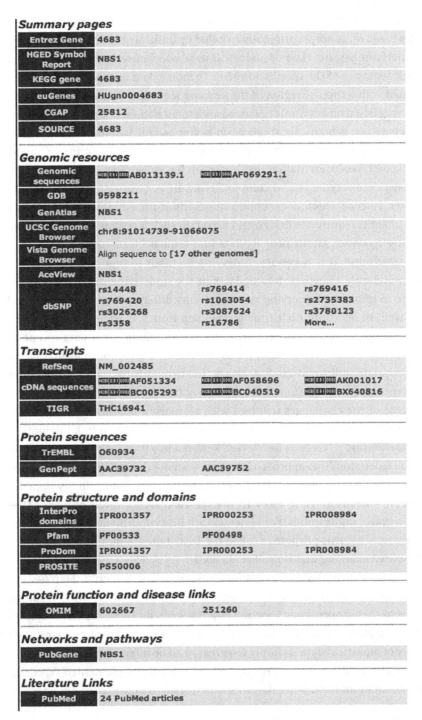

Figure 1.3 Link integration as implemented by Gene Cards. Only a small portion of the gene data presented by the Gene Cards web site is actually stored by Gene Cards. Most of the data are provided by hyperlinks to other databases.

in a uniform manner, and complex queries can be formed using keywords or key phrases based on concepts originally handled by multiple specialized databases. One particularly important class of biological data warehouse includes the underlying genomic sequence of the species. In such a "genome data warehouse," data are often associated with a specific region of the genome sequence. Consequently, in addition to querying based on keywords or database-element names, it is possible to formulate queries based on genomic locations or on genome-sequence similarity.

Although data warehousing has significant strengths in terms of integrated querying and data presentation, the approach has drawbacks as well. Genome data warehouses are very large and complex, typically consisting of multiple relational databases, each of which may contain hundreds of linked tables. A major software commitment is required to design and maintain such a system, as well as to develop programs to convert all the primary data into standard formats that can be handled by the warehouse. It is necessary to develop methods to regularly update the data in the warehouse to reflect data updates in the underlying databases. Developing software to implement querying across so many different types of data is challenging as well. In addition, each time a new data source is added to the warehouse, software needs to be developed to transform the new primary data to one of the standard warehouse data formats (or to create a new warehouse format), to create the appropriate additional database tables, and to modify the data manipulation and query-processing software to handle the new data.

Because of the challenges involved with regularly updating data warehouses, a third strategy for database integration – intermediate between link integration and data warehousing – has been developed. Referred to variously as view integration or database federation, this approach integrates remote data without the construction of local copies of the primary databases. To be practical, the primary databases in a database federation must agree to use specified formats and protocols that the view integrator understands. For biological applications, the principal protocol that has been used for this purpose is called the Distributed Annotation System, or DAS (http://www.biodas.org and Dowell et al., 2001).

In practice, current genome databases, such as the Ensembl, MapViewer, and UCSC Databases, are implemented using components of at least two and sometimes all three of these approaches. Ensembl, MapViewer, and UCSC are all primarily data warehouses. However, they all also include a link integration component; the total quantity of biological data is simply too great to all be integrated locally. In addition, Ensembl also includes a view integration component, implemented with DAS, by which annotations from other databases are displayed on the Ensembl Browser.

1.3.4 Builds and releases

In one way, our description of the generation of a genome database – namely, the sequencing of the genome, followed by genome assembly, followed by annotation generation, and finally leading to database construction – is rather misleading. In reality, these tasks are carried out iteratively and somewhat in parallel. Once a draft assembly has been constructed – and sometimes even earlier, when just a 1x or 2x coverage assembly for a new species is available – the tasks of transcript alignment and gene annotation and even the construction of a "skeleton" genome database are begun. Moreover, because even this limited data may be quite useful to some researchers, these incomplete and skeleton databases are released for public use at the genome database sites.

Subsequently, as the coverage of the genome increases, more complete sequence assemblies are produced by the groups responsible for sequence assembly. Similarly, each time a new genome assembly is released, the groups responsible for producing the genome annotations will update their annotations to reflect the changes introduced by the new assembly. In addition, even after the sequence assembly has become relatively complete (i.e., "finished"), the genome database maintainers will need to regularly update their databases as more, and more accurate, annotations become available. Consequently, maintaining a genome database involves balancing two often conflicting goals. On the one hand, one wants to have a stable data repository from which one can retrieve reproducible results. On the other hand, one wants the database to remain current with the continuing expansion of known genomic information.

The three major genome databases address these goals with somewhat different strategies. With Ensembl and NCBI, data in the databases are modified only as a part of new data "releases," which are indicated by changes in a database version or release number. For the NCBI system, each species database has its own version number, whereas Ensembl uses both species-specific and system-wide release numbers.[3] Importantly, between releases, the NCBI and Ensembl databases are stable, and repeated queries against them should provide identical results. (The only exceptions are Ensembl DAS tracks that are provided directly by external servers and that do not have their data stored in the Ensembl databases.) In contrast, in the UCSC system each species database only goes through a version number change – referred to as a database "build" – in response to a new sequence assembly of the species' genome. For example, the hg18 build of the UCSC database, released in March 2006, corresponded to the assembly changes from the November 2005 NCBI Build 36 of the human reference sequence.

Consequently, it is important to pay attention to what genome assembly number and what database release or build one is using when one uses data from a genome database. Similarly, when presenting one's own results (e.g., in a publication) on the basis of data extracted from a genome database, it is important to include information indicating exactly which sequence assembly and database build were used to carry out the analysis.

---

[3] See http://www.ensembl.org/info/using/api/versions.html for a complete description of the Ensembl release-numbering system.

1.4 A research query scenario

With this background, we are now ready to begin our description of the genome databases. We will begin with an example, the details of which are made up and are chosen for pedagogical purposes. However, the types of questions they bring up are quite real and will be familiar to readers experienced in molecular biology and genetics research. So let us imagine a scenario in which we have performed genetic association studies in an attempt to identify factors contributing to some central nervous system (CNS) disease. Let us further imagine that we have identified some genomic markers on the X chromosome that appear to be correlated with the disease (i.e., the markers and the presence of the disease among family members occur together more often than can be explained by chance). When this chromosomal region is resequenced in the affected family members, we observe a nucleotide that is C in the affected members and G in the unaffected members. This single nucleotide polymorphism (SNP) is at nucleotide 1905 of the CXorf34 mRNA. Because the function of CXorf34 is not known and the significance, if any, of the observed polymorphism is also unknown, it is unclear whether the observed variation is causally related to the occurrence of the disease.

In an attempt to determine whether our SNP contributes to the disease, there are many questions that we might want to answer. It is not important at this point that you understand these questions in detail. We will describe many of them more fully later. For now, the important point is just to illustrate the wealth of available information relevant to our question that exists within the various biology databases, and to indicate that locating this information is not always easy. Questions we might want to answer include (biological terms are defined in the glossary):

- Is the polymorphism in the dbSNP database, indicating that it has been previously identified?
- Does it overlap a known repeat sequence?
- Does the polymorphism overlap a CpG island?
- Has it been observed in any known EST?
- Is CXorf34 expressed in the CNS?
- Have any other genes, possibly including ones of known function, been identified as having similar expression patterns to CXorf34?
- Is the more common variant at the polymorphism site conserved in other vertebrates?
- Does CXorf34 have homologs in other species that could be manipulated experimentally or that might provide clues as to its function?
- Are there common, nearby SNPs that could be used as markers to identify other individuals who might have the new allele?
- What nearby regions would be appropriate to use as primers if one wanted to genotype individuals for the new allele using PCR?

To be sure, it is possible to at least partially answer these questions without a genome database. However, at a minimum, this approach requires identifying, becoming familiar with, and using multiple different resources (dbSNP, GenBank, dbEST, the GNF database, and so on), each with its own idiosyncrasies and learning curves. Moreover, some of the needed data – such as that displaying cross-species sequence alignments and evolutionary conservation – are not available from other types of databases at all. Determining this information without one of the genome databases would require actually carrying out one's own multiple sequence alignment with a tool such as ClustalW or PSIBLAST. I will not describe these procedures further here; suffice it to say that without the use of a genome browser, the task would be very tedious and time-consuming. In contrast, in the following chapters we will see how to efficiently address these questions in the context of the genome browsers.

1.5 The road ahead

In the following chapters, we will learn how to answer the questions posed in our scenario and to answer much more complex biological questions as well. To accomplish this, we will need to learn how genome databases are organized; how the data within them are stored, accessed, and can be manipulated; and how the data are presented in the genome browsers. We will begin in Chapter 2 with an introduction to the UCSC Genome Browser. Using the UCSC system as an illustrative example, the chapter focuses on the core genome browser features and how they are used to answer practical biological questions. The general concepts are illustrated using the research example described in the previous section. In Chapter 3, we introduce the Ensembl and MapViewer Genome Browsers, emphasizing the ways that they differ from the UCSC system, and revisiting our example research query from the perspective of the other browsers. We will also briefly introduce the other genome databases – for example, the single-genome databases and the prokaryote databases – in this chapter.

Chapter 4 introduces batch genome-database querying techniques for accessing data from multiple genomic locations in a single query. The chapter includes examples of biological questions that can be addressed by batch querying and presents tools for interactive batch querying of the major genome databases. These include web-based tools, such as Ensembl's BioMart and UCSC's Table Browser and Gene Sorter, as well as direct querying via the SQL query language.

Chapter 5 describes batch data post-processing. The chapter includes a detailed introduction to the Galaxy tool set, which enables post-processing to be performed interactively without the need for any computer programming. Chapter 6 introduces the techniques of programmed batch post-processing via SQL-based programming, the Taverna Toolkit, and an overview of the Ensembl and UCSC Application Programming Interfaces (API).

Chapter 7 describes Ensembl's Perl API for genome-database programmed querying. The chapter includes a brief review of the BioPerl package and illustrates how

BioPerl functionality is integrated into the Ensembl Perl API. The general discussion is accompanied by concrete programming examples that not only illustrate the power of this approach but can be readily modified by readers for their own applications. The chapter also includes a brief description of Ensembl's Java API.

Chapter 8 consists of more advanced techniques for programmatic access to data from Ensembl. These include methods for accessing data from Ensembl's comparative genomics database, tools for programmatic access to Ensembl's DAS data sources, and an overview of the procedures for installing a local mirror of Ensembl databases.

Chapter 9 introduces UCSC's C API, emphasizing UCSC capabilities that differ from those of Ensembl. The descriptions are again accompanied by sample working code. Chapter 10 continues the discussion of the UCSC API, focusing on important biological applications that can only be addressed by installing a local mirror of part of the UCSC databases. The chapter also includes a description of the steps required to install such a UCSC database mirror.

Chapter 11 describes methods for adding one's own data to existing genomic databases or creating one's own genome database. Methods covered include adding tables and tracks to the UCSC Genome Browser and database and using the Generic Model Organism Database (GMOD) construction tools. Finally, in Chapter 12 we speculate on the features of the genomic databases and browsers of the future. The chapter also points the reader toward tools and web sites that can be helpful for monitoring and anticipating these new features.

The book also contains several appendices. These include a glossary of biological and computer terms, a description of online files associated with the book, a bibliography of print references, a list of online resources, and descriptions of genome-database file and table formats, coordinate system conventions, and sequence alignment algorithms.

I have attempted to organize the book so that readers with specific interests are able to go directly to the topics that interest them, skipping some of the earlier chapters. In particular, the reader who is already familiar with genome browsing and wants to learn about batch querying should be able to start with Chapter 4, skipping much of Chapters 1 to 3. Readers with a good knowledge of interactive querying who are interested in programmed querying should be able to begin at Chapter 6. Also, Chapters 7 and 8 on Ensembl's API, and Chapters 9 and 10 on UCSC APIs, are relatively independent of each other and could be read independently. Lastly, Chapter 12 is of a somewhat more general nature, and a reader should be able to follow it after reading Chapters 1 through 3.

In addition to the printed text, an important component of the book is the associated web site at www.cambridge.org/9780521711326. The web site includes the complete code for all the programming examples and color reproductions of all of the browser screenshots shown in the text, and will include corrections of errors and misprints that have been identified in the text. An important component of the web site is the documentation of changes in web and application program interfaces that cause any of the examples in the text to not function properly; for example, changed web addresses and modified UCSC or Ensembl API function calls. If any example in this text does not work as indicated, the reader should check the web site to determine if a change in the interface has already been observed for this example and what modifications are necessary for the example to produce the results described in the text. If the web site does not contain such information, then the reader is encouraged to e-mail the author at schattner@alum.mit.edu.[4] I will use this feedback to update the web site so that other readers can find the necessary information easily. (Note that this information will be restricted to the specific examples in the text, as implemented on computer systems described in the book, e.g., Unix systems. Unfortunately, I do not have the resources to address queries beyond the examples described in the text.)

Chapter summary

- The amount of genomic sequence and annotation data available continues to increase at a very fast rate.
- Genome databases and browsers provide a means for accessing and querying this large body of data in a systematic and efficient manner.
- Building a genome database is a complex process including sequencing and assembling the genome sequence data, generating and organizing the annotation data, and developing effective data storage systems and user interfaces for storing and accessing the data.
- Because the state of the art for identifying and characterizing genomic annotations is continually improving, it is important for the genome database user to be aware of the capabilities and limitations of currently available genome database annotations.

Exercises

Note: Do not be discouraged if these exercises seem difficult. Without the use of a genome browser some of these questions *are* difficult. The purpose of these exercises is simply to motivate you to continue on to the next chapter, where you will see that by using the genome browsers, answering these questions is quite easy.

The example in the text concerns a putative polymorphism in the predicted gene, CXorf34. Without using any of the genome browsers can you determine:

1. How many exons does this gene have in all?
2. Is there any evidence for multiple isoforms of this gene from EST or cDNA data?

---

[4] It is a good idea to first try the example again after waiting a few hours or a day, since web sites and resources are sometimes "down" for short periods of time.

3. Do any ALUs, SINEs, or other repeat sequences overlap the polymorphism-containing exon of CXorf34?

4. Does CXorf34 have any mouse homologs?

5. How many ESTs overlap CXorf34?

6. To what extent is the exon containing the CXorf34 polymorphism conserved among mammals? Is the intronic sequence surrounding the exon conserved as well?

7. Are there any known SNPs in this exon?

2

Introduction to Genome Browsing with the UCSC Genome Browser

In this chapter, we introduce genome browsing using the UCSC Genome Browser as an example. We will emphasize the general layout and features of the user interface and indicate sources of detailed online documentation rather than attempting a complete survey of all of its annotations and features. We will then return to our polymorphism characterization example to illustrate some of these features and capabilities. In Chapter 3, we will revisit these topics, this time in the context of other genome browsers, particularly EBI's Ensembl and NCBI's MapViewer.

2.1 Introduction to the UCSC Genome Browser

The teams that have developed the genome databases and genome browsers have each had somewhat different project goals, and these differences are reflected in the tools and resources they provide. In the case of the UCSC Genome Browser Project, by far the most important objective has been to provide a comprehensive genome browser that is fast and easy to use. In contrast to other groups, such as NCBI or EBI, UCSC does not currently carry out either sequence assembly or *de novo* gene prediction or annotation. The only annotations that UCSC generates are genome-wide sequence alignments and a gene annotation track based on filtering Ensembl and NCBI gene annotations. On the other hand, UCSC does provide a vast array of annotations that it has integrated from multiple primary databases and that it presents in a uniform and standardized manner that many people find intuitive and natural. As just one example of the comprehensiveness of the UCSC annotations, analyses involving base-level sequence comparisons between mRNA or EST transcripts and the genome are currently easiest (and often solely possible) to carry out with the UCSC system.

So let us see how one actually uses a genome browser. The procedure consists essentially of three steps. First, one selects the species of interest and usually a specific genome assembly of that species as well. Second, one selects the genomic region in which one is interested. And finally, one chooses a set of annotations, variously called *tracks*, *views*, or *maps* in the different browsers, that contain the type of information that one wants to determine for the selected region. To execute these steps,

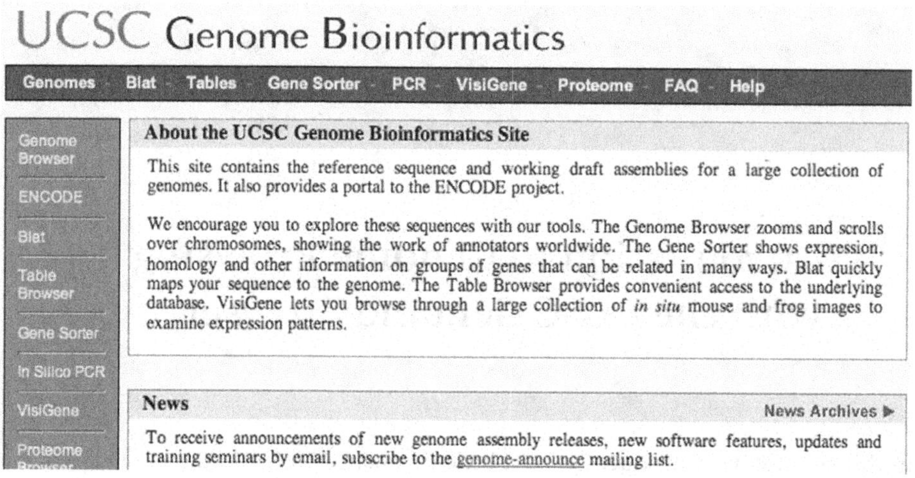

Figure 2.1 UCSC Genome Browser home page.

one generally begins at the home page for the browser, which is http://genome. ucsc.edu for the UCSC Browser. A screenshot of the UCSC Browser home page is shown in Figure 2.1.

2.1.1 Selecting a species and build

With the UCSC Browser, we specify species, assembly, and genomic region via the browser gateway page, which is reached by clicking on either the "Genomes" or "Genome Browser" buttons on the UCSC Browser home page or directly at http:// genome.ucsc.edu/cgi-bin/hgGateway. Parts of the gateway page are shown in Figure 2.2. On the gateway page, there are pull-down menus for selecting an organism and the desired genome assembly of that organism. Species choice is divided into separate menus for differing groups, or *clades*, of organisms to facilitate finding the desired species quickly.

For our examples, we will usually use the human genome. With the UCSC system, we next need to select a database build corresponding to a specific genome assembly. Often, simply selecting the most recent build is appropriate. However, sometimes the most recent database build may not yet include certain useful annotations. This is because in the UCSC system, the entire annotation database for the species must be reconstructed after each new sequence assembly. This involves recomputing the chromosomal coordinates of all annotations and reloading all the database tables. Moreover, transferring annotations to a new database build is currently only partly automated and, consequently, can take months to complete.

What can make this situation more confusing is the fact that some UCSC annotations are *never* transferred to the new build. Generally, this will be the case for annotations that are no longer considered important, for example, if the data has been moved within the database so that it can be accessed from an alternative annotation track. However, sometimes useful UCSC annotations simply disappear between

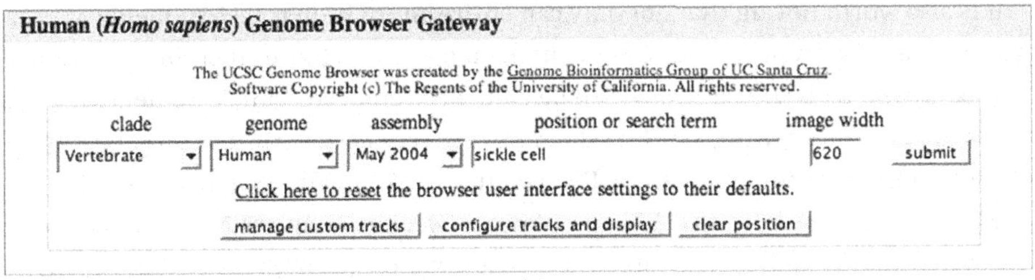

(a)

Sample position queries

A genome position can be specified by the accession number of a sequenced genomic clone, an mRNA or EST or STS marker, or a cytological band, a chromosomal coordinate range, or keywords from the GenBank description of an mRNA. The following list shows examples of valid position queries for the human genome. See the User's Guide for more information.

Request:	Genome Browser Response:
chr7	Displays all of chromosome 7
20p13	Displays region for band p13 on chr 20
chr3:1-1000000	Displays first million bases of chr 3, counting from p arm telomere
chr3:1000000+2000	Displays a region of chr3 that spans 2000 bases, starting with position 1000000
D16S3046	Displays region around STS marker D16S3046 from the Genethon/Marshfield maps. Includes 100,000 bases on each side as well.
RH18061;RH80175	Displays region between STS markers RH18061;RH80175. Includes 100,000 bases on each side as well. This syntax may also be used for other range queries, such as between cytobands and uniquely-determined ESTs, mRNAs, refSeqs, etc.
AA205474	Displays region of EST with GenBank accession AA205474 in BRCA1 cancer gene on chr 17
AC008101	Displays region of clone with GenBank accession AC008101
AF083811	Displays region of mRNA with GenBank accession number AF083811
PRNP	Displays region of genome with HUGO Gene Nomenclature Committee identifier PRNP
NM_017414	Displays the region of genome with RefSeq identifier NM_017414
NP_059110	Displays the region of genome with protein accession number NP_059110
pseudogene mRNA	Lists transcribed pseudogenes, but not cDNAs
homeobox caudal	Lists mRNAs for caudal homeobox genes
zinc finger	Lists many zinc finger mRNAs
kruppel zinc finger	Lists only kruppel-like zinc fingers
huntington	Lists candidate genes associated with Huntington's disease
zahler	Lists mRNAs deposited by scientist named Zahler
Evans,J.E.	Lists mRNAs deposited by co-author J.E. Evans

(b)

Figure 2.2 Gateway page for UCSC Genome Browser. (a) Top of gateway page, used for selecting species, assembly, and genomic location. (b) Bottom of gateway page, showing examples of available modes for specifying genomic location.

builds for no obvious reason. For example, tRNA genes could be located previously (e.g., in the hg16 build of the UCSC Human Genome Browser) via the "RNA Genes" track. But since the hg17 build, the "RNA Genes" track is no longer available, nor is there currently any other track for locating tRNAs in the UCSC Human Genome Browser.

Consequently, it is sometimes preferable to use an older UCSC build, especially if the differences in the underlying sequence assemblies and annotation databases are minor. In particular, because at the time of this writing, UCSC's hg17 Human Genome Database build (May 2004, corresponding to NCBI's human genome build 35) has several useful annotations that are not available in the hg18 build (March 2006), we will often choose the hg17 build for our examples.

It is also worth noting that not only can entire annotation tracks suddenly appear or disappear in a UCSC database without a change in build number, but individual records within a given track may also be added or removed. In fact, some datasets – which are rapidly growing and for which having very up-to-date data may be important – are updated quite frequently. For example, UCSC updates GenBank and RefSeq mRNA annotations daily and EST annotations weekly. If an mRNA is added to or removed from the RefSeq database, that change will be quickly reflected in the UCSC RefSeq annotation track. Although such changes are desirable for keeping UCSC's data up to date, they can be confusing if one is expecting completely reproducible results from a query of the UCSC database.

2.1.2 Choosing a genomic region

Next, we need to specify what region of the genome we want to query. There are several ways of doing this, including by explicit genomic coordinates, by gene name or description, or by locating a specified sequence in the genome with the BLAT sequence search tool. Some examples illustrating ways of specifying genome locations are presented on the gateway page itself (see Figure 2.2b).

For our first example, we want to obtain information pertaining to the gene whose mutated form is responsible for sickle cell anemia, the HBB human beta globin gene. To do this, we could enter "HBB" in the "Position or Search Term" field, or if we have forgotten the gene name, we can take advantage of the wide range of indexed keyword phrases available in the UCSC database and simply enter "sickle cell" into the "Position or Search Term" field.

After you enter "sickle cell" in the "Position or Search Term" field and press the "Submit" button, the browser returns a page with several genes and transcripts from which to select. The browser returned several choices both because the term "sickle cell" occurs in many places in the database and because some genes listed have more than one transcript isoform in the database. Because we are interested in human beta globin itself, we click on one of the links to the HBB gene, for example, the link for the accession M25079 under the "Human-aligned mRNAs" subheading. The result is a view of the genome browser – which we refer to as the "main browser display" – at the position of the HBB gene. The display will be similar to that shown in Figure 2.3, although the specific annotations displayed – which will either be the default annotations or the ones you specified the last time you used the UCSC Browser – are likely to be different than the ones shown here.

2.1.3 Tracks and track controls

The final step is to choose the specific annotations ("tracks") that you want to display from the track control list in the lower part of the browser screen. A portion of the available track controls is shown in Figure 2.4. It is probably a good idea to begin by selecting the "Hide All" button so that only annotations that you explicitly select are displayed. Because there are often a large number of possible tracks (for example, over

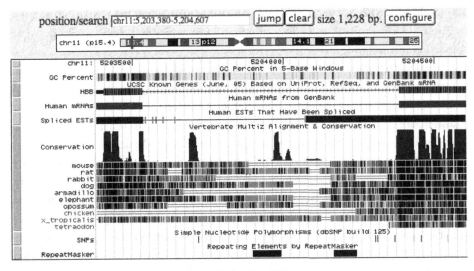

Figure 2.3 UCSC Genome Browser at the position of the HBB gene.

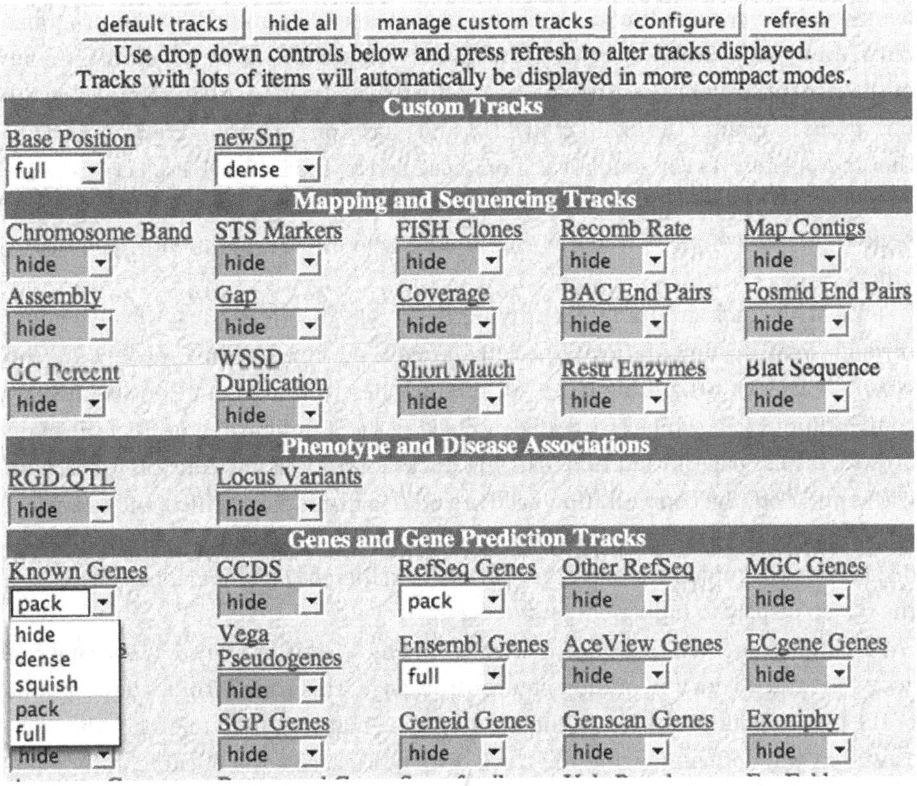

Figure 2.4 Portion of UCSC Browser track controls. Pull-down menu, shown for Known Genes track, shows available track density options.

100 tracks for hg17, or over 160 if one selects one of the genome regions annotated by the ENCODE Project), selecting among all the available tracks can initially be somewhat intimidating.

To make track selection easier, the various annotations are grouped together into categories, including chromosome descriptions, genes and gene predictions, expression data, comparative genomics, and sequence variations. Within the gene group, for example, are annotations of genes derived from manually curated gene annotations (e.g., RefSeq genes) as well as results from various gene prediction programs such as GENSCAN, N-SCAN, and ACESCAN. The mRNA/EST alignment group includes alignments of mRNAs and ESTs to the genome. The gene expression group includes tracks with microarray probe locations used in several widely used expression chips from Affymetrix, as well as tracks annotating expression levels from multiple human tissues that have been determined with these chips via the gene expression project of the Genomics Institute of the Novartis Research Foundation (GNF). The variation section includes tracks that annotate genetic variations such as single nucleotide polymorphisms, copy number polymorphisms, SNP linkage disequilibrium maps, and so on. The comparative genomics category includes alignments of the region with homologous regions in non-human vertebrate genomes as determined by the programs BLASTZ and MULTIZ. The comparative genomics annotations also include an estimate of the level of sequence conservation over the selected region, as determined by the phastCons program. (The BLASTZ, MULTIZ, and phastCons programs are described further in Appendix 4.) For genomic regions specified by the ENCODE Project, there are an additional group of 60 annotations, including tracks that locate promoter regions, regions identified by chromatin immunoprecipitation experiments, and many others.

Even with track grouping, identifying the tracks of interest may involve going through some documentation, especially if one is new to genome browsers. A good way to start is by examining the list of track descriptions found on the display configuration page, which is accessed by clicking the "Configure" button of the main browser display page. In addition to listing track descriptions and controls for turning tracks on or off, the configuration page includes controls from which one can select display parameters such as text size, overall screen image size in pixels, whether to display vertical guidelines, and other parameters, thereby enabling one to customize the screen display to individual preferences.

To obtain more detailed information about any single annotation track, one can access the track's documentation and controls page. This page includes descriptions of the type of data presented in the track and the methods used to obtain the data, as well as ways of configuring the track display. To access the documentation and controls page, click on the name of the annotation either on the display configuration page or in the track controls section at the bottom of the main browser display, or else click on the marker at the left-most end of the track's display, if the track is currently visible in the main browser display.

Display mode: |full ▾| |Submit|
Filter: ⊂ red ⊂ green ⊂ blue ⊂ exclude ⊙ include **Combination Logic:** ⊂ and ⊙ or

accession:	author:	library:	tissue:
BE256422			
cell:	keyword:	gene:	product:

description:

Color track by bases: |different mRNA bases ▾|

Figure 2.5 Portion of track documentation and controls page for spliced EST track. Filtering by accession number is selected so that only EST BE256422 will be displayed. Color track option is configured to highlight nucleotides that differ between the EST and the genomic sequence.

If we choose to display a track, we also can configure how to display the track. Each track control has its own pull-down menu that offers choices such as "Full," "Pack," "Squish," "Dense," or "Hide" (see Figure 2.4). As indicated by the name, full display mode presents the annotation in its most complete form. However, it also uses the largest amount of screen real estate and takes the longest time to display. Consequently, if one is looking at many annotations over a large region (e.g., if there are hundreds of ESTs in the region), using full annotations can be inconvenient. Pack, squish, and dense modes each display annotation data in a different condensed format. Dense mode condenses all annotations onto a single line. Squish mode shows each annotation at 50% height, whereas pack mode puts multiple non-overlapping annotations onto a single line. Consequently, it is often better to initially select a few tracks in squish, pack, or dense modes to see the genomic context of the region of interest and then zoom in to a smaller region where one may want to select a larger number of tracks, including some in full mode. In addition to controlling the density of the track display with the track control pull-down menu, we can obtain finer control of the track display via the documentation and controls page described previously. As an example, the available controls for the spliced ESTs track are shown in Figure 2.5.

Certain data are more easily visualized by changing the resolution of the genome display. This is most easily accomplished by using the "Zoom In" and "Zoom Out" buttons at the top of the main display page. By decreasing the genomic resolution (i.e., by "zooming out"), one can get a better sense of the genomic context of the feature or region being examined. In contrast, other features can only be observed by zooming to high resolution where individual base pairs are displayed. The easiest way of zooming to base-pair resolution is by clicking the "Base" zoom button at the top of the main browser display. When zoomed to base resolution, the "Base Position" track will display the genomic DNA and the three-frame translated amino acid sequences. An example of a genome display zoomed to base level is shown in Figure 2.6.

On the UCSC Browser, track annotations are shown separately for each of the two chromosomal strands. Because the genomic region of interest is often restricted to

28

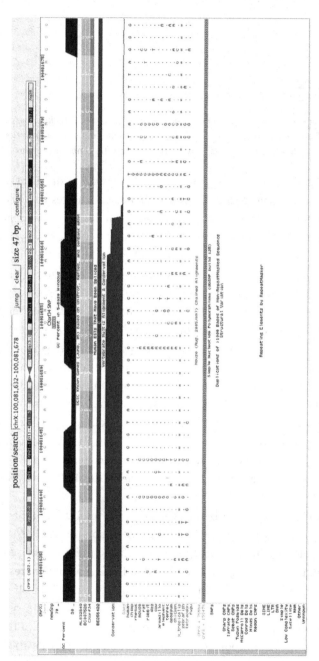

Figure 2.6 High-resolution UCSC Browser display of the region surrounding our polymorphism in the CXorf34 gene. The newSnp custom track, created by the file in Figure 2.10, shows the location of the polymorphism itself. The spliced EST track is configured to show the single EST (BE256422) that has a variant nucleotide at the polymorphism location. The conservation track shows conservation at polymorphism site in most mammals but not in opossum, tenrec, or chicken. See text for details.

one strand, displaying annotations for only a single strand at a time can lead to a less cluttered display. A toggle button in the upper left corner of the track display determines which strand of DNA sequence is displayed. Note that the strand initially displayed will be the default strand or the one that was most recently selected, and may not be the strand of the feature in which you are currently interested. Typically, the strand location of a feature is indicated by ">>>>" or "<<<<" markers in the annotation track, for example, as shown in the intron region of the HBB track in Figure 2.3.

However, displaying tracks for only a single strand can be confusing if one does not pay proper attention to the strand display toggle button. This is illustrated in Figure 2.7. Looking at the intronic region of Figure 2.7a, the (human) nucleotide sequence at the top of the display is seen to be identical to the human sequence in the multiple alignment. However, the sequence of amino acids shown in the exonic part of the multiple alignment does not match any one of the three genomic reading frames shown at the top of the screen. The reason is that the displayed amino acids are those of the aligned mRNA. In Figure 2.7, this mRNA aligns to the negative strand, but we might not be aware of this because the display is zoomed to a level where one does not see the intron strand arrows. In contrast, the *intron* bases are always aligned to the strand selected by the strand toggle button, which in Figure 2.7a is the positive strand. Clicking on the strand toggle (see Figure 2.7b) causes the entire display to show negative-strand data so that one can clearly see that the exon alignment does indeed match one of the three reading frames.

Of course, precisely which tracks you select for the main browser display page will depend on the question at hand. For our HBB gene example, we have selected the Known Genes, GenBank human mRNAs, spliced ESTs, SNP, repeats, GC%, and vertebrate conservation tracks. Selecting these tracks results in the display shown in Figure 2.3.

2.1.4 Other UCSC Genome Browser features

The UCSC Browser includes many other annotations and features that are not directly associated with a genomic region and hence are not represented as annotation tracks. For example, each annotation track has associated with it a specific "details" page. The details page is accessed by clicking on the name of the specific annotation, for example, clicking on "HBB" in the Known Genes track in Figure 2.3. (Note that individual annotation names are only displayed when the track itself is shown in full or pack mode.) Precisely what type of data is returned on a details page depends on the track type and varies for different species and genome assemblies as well. For example, the Known Genes details page provides information regarding the description, function, cellular location, expression data, protein structure, disease associations, and other characteristics of the gene and the protein that it encodes. A portion of the details page for the HBB gene is shown in Figure 2.8. In addition, the

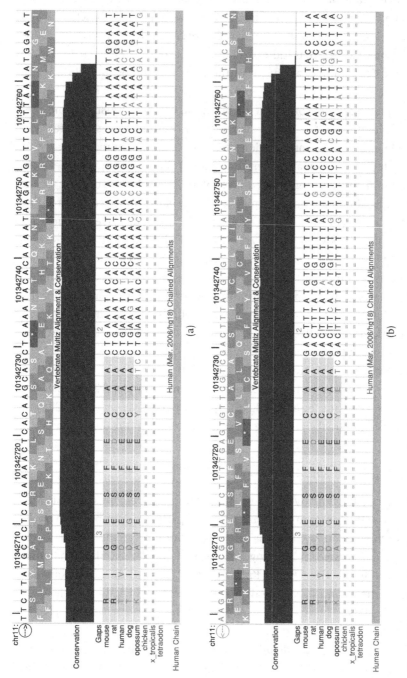

Figure 2.7 Display of negative-strand sequence data in the UCSC Genome Browser. Toggling the strand display control – the small horizontal arrow within the circle in the upper left of the figure – switches the strand of the genomic sequence data displayed as well as the sequence data in the intronic part of the multisequence alignment. However, the amino acid sequence within the multiple alignment is not switched. In (a), sequence data from the opposite strand as that of alignment has been selected. In (b), the strand selected agrees with the strand of alignment. See text for more details.

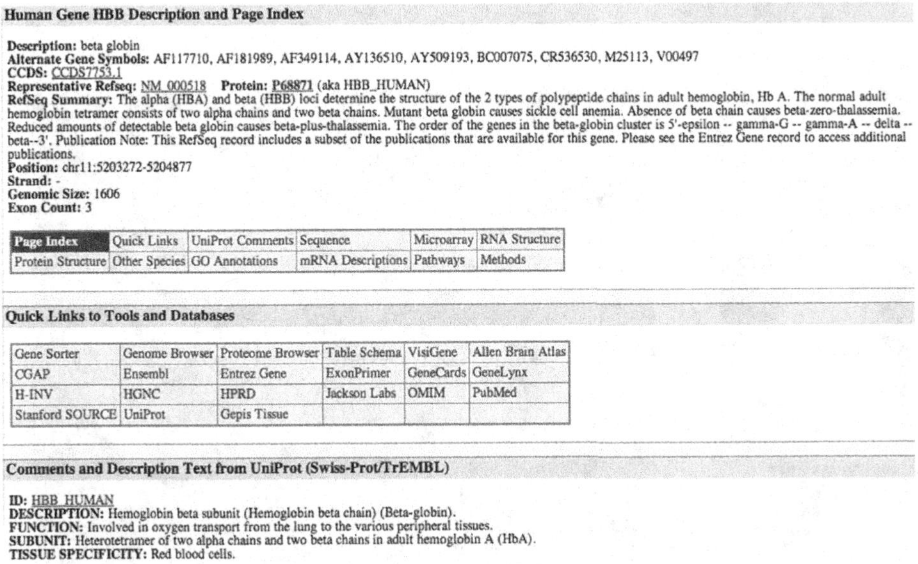

Human Gene HBB Description and Page Index

Description: beta globin
Alternate Gene Symbols: AF117710, AF181989, AF349114, AY136510, AY509193, BC007075, CR536530, M25113, V00497
CCDS: CCDS7753.1
Representative Refseq: NM_000518   **Protein:** P68871 (aka HBB_HUMAN)
RefSeq Summary: The alpha (HBA) and beta (HBB) loci determine the structure of the 2 types of polypeptide chains in adult hemoglobin, Hb A. The normal adult hemoglobin tetramer consists of two alpha chains and two beta chains. Mutant beta globin causes sickle cell anemia. Absence of beta chain causes beta-zero-thalassemia. Reduced amounts of detectable beta globin causes beta-plus-thalassemia. The order of the genes in the beta-globin cluster is 5'-epsilon -- gamma-G -- gamma-A -- delta -- beta--3'. Publication Note: This RefSeq record includes a subset of the publications that are available for this gene. Please see the Entrez Gene record to access additional publications.
Position: chr11:5203272-5204877
Strand: -
Genomic Size: 1606
Exon Count: 3

Page Index	Quick Links	UniProt Comments	Sequence		Microarray	RNA Structure
Protein Structure	Other Species	GO Annotations	mRNA Descriptions	Pathways	Methods	

Quick Links to Tools and Databases

Gene Sorter	Genome Browser	Proteome Browser	Table Schema	VisiGene	Allen Brain Atlas
CGAP	Ensembl	Entrez Gene	ExonPrimer	GeneCards	GeneLynx
H-INV	HGNC	HPRD	Jackson Labs	OMIM	PubMed
Stanford SOURCE	UniProt	Gepis Tissue			

Comments and Description Text from UniProt (Swiss-Prot/TrEMBL)

ID: HBB_HUMAN
DESCRIPTION: Hemoglobin beta subunit (Hemoglobin beta chain) (Beta-globin).
FUNCTION: Involved in oxygen transport from the lung to the various peripheral tissues.
SUBUNIT: Heterotetramer of two alpha chains and two beta chains in adult hemoglobin A (HbA).
TISSUE SPECIFICITY: Red blood cells.

Figure 2.8 Portion of UCSC Genome Browser details page for the HBB gene.

details page includes links to dozens of other databases with additional annotation information for the gene.

Other useful tools provided by the UCSC Browser include the Gene Sorter tool for identifying "similar" genes (which is described later), the LiftOver tool for genome-assembly coordinate conversion illustrated in Chapter 4, the *in silico* PCR tool for primer design (http://genome.ucsc.edu/cgi-bin/hgPcr?command=start), and the Proteome Browser for protein analysis (Hsu et al., 2005). For visual images of mRNA expression patterns, UCSC's VisiGene (http://genome.ucsc.edu/cgi-bin/hgVisiGene) provides a large collection of *in situ* gene expression images that are indexed by gene. Figure 2.9 is an example of a VisiGene display showing the *in situ* expression of the hoxA gene in the embryonic mouse nervous system. In addition to these, there are dozens of other annotation tracks and useful features available on the UCSC Genome Browser that we have not mentioned and will not describe further here. They are all well documented at the UCSC web site as well as in the excellent introductory tutorials available at the OpenHelix web site (http://www.openhelix.com/tutorials.shtml). However, there is one tool, the Custom Track tool, that is of such general utility and importance that we describe it next.

2.1.5 UCSC custom tracks

Often one wants to combine one's own data with annotations from a genome database. For example, in our polymorphism characterization example, we might want to add an annotation track showing the exact location of the polymorphism. Or if we have identified a set of highly expressed transcripts in a microarray experiment, we might want to determine how the locations of these transcripts relate to

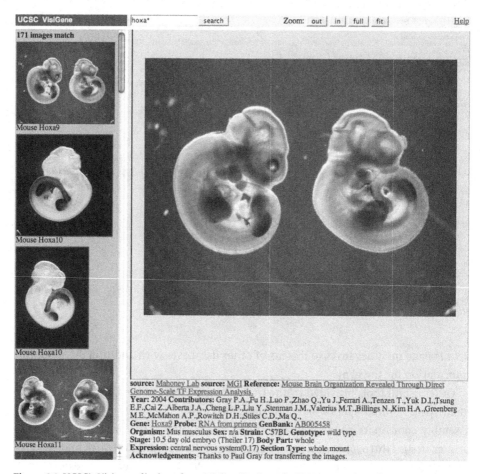

Figure 2.9 UCSC's Visigene display of spatial distribution of mRNA expression of hoxA gene in the mouse.

known genomic features such as gene predictions or pseudogenes. Or perhaps we have sequenced some ESTs and want to compare them with ESTs that have been identified previously.

Integrating custom data into the UCSC Genome Browser display is accomplished by using a custom track. A *custom track* is essentially a list of genomic coordinates along with formatting information for displaying them on the browser. The coordinates can be specified in one of several coordinate formats including BED, PSL, or GFF. These formats are described in detail in Appendix 2. In order to be displayed in the UCSC Browser, the custom track data must be stored in a file (the custom track file). It is possible to simultaneously display multiple custom tracks, including ones with different data formats (e.g., some BED tracks and some PSL tracks).

Custom track files are stored on the UCSC server for eight hours from the time they were last accessed. While the custom tracks are on the UCSC Browser, they can only be viewed from a computer with the same IP address as the one that originally uploaded the track. In this way, one's custom data remains private, even though the

```
browser position chrX:100081640-100081670
track name=newSnp description="CXorf34 SNP" color=255,0,0
chrX 100081653 100081654 CXorf34    0    -
```

Figure 2.10 Sample UCSC custom track file for single genomic location. Lines one and two are header lines specifying the initial browser location to be chosen and the name and coloring of the custom track. The single location in the track itself is in line three. It is one nucleotide in length and specifies the location of the polymorphism in our example.

data has been transferred to the main public UCSC site. Once the custom track data has been removed from the UCSC site, one must re-upload the custom track file to view it again. You can share your custom data with other users by sending them a copy of your custom track file, which they can then upload to UCSC for viewing from their own computers.

There are two basic ways of creating a custom track, via direct upload or by using the Table Browser tool. We will discuss the Table Browser in Chapter 4. Here, we will look at the direct upload approach. Figure 2.10 shows a simple custom track file consisting of a single custom track containing a single annotation, specifically, the location of the CXorf34 polymorphism in our example. The location is specified in "BED" format consisting of the chromosome name, the feature start and end positions, the strand on which the feature is located, and the name of the feature. Our custom track file also includes two header lines that specify the color and track name to be used by the browser and the genomic region to be displayed when the custom track is initially selected.

Once the custom track file is completed, the file is uploaded by clicking on the "Add Custom Track" or "Manage Custom Tracks" buttons on the main browser display or gateway pages (see Figures 2.2 and 2.4) and following the on-screen instructions for selecting the custom track file for uploading. After this step is completed, the custom track can be selected just like any other annotation track and the custom annotations are visible at each of the locations in the custom track file. Detailed information on creating and displaying more complex custom tracks can be found in the online UCSC Browser User's Guide at http://genome.ucsc.edu/goldenPath/help/hgTracksHelp.html#CustomTracks.

2.2 Returning to the disease polymorphism example

Having introduced the basics of using the UCSC Browser, let us return to our scenario from the previous chapter. We recall that we had identified a G → C polymorphism in the CXorf34 gene and suspect that this polymorphism may predispose an individual to disease. To explore this hypothesis, we want to ask several questions: Is the polymorphism in dbSNP? Is it in a genomic region that has previously been linked to a known disease? Does it occur in any known EST? Is the site conserved in other vertebrates? Does the polymorphism site overlap any known repeats? Can we get any clues as to the function of this uncharacterized gene? Can we identify candidate mouse homologs?

Starting at the genome browser gateway page, we select the hg17 (May 2004) build of the human genome. Next, we need to choose the genome region. One possibility would be to enter "CXorf34" in the "Position and Search Term" field. However, doing so would return the region corresponding to the entire 42 KB of the CXorf34 gene. Because we are primarily interested in the short region immediately surrounding the polymorphism, we will instead use the BLAT sequence search tool. Selecting the "BLAT" button from the top menu brings us to the BLAT input page. Now we can cut and paste[1] a 60 nt sequence, (GTCCTTGGGATTGAATTGTTGGAGCAGGCACTGGAGGAT GCAAGATGGACTGCAGCCTTC) centered around the polymorphism (underlined), into the input field and select the (default) hyperlink output option. BLAT then returns its results, indicating that there is a single place in the genome that matches our input at the 98.4% level over its entire length (the match will not be 100% because of the mismatch coming from the polymorphism itself). Clicking on the "Browser" link from the BLAT results page returns a view of the 60 nt region surrounding the polymorphism site.

Next, we need to select the appropriate annotation tracks to answer our biological questions. First, we click "Hide All" to remove any previously selected tracks. We specify the locus variant track to see if the polymorphism site is near any regions associated with known diseases. We choose the Repeatmasker and Segmental Duplications tracks to see if the site overlaps any known genomic repetitive elements. We will select the SNP and structural variation tracks to look for any known human variations at the site.

To determine whether our allele variant has been observed in an EST or mRNA, we select the human mRNA and spliced EST tracks in the full display mode. Because we are interested in identifying places where the mRNA or EST differ from the genome, we select the "Different mRNA bases" option from the "Color Track by Bases" pull-down menu in the mRNA and spliced EST track configuration pages (see Figure 2.5). We select the conservation track to see to what extent the site is conserved among vertebrates, and because we want to search for candidate mouse homologs, we specify the mouse chain track as well. To facilitate visualizing sequence conservation at the polymorphism site, we will select "Display bases identical to reference as dots" and "No codon translation" on the conservation track configuration page.

We will also select the base track to display the actual sequence, if necessary toggling the strand select switch so that the displayed sequence is from the same strand as the transcribed gene (the negative strand, in this case). Finally, to more easily visualize where the polymorphism is in the gene structure, we will select the

---

[1] For cut-and-paste convenience, this sequence can be found in the file CXorf34Poly.fa in the data directory of the tar file available from the book web site.

Known Genes track and add the custom track, shown in Figure 2.10, that annotates the precise position of the polymorphism site.

Clicking "Refresh" displays the selected annotations. The resulting display has many tracks, including over 40 mRNA and EST tracks. However, with the mRNA and spliced EST tracks configured to only display different bases, we can quickly determine that no mRNAs and only a single spliced EST (BE254622) differ from the genome at the polymorphism site. Moreover, we see that the variant is a G → C transversion, which is the same variation as in our allele. From the translated genomic sequence, we learn that this single nucleotide change results in an amino acid change as well because the affected codon changes from "GTG," which codes for valine, to "CTG," which codes for leucine. Having identified this EST, we simplify our display by hiding the mRNA track and reconfiguring the spliced EST track, as illustrated in Figure 2.5, so that only BE254622 is displayed. Refreshing the main browser display then yields Figure 2.6.

From Figure 2.6, we see that the variation is not in dbSNP nor are there any other known mutations or variations or any repetitive elements at this location. From the vertebrate alignment in the figure, we see that the same nucleotide ("G") is present at the homologous sites in twelve other vertebrate genomes but that there is a "C" in opossum and tenrec (as in our new allele), whereas in chicken there is an "A." Both of these nucleotide changes result in different amino acids as well.

We next want to identify candidate homologous mouse genes. From the mouse chain track of Figure 2.6, we see that BLASTZ genome-wide alignment predicts the homologous region in the mouse genome of our polymorphism site is located on mouse chromosome X near position 19.5 MB. Examining the mouse genome at this location, we find that this is located within the mouse 4732479N06Rik gene. For protein-coding regions, such as CXorf34, an alternative approach to identifying homologous regions is by determining the most similar translated protein sequence using the reciprocal blastp method. The best reciprocal blastp hit can be identified with the UCSC Browser from the "Orthologous Genes" section of the Known Gene details page. Performing this operation to determine the blastp best hit to CXorf34 in mouse again yields 4732479N06Rik, showing that, in this case, both BLASTZ and reciprocal blastp predict the same mouse homolog.

Next, we want to see if we can get some clues as to the function of this uncharacterized gene. To this end, we will use the Gene Sorter, which is one of several auxiliary genome analysis tools that is incorporated with the UCSC Genome Browser. The Gene Sorter searches for genes that are in some way "similar" to a specified query gene. The idea behind its use in our present application is that if the most similar genes to our query gene all belong to a certain pathway or functional class, then there is a reasonable chance that the uncharacterized query gene may belong to that class as well. Clicking on the "Gene Sorter" button at the top of the main display page returns the Gene Sorter input screen. With the Gene Sorter, we can search for

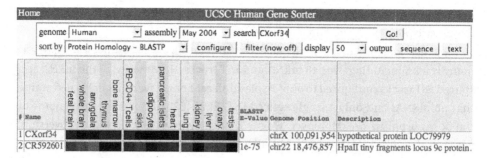

Figure 2.11 Gene Sorter display. Gene Sorter is used to find genes that are similar to a query gene in terms of mRNA expression patterns, blastp homology, protein-protein interactions, and so on. With CXorf34 as input, the only related gene identified is CR592601, which, as indicated in the display, is homologous to CXorf34 using blastp comparison.

"similar" genes on the basis of BLAST sequence similarity, expression pattern similarity, proximity in protein interaction networks, shared Pfam protein domains, and so on. Using CXorf34 as input, we find, for example, that CXorf34 has significant BLAST similarity to the partially annotated gene CR592601, suggesting that CXorf34 may have a related function to CR592601 (see Figure 2.11).

Finally, we would like to be able to search for the new allele in other individuals. One approach would be to design primers specific to our region of interest for use in a PCR assay. UCSC's *in silico* PCR design tool (http://genome.ucsc.edu/cgi-bin/hgPcr) enables one to do this easily. An alternative approach would be to look for other nearby genetic variations that might be inherited in tandem with the new polymorphism. This method of looking for evidence of variations based on correlated genetic inheritance, referred to as genetic markers being in *linkage disequilibrium*, is possible via the HapMapLD track available on some UCSC genome builds (e.g., hg17). We will illustrate this approach in more detail in the context of Ensembl's LDView, described in the next chapter.

Chapter summary

- The three main genome browsers – Ensembl, MapViewer, and UCSC – contain similar data from many of the same species. However, the precise species and annotations contained by each do differ and some tasks are more easily carried out in one browser than another.
- The UCSC Genome Browser provides perhaps the most comprehensive collection of genomic annotations available from many primary databases.
- UCSC itself produces only a small number of genomic annotations, primarily in the form of genomic sequence alignments.
- Using a genome browser, such as the UCSC Genome Browser, generally involves three steps:

  a. Choose a species and, in some cases, a genome assembly.

   b. Choose a genomic location.
   c. Select the annotation tracks or maps corresponding to the data in which one is interested.
- Most of the UCSC tracks and display tools have numerous configuration options, enabling users to customize the display to the needs of their specific applications.

Exercises

1. Many vertebrate genes may be spliced in multiple forms. One such gene is DRD2, a dopamine receptor gene. By locating this gene in the UCSC Browser, determine how many exons each of its two principal isoforms contain.
2. In some cases, mRNA expression data can also be obtained from genome browsers. Locate the DRD2 gene in the UCSC Browser. Determine whether there are annotation track(s) for mRNA expression in different tissues. In which tissues is DRD2 most highly expressed? Are there tissues in which the short isoform of DRD2 is highly expressed but the long isoform is not?
3. The gene PALB2 (previously known as FLJ21816) has recently been implicated in some forms of cancer through its association with BRCA2. In which tissues is PALB2 most highly expressed? Does there appear to be a different level of expression of PALB2 in cancer cell lines than in normal cells?
4. Use the UCSC Gene Sorter to identify any genes that are similar to or related to PALB2 on the basis of sequence homology, expression patterns, protein domains or families, or protein-protein interactions.
5. Custom tracks are not only useful for integrating custom data but for collecting regions with similar or related known annotations. Make a UCSC custom track of all genes whose descriptions include the term "Fanconi anemia" (e.g., by using the "Position and Search Term" input on the UCSC Browser).

3

Browsing with Ensembl, MapViewer, and Other Genome Browsers

In Chapter 2, we began our study of genome browsers by examining the UCSC Genome Browser. In the present chapter, we continue our examination of genome browsers by looking at EBI's Ensembl and NCBI's MapViewer tools. We will focus on the ways that Ensembl and MapViewer differ from the UCSC Genome Browser, indicating some of the situations in which one tool may be more appropriate than the others. We will revisit our example involving the putative disease-causing polymorphism and see how we could study the properties of this polymorphism using these tools. Finally, we will introduce the other genome browsers (e.g., Gramene, the prokaryote browsers, and the single-organism browsers) and indicate when we might want to use these tools as well.

3.1 Introduction to the Ensembl Browser

As with the UCSC Browser, the Ensembl Browser, from the European Bioinformatics Institute (EBI), has been designed to be a comprehensive, integrated genome data resource with a fast and intuitive user interface. However, EBI has had additional objectives for the Ensembl project. The Ensembl system is also designed to provide an automated pipeline for gene annotation as well as a well-supported programming interface to enable querying of the Ensembl databases by researchers outside of EBI. In addition, Ensembl is designed to facilitate the direct incorporation of annotations from outside data providers into the Ensembl display via DAS-based "view integration." Consequently, because of Ensembl and UCSC's somewhat different objectives, making direct comparisons between the UCSC and Ensembl systems is difficult and not always useful.

3.1.1 Ensembl views

To provide its comprehensive set of annotations, Ensembl uses some thirty different types of displays or *views*, each one customized for a specific type of annotation. One advantage of Ensembl's multiple views is that the screen display can be less cluttered, as it is not filled with multiple controls and features that are not needed for the query at hand. Moreover, multiple views provide Ensembl with the flexibility to present browser displays not found in other systems.

However, there are drawbacks to Ensembl's multiplicity of views as well. One disadvantage is that it is sometimes not obvious just what annotations are even available. If a desired type of annotation is not visible on the menu in one view, you do not know whether it might be available in one of the other thirty views. For example, if one is interested in coding-exon SNPs, should one look for annotations in SNPView or in GeneSNPView? Or perhaps in TranscriptSNPView? For annotations stored locally at Ensembl, the answer can usually be found via the Ensembl HelpView pages at http://www.ensembl.org/common/helpview. However, if the annotations of interest are not stored locally (i.e., if they are DAS tracks), then determining the available annotations from the sometimes cryptic names in Ensembl's DAS-Sources menu can be difficult. Additional information about DAS tracks can be found at the DAS registry site at http://www.dasregistry.org.

It is also sometimes not obvious *how* to switch views, for example, between SNPView and GeneSNPView or TranscriptSNPView. In such cases, it may be necessary to refer to the Ensembl Sitemap page (http://www.ensembl.org/ sitemap.html) and link to the species-specific site map, from which the desired view can be selected.

Another disadvantage of multiple views is that one can only simultaneously display annotations if all of them are available within a single view. If the desired annotation data is distributed over multiple views, one needs to switch back and forth between views. In particular, Ensembl custom tracks cannot be displayed in many views where they could be useful, such as AlignSliceView or LDView. For another example, one might want to look at both a base-level multiple sequence alignment as well as at the occurrence of overlapping ESTs. In this case, we would again need to alternate views because the ESTs can be viewed only in ContigView, whereas the base-level alignment is only seen in AlignSliceView.

3.1.2 Ensembl ContigView

Among the various Ensembl views, one of the most important is Ensembl's ContigView. ContigView is similar to UCSC's main browser display; however, they have some significant differences. First, as shown in Figure 3.1, Ensembl ContigView actually consists of four separate displays (Chromosome, Overview, Detail, and Basepair), each one displaying the same region at a different resolution. This is a convenient feature. Because any combination of the four resolutions can be displayed simultaneously, it is possible, for example, to view the nucleotide sequence of an intron-exon splice junction together with the gene context in which the splice junction occurs (Figure 3.1b).[1] A second difference between the Ensembl ContigView and UCSC

---

[1] Of course, it is possible to visualize both a genomic feature and its context in the UCSC display as well by successively increasing or decreasing the display resolution. However, it is currently not possible to view such multiple resolutions simultaneously without opening multiple windows or performing other cumbersome manipulations.

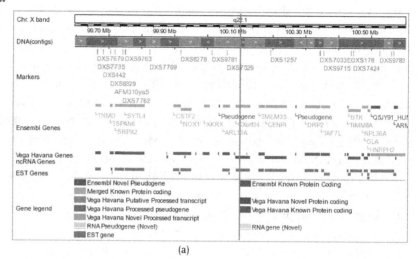

(a)

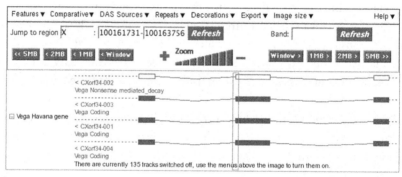

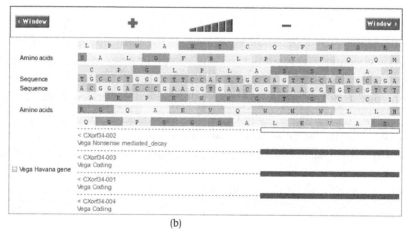

(b)

Figure 3.1 Ensembl ContigView displays. Portion of Ensembl display showing a splice junction in CXorf34 together with its the genomic context: (a) ChromosomeView and Overview (b) Detail and Basepair Views. Detail View shows entire intron and exon. Open rectangle indicates region immediately surrounding splice junction, which is displayed at the nucleotide level in the Basepair View shown in the lower part of the figure.

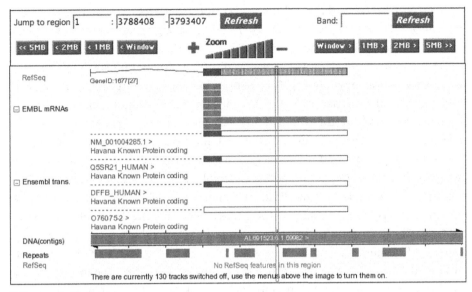

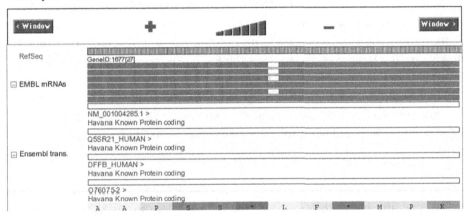

Figure 3.2 Ensembl (and MapViewer) display only a subset of overlapping mRNAs and ESTs. Ensembl Detail View (upper display) appears to indicate that only a single mRNA overlaps the region of interest, which is indicated by the transparent rectangle. However, by opening Basepair View (lower display), we see that at least six mRNAs actually intersect the region.

displays is that with Ensembl, the nucleotide and codon sequence data for both genomic strands are included in the same display. As described in Chapter 2, this may make the display somewhat more cluttered but it is also somewhat less ambiguous.

Ensembl ContigView and the UCSC main browser display also differ in that in Ensembl, each track is simply selected or not selected. There are no track display options, such as UCSC's "dense," "pack," or "squish" to configure. Although having only "on" and "off" modes for track displays makes the user track controls simpler, it may lead to confusion, as illustrated with the Ensembl mRNA track in Figure 3.2. In the Detail View screen image, one sees seven mRNAs. However, this is *not* because there are only seven mRNAs in this region, but rather because Ensembl only displays seven mRNAs at a time. Moreover, the seven displayed mRNAs in Detail View are not the same mRNAs as those shown in Basepair View. As long as you know what is going on, this is not a problem because you can "mouse over" each mRNA to identify which mRNA it is, and can click on the screen to display additional mRNAs. However, for the unwary this can be confusing. For example, looking only at the open rectangle in the upper (Detail View) part of Figure 3.2, one might have deduced (incorrectly) that just one mRNA overlaps the region of interest. However, looking at Basepair View in the lower part of Figure 3.2 shows that, in fact, multiple mRNAs overlap the region.

Another important (and sometimes confusing) difference between the Ensembl and UCSC system of annotations is that Ensembl and UCSC have different (implicit) definitions of what is a "gene."[2] In Ensembl, a "gene" is a collection of overlapping transcripts. For example, the BRCA1 gene has the Ensembl gene ID of ENSG00000012048, and the many isoforms of BRCA1 each have their own Ensembl transcript ID beginning with the letters "ENST" (e.g., ENST00000351666). As a result, an Ensembl gene does not have exons or introns *per se* that can be displayed in a browser. Instead, selecting the Ensembl "gene" track in the Ensembl Browser results in the display of the exon-intron structure of multiple Ensembl *transcripts*, each with a different Ensembl transcript ID.

In contrast, for UCSC, each transcript isoform is a separate "gene." For example, looking on the Ensembl Gene track in the BRCA1 region with the UCSC Genome Browser shows many "genes," one corresponding to each Ensembl transcript (e.g., ENST00000351666). Ensembl's BRCA1 gene – ENSG00000012048 – does not have a separate entry in the UCSC database at all, even though ENSG00000012048 is in the UCSC database *index* and points to the various Ensembl BRCA1 transcripts.

3.1.3 Ensembl features

Ensembl has many powerful features, such as custom tracks and multispecies alignments, that are similar to features we have already seen in the UCSC Browser. In addition, Ensembl provides some important tools and annotations not available in the UCSC system. For example, later in this chapter, we will see the usefulness of the Ensembl GeneTreeView (Figure 3.3) and TranscriptSnpView (Figure 3.4) for displaying gene-level phylogenetic trees and SNP variations across mouse strains, respectively. Similarly, we will see the utility of Ensembl's MultiContigView (illustrated in Figure 3.5) for simultaneously visualizing aligned and annotated regions from multiple species when performing syntenic comparisons.

Ensembl also provides two sequence-searching tools (SSAHA and BLAST) for locating one's sequence in the genome (UCSC only provides one, BLAT). SSAHA is a fast sequence similarity tool, similar to BLAT. However, the speed of BLAT and SSAHA

---

[2] Actually, defining what is meant by a gene is not trivial, especially for organisms that are capable of alternative splicing or other means for producing multiple transcripts from a single genomic location. For a detailed discussion of this topic, see, for example, Gerstein et al., 2007.

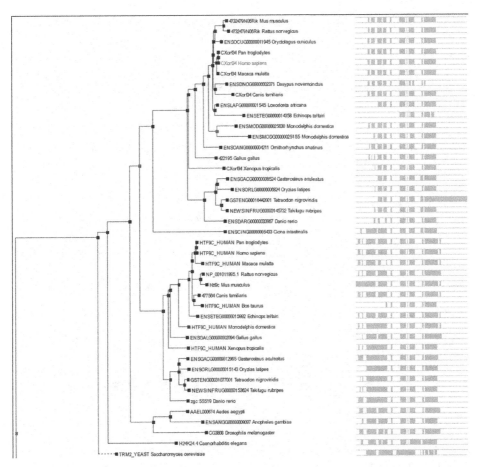

Figure 3.3 Ensembl GeneTreeView showing phylogenetic relationships among CXorf34 homologs in more than forty different species.

comes at a price. Specifically, SSAHA (and BLAT) require that at least a part of the query exactly matches part of the genome sequence. Consequently, if the query sequence has only limited similarity to the target – for example, if the two sequences are from different species – the query sequence may not be located at all. If the query is a protein-coding sequence, this problem can be minimized by first translating both the query and the genome sequences, and each of the browser sequence-search tools have options for performing such translated searches. However, translated searches cannot be used for non-protein-coding sequences. Moreover, even for protein-coding sequences, sequence-search tools requiring extended exact matches may still not find any "hits" in the genome if the query and genome sequences are not sufficiently similar. Consequently, if one needs to find more distantly related regions of homology in a genome, BLAST can be useful, even though BLAST is much slower than BLAT or SSAHA.

Ensembl has many other useful features that we will not cover here. They are well described in the Ensembl documentation. If you are interested in exploring

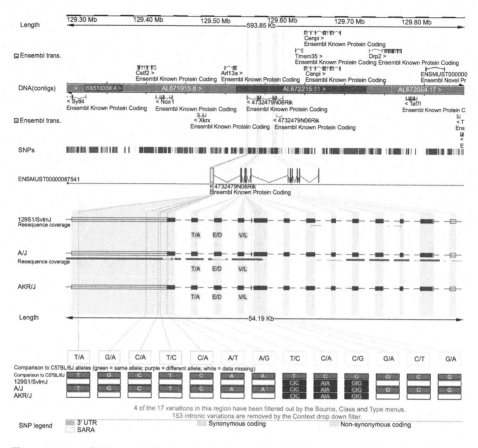

Figure 3.4 Ensembl TranscriptSnpView displaying variations in the 4732479N06Rik gene in different mouse strains.

more of these features, a good place to start is the main Ensembl Help page at http://www.ensembl.org/common/helpview.

3.2 Revisiting our polymorphism example in Ensembl

We now return to our CXorf34 polymorphism scenario, this time using Ensembl. The general approach is quite similar to the one we used with the UCSC Bbrowser: we first select the genome and database, then locate the sequence in the genome using a BLAST-like sequence search tool (SSAHA or BLAST), and finally choose the annotations to display using Ensembl's view and track selections.

We begin at the Ensembl home page (http://www.ensembl.org, and shown in Figure 3.6) and select the human genome from the current release of the Ensembl database. We note that as with the UCSC database, it is possible to view previous releases of the database either by selecting the "View previous release of page in Archive!" link or else directly, for example, via http://oct2006.archive.ensembl.org/index.html. However, in contrast to the UCSC databases, because Ensembl includes essentially all annotations

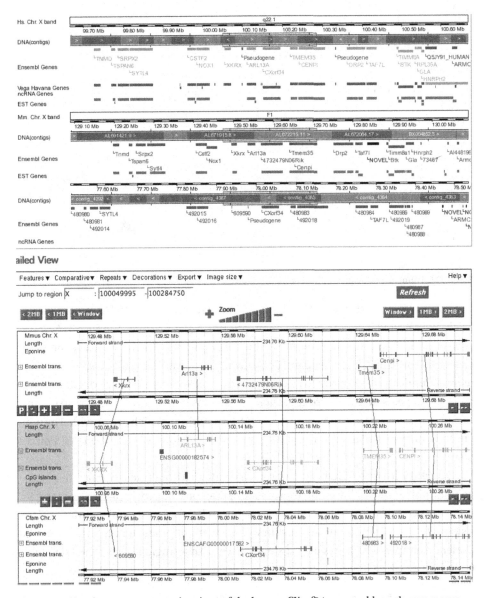

Figure 3.5 Simultaneous comparative views of the human CXorf34 gene and homologous mouse gene, 4732479N06Rik, using Ensembl's MultiContigView.

from the previous release as soon as a new version becomes available, there is rarely any reason not to use the most recent Ensembl release.

Next, we select the SSAHA tool from Ensembl's BlastView page and enter the 60 nt sequence surrounding our polymorphism (see Figure 3.7). We again receive a single "hit." Selecting the location of the hit on the X chromosome as shown on the Ensembl BLAST output page, we are returned a display in Ensembl's ContigView mode that again brings us to the CXorf34 gene.

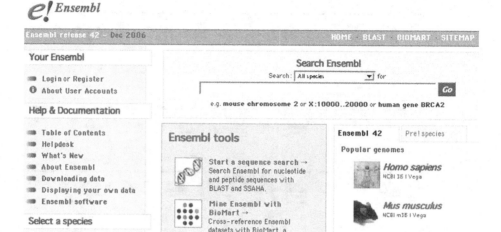

Figure 3.6 Home page for Ensembl Genome Browser.

Finally, we need to select the Ensembl annotations. In ContigView, annotation tracks are selected from the pull-down menus immediately above the Detail View display section. To address our questions about our polymorphism, we select the SNPs, Ensembl Genes, ESTs, and human cDNAs from the "Features" pull-down menu, segmental duplications (SegDup WashU) from the DAS Sources menu, and "All repeats" from the Repeats menu. Currently, there are no Ensembl annotations for structural variants or disease mutations. Note that DAS annotations, which are provided by non-Ensembl data providers, are listed in a separate pull-down menu, the DAS Sources menu.

As we did with the UCSC Browser, we will create a custom track to mark the precise polymorphism site. In Ensembl, there is more than one way to display custom tracks. Perhaps the easiest approach, which we describe here, uses Ensembl's internal DAS server.[3] To use this method, we must first list our annotations in DAS format. For the present example, which has only a single track with a single feature, the DAS input file is only one line long:

```
.  CXorf34Snp  .  .  X  100162165  100162165  -  .  0
```

The fields must be tab-delimited and the periods (".") are required for fields with no explicit data. The DAS format is described in more detail in Appendix 2 and at http://biodas.org/servers/LDAS.html. To upload this custom track, we select "Manage sources" under the DAS Sources menu, then select "Upload your data" from the returned DasConfView page and follow the three-step instructions for uploading data. Once the data has been uploaded, a new track, which we named CXorf34Snp, can be selected from the DAS Sources menu. We note that the coordinates in the Ensembl

---

[3] Other ways of creating Ensembl custom tracks are described at http://www.ensembl.org/info/using/external_data.

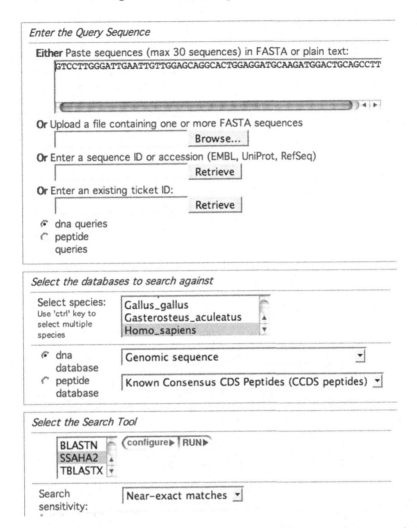

Figure 3.7 Ensembl BlastView with the sequence surrounding the polymorphism site entered as the query sequence. SSAHA with near-exact matching is selected because the query sequence differs from the reference genome by only a single nucleotide.

DAS record are different from those in the UCSC custom track of Figure 2.10. The differences are partly the result of our having used different genome assemblies and partly the result of different coordinate numbering conventions used by UCSC and Ensembl (genome coordinate-system numbering conventions are described in Appendix 1).

Once we have made our track selections, Ensembl returns the display shown in Figure 3.8. From the display, we again see that our site does not overlap any known SNPs, repeats, or disease regions. We again see that numerous cDNAs and ESTs overlap our polymorphism site. However, there are differences between Ensembl's EST display and what we saw with the UCSC Browser. First, the number of transcripts shown is often different – and generally smaller – than the number of ESTs returned in

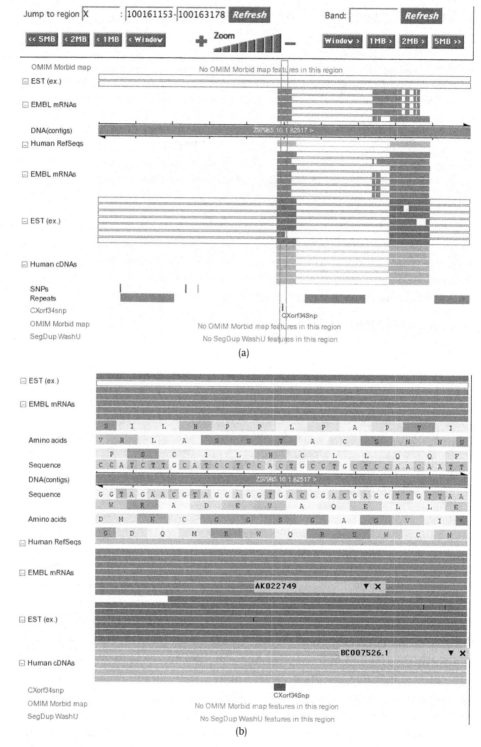

Figure 3.8 (a) Ensembl Detail View of the region surrounding our CXorf34 polymorphism. Position of polymorphism is shown via custom "CXorf34snp" track. (b) Basepair View at the polymorphism site. Overlapping mRNAs and ESTs can be identified by positioning the mouse over tracks, as shown. Only a subset of overlapping mRNAs and ESTs will appear in the display.

Homo sapiens	A	T	C	T	T	G	C	A	T	C	C	T	C	C	A	C	T	G	C	C	T
Mus musculus	A	C	C	T	G	G	C	A	T	C	C	T	C	T	A	C	T	G	C	C	T
Canis familiaris	A	C	C	T	G	G	C	A	T	C	C	T	C	T	A	C	T	G	C	C	T
Monodelphis do ...	.	.	.	.	.	.	.	.	.	.	.	.	.	.	.	.	.	.	.	.	.
Rattus norvegicus	A	C	C	T	G	G	C	A	T	C	C	T	C	T	A	C	T	G	C	C	T
Bos taurus	.	.	.	.	.	.	.	.	.	.	.	.	.	.	.	.	.	.	.	.	.
Gallus gallus	A	C	C	T	G	G	C	A	T	C	C	T	C	T	A	T	T	G	C	T	T
Pan troglodytes	a	t	c	t	t	g	c	a	t	c	c	t	c	c	a	c	t	g	c	c	t
Macaca mulatta	A	T	C	T	T	G	C	A	T	C	C	T	C	C	A	C	T	G	C	C	T

Figure 3.9 PECAN-generated multispecies alignment surrounding the polymorphism site shown in Ensembl AlignSliceView. The rectangle, which is not part of the AlignSliceView display, indicates the polymorphism location in the alignment.

UCSC's "full" display mode for ESTs. This is because Ensembl displays only part of the complete set of ESTs to keep the display from becoming too crowded. To see which ESTs are being displayed, one can place the mouse over any of the EST tracks and the name of the EST will be displayed. Refreshing the screen displays other ESTs. A second and more significant limitation of the Ensembl EST display is that there is no way to determine if any of these sequences have sequence variants at the polymorphism site because the Ensembl database does not store the actual mRNA/EST/cDNA sequences. Consequently, we cannot tell whether our variant has been observed in any mRNA, EST, or cDNA transcript.

To further characterize our polymorphism site, we need to use other Ensembl views besides ContigView. To locate these related views, we can use the pull-down menus from the menu list on the left side of the ContigView display. For example, to determine the level of sequence conservation at the polymorphism site, we can select "10 amniota vertebrates – PECAN" from the "View alignments" menu. This will switch the display to AlignSliceView. If we switch to AlignSliceView using the region immediately surrounding our polynomial site (chrX:100,162,136-100,162,195 in NCBI assembly 36 coordinates, currently used by Ensembl), we see an alignment similar to that shown in Figure 3.9.

We observe that Ensembl's alignment differs from the UCSC alignment we saw in Figure 2.6. First, because we cannot display our custom track in AlignSliceView, we need to note the precise coordinate of the polymorphism so that we can identify its location in the alignment. We have done this by manually superimposing a transparent rectangle at the position of the polymorphism in the alignment. Second, an Ensembl multisequence alignment can only be displayed relative to the positive genome strand. In contrast, with the UCSC Browser, we can choose the strand to use for orienting multisequence alignments. In Figure 2.6, we chose to display the conservation alignment relative to the negative strand because CXorf34 is a negative-strand gene. The UCSC and Ensembl alignments also differ because, currently, UCSC's alignments include more species than Ensembl's. For the species included in both alignments, the observed nucleotides at the polymorphism site are the same, with (positive strand) "C" being found in the mammals and "T" found in chicken. However, such

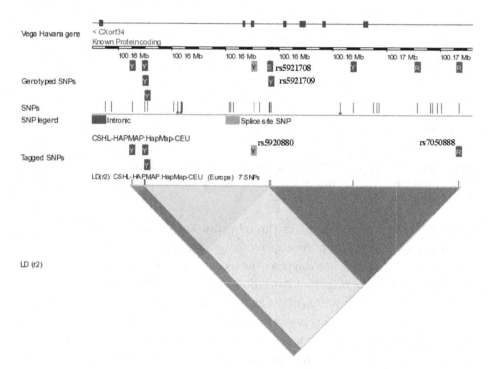

Figure 3.10 CXorf34 in Ensembl's Linkage Disequilibrium View (LDView). Note that SNP IDs have been manually added in the figure. One needs to "mouse" over the SNPs to identify them in the actual display.

agreement between UCSC and Ensembl alignments is not necessarily the case, as UCSC and Ensembl use different multiple-sequence alignment algorithms. Differences between UCSC's and Ensembl's alignment algorithms are described in Appendix 4.

Other Ensembl views provide additional information on our region of interest. For example, we can use Ensembl's LDView to search for common, nearby SNPs that may be in linkage disequilibrium with our new allele. Figure 3.10 shows Ensembl's LDView display in the neighborhood of our polymorphism site in CXorf34. (You may need to go to the CXorf34 page in "GeneView" – which can be reached by entering "CXorf34" in the main search box – and then link to LDView from there.) The display also includes a track showing which common SNPs in this region have been extensively genotyped for different human populations and which have been "tagged" by the HapMap project as markers of population variation. The figure shows that for European populations, our polymorphism site is within a region of strong linkage disequilibrium, indicated by the dark triangle. Consequently, by genotyping SNPs rs5921708, rs5921709, or rs5920880, as well as SNP rs7050888, in the individual with our novel allele, it may be possible to develop a marker to help identify other individuals who possess the new allele. For more details about the use of linkage disequilibrium, the HapMap project, and the use of triangle-type linkage-disequilibrium displays like Figure 3.10, see Barrett et al. (2005) and Frazer et al.

(2007) and references therein. We note that some UCSC Browser builds (e.g., hg17) include linkage-disequilibrium displays, as well; however, UCSC does not currently include tracks that explicitly indicate genotyped and tagged SNPs.

GeneTreeView, MultiContigView, TranscriptSnpView, and FamilyView also contain information that is useful for our example. GeneTreeView provides a phylogenetic tree at the individual gene level; the GeneTreeView for CXorf34 is shown in Figure 3.3. MultiContigView enables us to compare annotations for CXorf34 and each of its homologs together in a single display. An example of a MultiContigView display between CXorf34 and its mouse homolog is shown in Figure 3.5. If we anticipate studying CXorf34 further in the mouse, selecting TranscriptSnpView (see Figure 3.4) provides a SNP comparison of CXorf34-mouse homolog sequences in different mouse strains, which may be useful in selecting appropriate mouse strains for experimental studies. Finally, although there is no equivalent of UCSC's Gene Sorter to identify "similar genes" on the basis of gene expression or protein-interaction network proximity, Ensembl's FamilyView does allow one to search for genes that are similar on the basis of protein-family grouping.

3.3 Genome browsing with NCBI MapViewer

NCBI's MapViewer is the third major multispecies genome browser. Because of the significant differences in genome-database design objectives of NCBI as compared to UCSC or EBI, directly comparing MapViewer with the Ensembl or UCSC Browsers is not particularly instructive. Rather than being a comprehensive stand-alone genomic database resource, MapViewer is one component of a suite of database resources and tools offered by NCBI. NCBI database resources include archival databases such as GenBank, dbEST, dbSNP, and the genomic sequence trace databases as well as the curated databases, such as RefSeq, UniGene, and Homologene. NCBI genome-analysis tools include *in silico* PCR tools for primer design (http://www.ncbi.nlm.nih.gov/sutils/e-pcr), tools for testing putative novel gene isoforms from known transcript data (http://www.ncbi.nlm.nih.gov/mapview/static/ModelMakerHelp.html), and genome-wide sequence-searching tools such as Genomic BLAST. In particular, no matter which browser you use, NCBI Genomic BLAST can be useful for searching for sequences within a genome, even if the genome is not yet available in any of the genome browsers. Genomic BLAST can be reached via http://www.ncbi.nlm.nih.gov/sutils/genom_table.cgi.[4]

As its name suggests, MapViewer is particularly designed for facilitating the construction of genomic maps. This objective is reflected in the fact that the MapViewer display is oriented vertically. Such vertical displays facilitate the simultaneous viewing of multiple genomic coordinate systems. For example, with MapViewer one can

---

[4] For eukaryotes, there is an alternate interface accessible via the single-organism BLAST buttons on the NCBI Genome page at http://www.ncbi.nlm.nih.gov/Genomes.

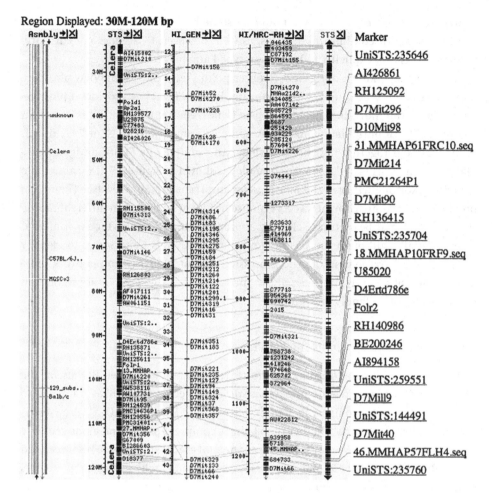

Figure 3.11 Chromosomal inversion on NCBI MapViewer. Region displayed is from 30 to 120 megabases on mouse chromosome 7. Note the large chromosomal inversion, near 70 megabases, between the radiation hybrid data (map WI/MRC-RH) and the STS map, indicated by the large number of crossed lines between corresponding landmarks on the two maps.

compare two complete genome assemblies (e.g., the reference and Celera human genome assemblies), as well as partial assemblies and non-sequence-based genetic maps. Such comparisons are more difficult or impossible to perform with the UCSC or Ensembl systems. An example of map comparison with MapViewer is shown in Figure 3.11. The figure shows linkage and radiation hybrid maps along with the mouse reference assembly, illustrating a large inversion of the radiation hybrid mapping relative to the reference assembly in the middle of the chromosome.

Another useful feature of MapViewer is the support of three tools (MegaBLAST, discontiguous MegaBLAST, and BLAST) for sequence searching. MegaBLAST is fast but requires longer regions of exact matches between the query and genome sequences (similar to BLAT and SSAHA). In contrast, discontiguous MegaBLAST is more sensitive

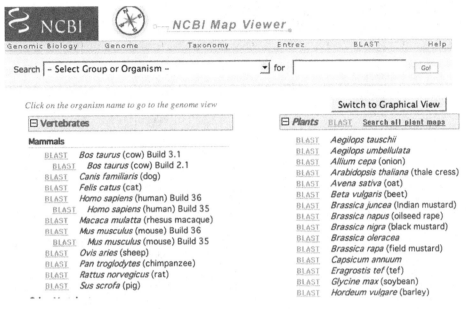

Figure 3.12 Home page for MapViewer.

than BLAT, SSAHA, or conventional MegaBLAST, and yet is much faster than conventional BLAST.

However, despite these features, for general genome querying, MapViewer has some significant limitations. First, because of MapViewer's vertical orientation, it would be awkward to "zoom in" to a base-pair level view of the chromosome and, in fact, MapViewer does not currently provide base-level viewing of the genome sequence at all. Similarly, MapViewer currently does not provide base-level, cross-species, pairwise, or multiple-sequence alignments (this capability is currently under development at NCBI), nor does MapViewer offer custom tracks or direct querying of its database, either interactively or via computer programs. Consequently, for applications requiring these capabilities, one needs to use the Ensembl or UCSC systems. For a more detailed introduction to MapViewer's capabilities, the reader is referred to the MapViewer online documentation and tutorials listed in Appendix 7.

3.4 Polymorphism characterization with MapViewer

We now consider MapViewer in the context of our polymorphism-characterization example. Starting at the MapViewer home page (http://www.ncbi.nlm.nih.gov/mapview), we select the most recent build of the human genome and BLAST search input (see Figure 3.12). From the BLAST search page, we select MegaBLAST for searches with query sequences that are very similar to the underlying genome. We also restrict the search to the "reference" human genome. Inserting our 60 nt sequence and running MegaBLAST, we receive a BLAST report that includes a "Genome View" link,

Region Displayed: 100,162,088-100,162,242 bp

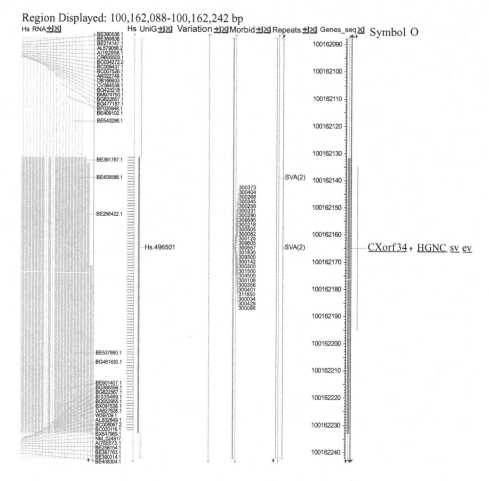

Figure 3.13 MapViewer display for the genomic region surrounding CXorf34.

enabling us to view the matching region in MapViewer. The result will be similar to that seen in Figure 3.13, though the selected annotations may differ.

The most obvious difference from the UCSC and Ensembl displays is that the MapViewer display is oriented vertically. One advantage of MapViewer's vertical view is that it enables a convenient overview of regional gene descriptions. For example, if a region of a chromosome contains a cluster of genes all associated with a specific molecular function or pathway (as is often the case in prokaryotes), this fact will be immediately apparent. Such an immediate functional overview of a genomic region is not easy to display with horizontally oriented annotation tracks.

We now need to select the appropriate annotations. In MapViewer, annotations are selected via the "Maps and Options" button. Descriptions of the various maps available for each genome are found in MapViewer's species-specific help pages. For the human genome, this documentation is located at http://www.ncbi.nlm.nih.gov/mapview/static/humansearch.html.

For our polymorphism example, we select the repeats, rnaHs (GenBank mRNAs), ugHs (UniGene EST clusters), Variation, Morbid/Disease, and Gene maps from the "Maps and Options" menu. As with Ensembl, MapViewer displays annotations for both strands and the user only selects whether a track is "on" or "off." In choosing the "Gene" map (which is called Genes_seq on the MapViewer display), we should note that NCBI has yet a third implicit definition of a "gene" that is slightly different from both UCSC and Ensembl's definitions, described previously. Specifically, MapViewer's "Gene" map displays a "flattened view" of the entire set of transcripts associated with the gene. That is, MapViewer's Gene map displays all exons that exist in *any* transcript of the gene set.

We will make the Gene Map MapViewer's "master map" so that the Gene Map links and annotations are included in the MapViewer display. By clicking on the links in the Gene Map display, one is led to other tools and databases within the NCBI site that provide data and functionality similar to that provided by the "Details" pages at UCSC or the alternate "Views" at Ensembl. For example, clicking on the "SNP" link would take us to any SNP annotations in dbSNP in the currently selected region. MapViewer does not currently provide multispecies alignments, so we cannot determine whether our polymorphism site is conserved among vertebrates. Also, MapViewer does not support custom tracks, so we cannot highlight the specific polymorphism site as we did previously.

Upon refreshing the screen, we see a display similar to that shown in Figure 3.13. No SNPs or other variations are annotated in the region. Two repetitive regions are annotated; however, without a custom track or other method to specify exactly where in the display our polymorphism is located, it is difficult to ascertain that the repeat elements do not actually overlap the polymorphism site. Again, we see that numerous ESTs and mRNAs overlap our polymorphism site. For the mRNAs, we could determine whether any of them have mismatches with the genomic sequence by selecting the "ev" EvidenceViewer tool, shown at the extreme right of Figure 3.13. However, we would not be able to directly determine whether any ESTs have our sequence variant at the polymorphism site from the browser display. (We could return to the main BLAST search page and BLAST our sequence against NCBI's EST database to determine whether our variant had been previously observed.) On the other hand, NCBI offers a useful trace searching utility (http://www.ncbi.nlm.nih.gov/blast/mmtrace.shtml) with which we could check whether our allele had been deposited in any of the trace archives. (Ensembl also has a trace searching facility at http://trace.ensembl.org/cgi-bin/tracesearch but, currently, Ensembl's site only includes traces for a limited number of species.)

3.5 Using other browsers

All three "major" genome browsers – Ensembl, MapViewer, and UCSC – have quite good coverage of animal genomes. In addition, MapViewer includes some 25 plant

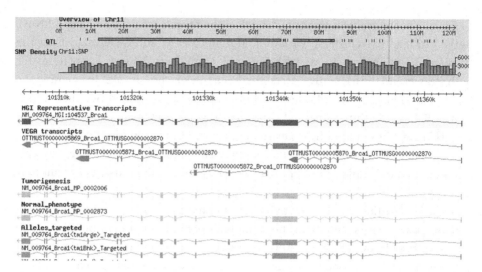

Figure 3.14 GBrowse screenshot of BRCA1 gene in the MGI genome browser.

and 14 fungal species. However, for other species and for more detailed annotations of some species, other genome browsers may be useful. For example, the Gramene genome browser (http://www.gramene.org/genome_browser) includes four important plant species. Genome browsers for individual model organisms are available via the single-genome databases, such as the Saccharomyces Genome Database (SGD) (http://www.yeastgenome.org), the Mouse Genome Database (http://www.informatics. jax.org), WormBase (http://www.wormbase.org), FlyBase (http://www.flybase. org), BeetleBase (http://www.bioinformatics.ksu.edu/BeetleBase), DictyBase (http:// dictybase.org/db/cgi-bin/ggb/gbrowse/dictyBase), the Sigenae farm animal browser (http://public-contigbrowser.sigenae.org:9090), and the cat database (http://lgd.abcc. ncifcrf.gov/cgi-bin/gbrowse/cat). In addition, although none of the major browsers officially cover prokaryotes, UCSC, Ensembl, and NCBI do have separate web sites for microbial genomes. For UCSC, prokaryote genomes are accessed via the microbial genome portal at http://archaea.ucsc.edu, whereas Ensembl and NCBI use the Genome Reviews web browser (http://www.ebi.ac.uk/GenomeReviews) and the NCBI microbial browser (http://www.ncbi.nlm.nih.gov/genomes/MICROBES/microbial_taxtree.html), respectively.

Using these other genome browsers is straightforward and quite similar to using Ensembl, MapViewer, or the UCSC Browser. In fact, several of them use the same architecture and user interface as one of these main genome browsers. For example, UCSC, EBI, and NCBI's prokaryote databases each use the same interface as the main genome browser with which it is associated, and the Gramene browser uses the Ensembl interface. In contrast, most of the single-genome databases use the GBrowse genome browser, which we have not yet described. However, the GBrowse interface is quite similar in spirit to the interfaces of the main genome browsers (especially Ensembl), and it is quite easy to learn if one is already familiar with the interfaces

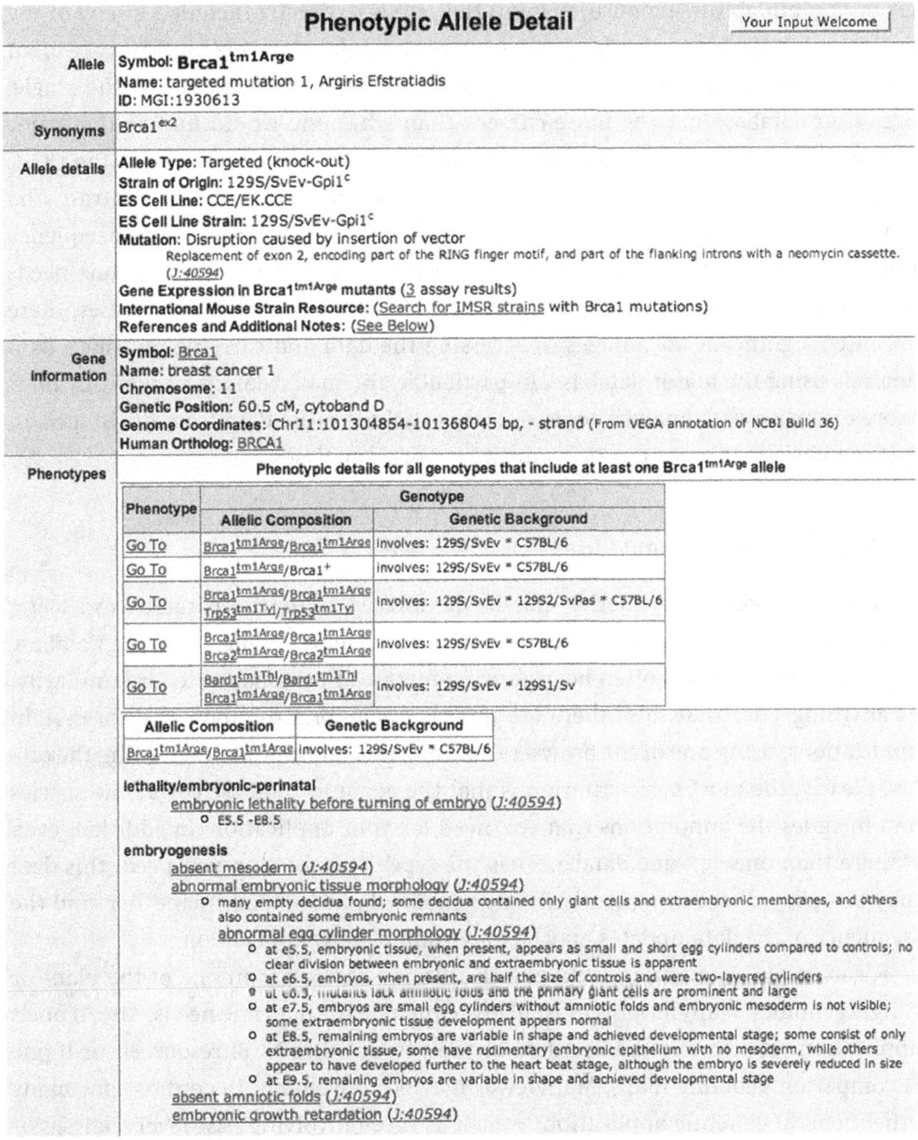

Phenotypic Allele Detail | Your Input Welcome |

Allele	Symbol: **Brca1**^{tm1Arge} Name: targeted mutation 1, Argiris Efstratiadis ID: MGI:1930613
Synonyms	Brca1^{ex2}

Allele details	**Allele Type:** Targeted (knock-out)
	Strain of Origin: 129S/SvEv-Gpi1^c
	ES Cell Line: CCE/EK.CCE
	ES Cell Line Strain: 129S/SvEv-Gpi1^c
	Mutation: Disruption caused by insertion of vector
	    Replacement of exon 2, encoding part of the RING finger motif, and part of the flanking introns with a neomycin cassette.
	    (J:40594)
	Gene Expression in Brca1^{tm1Arge} **mutants** (3 assay results)
	International Mouse Strain Resource: (Search for IMSR strains with Brca1 mutations)
	References and Additional Notes: (See Below)

Gene information	**Symbol:** Brca1 **Name:** breast cancer 1 **Chromosome:** 11 **Genetic Position:** 60.5 cM, cytoband D **Genome Coordinates:** Chr11:101304854-101368045 bp, - strand (From VEGA annotation of NCBI Build 36) **Human Ortholog:** BRCA1
Phenotypes	Phenotypic details for all genotypes that include at least one Brca1^{tm1Arge} allele

Phenotype	Genotype	
	Allelic Composition	**Genetic Background**
Go To	Brca1^{tm1Arge}/Brca1^{tm1Arge}	involves: 129S/SvEv * C57BL/6
Go To	Brca1^{tm1Arge}/Brca1$^+$	involves: 129S/SvEv * C57BL/6
Go To	Brca1^{tm1Arge}/Brca1^{tm1Arge} Trp53^{tm1Tyj}/Trp53^{tm1Tyj}	involves: 129S/SvEv * 129S2/SvPas * C57BL/6
Go To	Brca1^{tm1Arge}/Brca1^{tm1Arge} Brca2^{tm1Arge}/Brca2^{tm1Arge}	involves: 129S/SvEv * C57BL/6
Go To	Bard1^{tm1Thl}/Bard1^{tm1Thl} Brca1^{tm1Arge}/Brca1^{tm1Arge}	involves: 129S/SvEv * 129S1/Sv

Allelic Composition	**Genetic Background**
Brca1^{tm1Arge}/Brca1^{tm1Arge}	involves: 129S/SvEv * C57BL/6

lethality/embryonic-perinatal
    embryonic lethality before turning of embryo (J:40594)
        ○ E5.5 -E8.5

embryogenesis
    absent mesoderm (J:40594)
    abnormal embryonic tissue morphology (J:40594)
        ○ many empty decidua found; some decidua contained only giant cells and extraembryonic membranes, and others
          also contained some embryonic remnants
    abnormal egg cylinder morphology (J:40594)
        ○ at e5.5, embryonic tissue, when present, appears as small and short egg cylinders compared to controls; no
          clear division between embryonic and extraembryonic tissue is apparent
        ○ at e6.5, embryos, when present, are half the size of controls and were two-layered cylinders
        ○ at e6.5, mutants lack amniotic folds and the primary giant cells are prominent and large
        ○ at e7.5, embryos are small egg cylinders without amniotic folds and embryonic mesoderm is not visible;
          some extraembryonic tissue development appears normal
        ○ at E8.5, remaining embryos are variable in shape and achieved developmental stage; some consist of only
          extraembryonic tissue, some have rudimentary embryonic epithelium with no mesoderm, while others
          appear to have developed further to the heart beat stage, although the embryo is severely reduced in size
        ○ at E9.5, remaining embryos are variable in shape and achieved developmental stage
    absent amniotic folds (J:40594)
    embryonic growth retardation (J:40594)

Figure 3.15 Example of detailed allele data available from the MGI genome browser.

of the major genome browsers. An example of a GBrowse genome browser display is shown in Figure 3.14, which shows a screenshot of part of mouse chromosome 11 from the MGI Mouse Genome Database's web site. We will return to GBrowse in Chapter 11 in the context of building one's own custom genome database.

If a species has its own single-genome database, there may be reasons to check it even if the species is also covered by one of the major databases. First, some of the more detailed annotations from the single-organism database may not be integrated into any of the major genome browsers. For example, Figure 3.15 shows mouse allele data from the genome browser at the Mouse Genome Database (http://gbrowse.informatics.

jax.org/cgi-bin/gbrowse/mouse_current) that are not directly included in any of the multispecies databases. In addition, because the single-organism databases are often the primary data repositories for data on that organism, the data in the single-organism database may be more current than what one would find in the major genome databases. For example, the genomic sequence for *S. cerevisiae* in the UCSC Genome Browser is from the October 2003 sequence assembly; in contrast, the sequence available from the SGD site has numerous updates to the genomic sequence that have been incorporated since 2003. On the other hand, if the data one needs are available and sufficiently current on one of the major genome databases, there are often significant advantages to accessing the data and carrying out one's data analysis using the major database. In particular, the major databases typically offer more extensive data analysis tools, and they include more data from other species, making it easier to carry out comparative genomic analyses.

3.6 Which browser should I use?

Many genome-database queries can be performed with comparable ease using Ensembl, MapViewer, or the UCSC system. Consequently, which genome database or browser you use will often be as much a matter of personal taste, or familiarity, as anything else. That said, there are differences among the tools and for certain applications, using one of the browsers may be more appropriate than using the others. Clearly, the most basic criterion is that the genome database covers the species and includes the annotations that you need for your application. In addition, even if more than one genome database has the type of annotation you need, this data may be updated more frequently in one genome database than in another, and the frequency of the data updates may be important to your application.

For example, as mentioned previously, if one's interest is in one of the plant or fungal genomes, MapViewer is most likely to include the data one needs. Also, if one's application requires integrating data or tools from other NCBI resources, or if one is comparing genomic maps, MapViewer may be convenient. In contrast, for many other general genomic applications – such as those involving base-level analyses or multigenome alignments, or ones that involve batch querying (described in Chapters 4 through 11) – UCSC or Ensembl are more appropriate choices.

Assuming that more than one of the browsers include the species and sufficiently current annotations for your application, then your tool selection criteria are likely to be more subjective. Do you prefer horizontal or vertical displays? Do you prefer a single integrated view or do you like multiple views? Is there some special tool or feature provided by one of the browsers that would be helpful for your application?

Last, but not least, browser convenience and utility are affected by browser speed, that is, how long it takes for the browser to display the desired annotations after performing a mouse click. One might not initially think it important whether a browser takes one second or ten seconds to refresh the screen after changing genome location or adding an annotation track. However, the frustration that accompanies waiting several seconds for a screen refresh is a major drawback for many users.

In the past, system performance could vary quite noticeably among the three major browsers, especially when searching for a sequence fragment in the genome. However, performance improvements in sequence-searching programs have resulted in the speed differences among the browsers to decrease dramatically. At present, short of performing careful benchmarking experiments, it is difficult to make definitive statements regarding the relative performance of the browsers for comparable queries. That said, at least my anecdotal experience suggests that sometimes there are still performance differences among the browsers (and that the UCSC Browser has the shortest average response time – but then again, I am generally accessing the browsers from a California location). Perhaps the best that can be said in this regard is that if you have a set of tasks that will need to be performed repeatedly and could be accomplished on any of the browsers, then it might be worthwhile to try them with more than one system and make your own comparison.

Chapter summary

- Like UCSC, Ensembl also provides a comprehensive collection of genomic annotations from many databases.
- In contrast to UCSC, an important goal of Ensembl is also to produce genomic annotations – in particular, predictions of genes and transcripts – and Ensembl's own annotations play a central role on the Ensembl Genome Browser.
- Ensembl provides over thirty different types of customized annotation displays, in contrast to UCSC and MapViewer's more integrated display strategy.
- MapViewer is especially designed for comparing genetic and genomic maps and integrating data from multiple NCBI data resources including RefSeq, Unigene, NCBI BLAST, dbSNP, dbEST, OMIM, and so on.
- MapViewer uses a vertical display and lacks certain capabilities, such as single-nucleotide resolution displays, multiple-sequence alignments, custom tracks, and batch-querying, which are not needed for manipulating genomic maps or utilizing NCBI's genomic resources.
- Although genomes and annotations for many species can be found on all three browsers, some classes of organisms are better represented on one browser than another, or may be available only on one of the more specialized genome browser web sites.

Exercises

1. Repeat Exercise 2.1 using the Ensembl and MapViewer Browsers. Do you find one or another of these browsers easier or more useful for determining transcript isoform data?

2. Can you repeat Exercise 2.2 using the Ensembl or MapViewer Browsers? Does one of these browsers have more, or more easily accessible, gene-expression data than the other?

3. Can you find information regarding genes that may be similar to, or related to, PALB2 using Ensembl or MapViewer? You may want to use Ensembl's FamilyView, NCBI's blastp, or other Ensembl or NCBI tools.

4. Make an Ensembl custom track of the genes associated with "Fanconi anemia."

5. This exercise explores the effectiveness of browser sequence searching tools when only short search sequences are available. In our polymorphism example, we used a long, 60-nt region (GTCCTTGGGATTGAATTGTTGGAGCAGGCACTGGAGGATGCAAGAT GGACTGCAGCCTTC) centered around the polymorphism site.

   a. Can you locate the site in the UCSC Browser using BLAT if you only use the central 30 nt of the query sequence?

   b. Can you locate the sequence if only the central 20 or 15 nt are available?

   c. Can the site be located in Ensembl using SSAHA with the shorter query sequences? If not, can it be found by changing the search parameters passed to SSAHA or using Ensembl's BLAST tool?

   d. Similarly, can the polymorphism site be located with MapViewer's search tools (i.e., MegaBLAST, discontiguous MegaBLAST, and BLAST) using shorter (e.g., 15 to 30 nt) query sequences?

4

Interactive Genome-Database Batch Querying

For all its power and convenience, genome browsing has limitations. In particular, genome browsing only enables one to query a single genomic locus at a time and, hence, is cumbersome for applications involving multiple genomic regions. Querying each region interactively is tedious, time-consuming, and, because of all the required manual mouse manipulations, potentially prone to error.

With genome batch querying, we can obtain data from multiple related locations (i.e., a "batch" of genes or genomic locations) via a single query. Genome-database batch querying can be performed either interactively – by direct database querying or via a web-based interface – or by programmed access.

In the present chapter, we introduce interactive genome batch querying. We begin with examples of the kinds of biological questions that can be addressed by batch querying. Then we describe three general strategies – SQL querying, data marts, and direct table access – used to implement interactive batch querying in the genome databases. Next, we turn our focus to BioMart, the Table Browser, and the Gene Sorter, web-based tools that Ensembl and UCSC provide for interactive batch querying, and show how they can be used in typical batch querying applications. Finally, we briefly describe interactive batch querying using the standard SQL querying language.

4.1 Batch querying applications

Batch querying applications are queries in which a single set of biological questions needs to be answered for many genes or genomic regions. These questions may be as simple as finding the RefSeq IDs for a list of genes for which we have Ensembl IDs, or determining what genes have annotations for a specific disease. Or they can be more complex, such as determining which regions in a list of highly expressed regions from a tiling-array experiment have already been observed as ESTs, or identifying places where repetitive elements are found within gene coding regions. In the remainder of this chapter and those following, we will consider these examples and numerous others that indicate the range of biological questions that can be addressed by genomic batch querying. However, first we briefly consider, in more general terms, the types of interactive batch-querying facilities provided by the genome databases.

4.2 Database architectures and batch querying

Genome databases are generally implemented as relational databases, with their data stored in tabular format. Consequently, the most direct way that the data in a genome database can be accessed is via the Structured Query Language (SQL). However, this approach, which we illustrate at the end of this chapter, requires that users are familiar with the SQL language and that they understand the descriptions of the tables used in the database.

To enable batch querying by users not familiar with SQL, a database system may include a user interface that takes user input and translates it into SQL queries of the underlying database tables. This is the approach used by the UCSC Table Browser, which is essentially a "user friendly" interface to SQL querying of the UCSC genome database. However, the Table Browser user still needs to have some understanding of the UCSC database table definitions.

Unfortunately, the way data are organized in relational database tables is not always very natural to humans. This is especially true in the case of hierarchical data, where one piece of data is, in some sense, a "part" of another piece of data. For example, a gene might naturally be considered as a collection of transcripts, each of which consists of a set of exons. However, in a relational database, the gene, transcript, and exon data for a single gene may well be stored in three separate tables.

Consequently, both Ensembl and UCSC provide interactive batch-querying tools to combine related data from multiple database tables. Such tools are typically called *data marts*. A data mart reorganizes data from the main database into a collection of fundamental software objects, variously called datasets or data foci, in a separate auxiliary database (see Figure 4.1). As a result, the data mart user does not need to know anything about the actual table structure in the database. UCSC's Gene Sorter, which we have encountered in Chapter 2, is a data mart in which the data focus is always the gene. Ensembl's data mart uses the BioMart software package, previously called EnsMart (Kasprzyk et al., 2004). Ensembl BioMart lets the user select different kinds of data foci, such as genes, SNPs, or alignments. BioMart is developed separately from Ensembl, and is incorporated in several other genome databases besides Ensembl. In particular, several of the GMOD databases, described in Chapter 11, also incorporate the BioMart software for interactive batch querying. More information on the BioMart software and its applications is available on the BioMart web site (http://www.biomart.org).

With the data mart approach, rather than querying arbitrary combinations of database tables, one is restricted to querying data related to one of the foci. If the foci are appropriately selected by the data mart designers to reflect the needs of system users, then the system can be streamlined, modularized, and made expandable

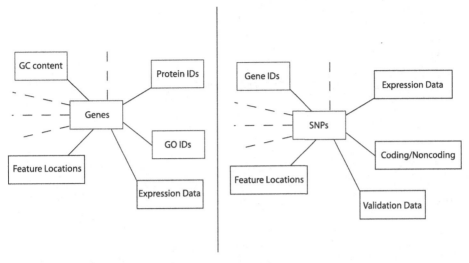

Figure 4.1 Schematic view of part of Ensembl's BioMart data mart architecture, showing two data foci – genes and SNPs – and five of the auxiliary data sets linked to each data focus.

(by adding more foci) – and will still be able to respond to nearly all queries that users may make. As a result, batch querying with BioMart or the Gene Sorter is, in some ways, easier than using the UCSC Table Browser. However, as we will see later, in some cases the table-based approach offers somewhat more flexibility, enabling one to address some questions with the Table Browser that could not be answered with a tool like BioMart or the Gene Sorter.

4.3 Ensembl BioMart and MartView

Ensembl's browser interface to BioMart is called MartView. Using MartView, inter active batch querying is carried out with a three-step procedure. Starting from the MartView input page (http://www.biomart.org/biomart/martview), one first chooses a genome and a "focus." Currently, the focus can be either manually curated genes from the VEGA project, Ensembl genes and gene predictions, SNPs, pairwise or multiple-sequence alignments, or homologies. In the second step, one selects the desired "attributes" of the data. Numerous attribute options are available, including SNP or gene annotations, accession IDs from many databases, sequence data, gene structure, and so on. Finally, in the third step, one limits the amount of data to be returned by specifying a single genome region and/or by restricting the annotations the focus (e.g., the selected genes) may have. Genes can be filtered on the basis of gene IDs, GO descriptions, Pfam protein domains, as well as many other annotations. It is even possible to filter genes on the basis of properties of a gene's homologs in other species using the second-database and intersection options on the MartView input page. Figure 4.2a shows a screenshot of part of the MartView input page for a query to convert a list of RefSeq accession numbers to Ensembl accessions.

» **Dataset:**
Homo sapiens genes
(NCBI36)
  » Attributes (Features)
    Ensembl Gene ID
    Ensembl Transcript ID
    RefSeq DNA ID
    Chromosome Name
    Start Position (bp)
    End Position (bp)
    Strand
  » Filters
    Refseq DNA ID(s): [ID-list
    specified]

» **Dataset:**
[None Selected]

◉ Features ○ SNPs
○ Structures ○ Sequences

⊞ REGION:

⊟ GENE:

Ensembl Attributes

☑ Ensembl Gene ID   ☐ Ensembl Peptide length
☑ Ensembl Transcript ID   ☐ Transcript count
☐ Ensembl Peptide ID   ☐ % GC content
☐ External Gene ID   ☐ Description
☐ External Gene DB   ☐ Biotype
☐ Ensembl CDS length   ☐ Source
☐ Ensembl cDNA length   ☐ Status

GO Attributes

☐ GO ID   ☐ GO evidence code
☐ GO description

External References (max 3)

☐ CCDS ID   ☐ RefSeq Predicted DNA ID
☐ Codelink ID   ☐ RefSeq Peptide ID
☐ EMBL ID   ☐ Rfam ID

(a)

Ensembl Gene ID	Ensembl Transcript ID	RefSeq DNA ID	Chromosome Name	Start Position (bp)	End Position (bp)	Strand
ENSG00000113525	ENST00000231454	NM_000879	5	131905035	131907113	−1
ENSG00000100296	ENST00000358079	NM_001002879	22	28231868	28279703	−1
ENSG00000172531	ENST00000312989	NM_001008709	11	66922228	66925978	−1
ENSG00000204075	ENST00000372866	NM_001008739	6	42966199	42966504	−1
ENSG00000146955	ENST00000275874	NM_001008749	7	139753916	139772419	1
ENSG00000163806	ENST00000296122	NM_001008779	2	28828118	28926981	1
ENSG00000171649	ENST00000307468	NM_001010879	19	62787320	62795570	1
ENSG00000105146	ENST00000302804	NM_001015879	19	62434240	62438727	1
ENSG00000127241	ENST00000337774	NM_001879	3	188418632	188492446	−1
ENSG00000003402	ENST00000309955	NM_003879	2	201689135	201737246	1

(b)

Figure 4.2 RefSeq to Ensembl ID conversion using BioMart. (a) Screenshot of a portion of the input screen. (b) Data output showing RefSeq and Ensembl IDs as well as genomic coordinates.

4.3.1 Applying BioMart

Let us now apply BioMart to some batch querying problems. For our first example, let us convert the IDs of a set of genes from Ensembl IDs to RefSeq IDs. We begin by selecting a focus of Ensembl genes for the human genome database of, say, Ensembl release 42. Next, we select the desired output fields – that is, Ensembl IDs and RefSeq IDs – from the Ensembl attributes and external references sections, as shown in Figure 4.2a. Finally, on the filter page, we paste or upload our list of Ensembl IDs into the "ID list limit," making sure to specify that the gene ID type is "Ensembl IDs." Selecting "Results" returns a list of RefSeq IDs and associated Ensembl IDs, as shown in Figure 4.2b.

For our second example, we will find genes that have a specific GO annotation, for example, synaptic transmission. To obtain these genes, we will need to know the GO code for the annotation, or else we will need to look up the GO code in the GO

Ensembl Gene ID	Description
ENSG00000171189	Glutamate receptor, ionotropic kainate 1 precursor (Glutamate receptor 5) (GluR-5) (GluR5) (Excitatory amino acid receptor 3) (EAA3). [Source:Uniprot/SWISSPROT;Acc:P39086]
ENSG00000128245	14-3-3 protein eta (Protein AS1). [Source:Uniprot/SWISSPROT;Acc:Q04917]
ENSG00000160307	Protein S100-B (S100 calcium-binding protein B) (S-100 protein beta subunit) (S-100 protein beta chain). [Source:Uniprot/SWISSPROT;Acc:P04271]
ENSG00000169862	Catenin delta-2 (Delta-catenin) (Neural plakophilin-related ARM-repeat protein) (NPRAP) (Neurojungin) (GT24). [Source:Uniprot/SWISSPROT;Acc:Q9UQB3]
ENSG00000144619	Contactin-4 precursor (Brain-derived immunoglobulin superfamily protein 2) (BIG-2). [Source:Uniprot/SWISSPROT;Acc:Q8IWV2]
ENSG00000183454	Glutamate [NMDA] receptor subunit epsilon-1 precursor (N-methyl D- aspartate receptor subtype 2A) (NR2A) (NMDAR2A) (hNR2A). [Source:Uniprot/SWISSPROT;Acc:Q12879]
ENSG00000174775	GTPase HRas precursor (Transforming protein p21) (p21ras) (H-Ras-1) (c-H-ras) (Ha-Ras). [Source:Uniprot/SWISSPROT;Acc:P01112]
ENSG00000156642	Neuroplastin precursor (Stromal cell-derived receptor 1) (SDR-1). [Source:Uniprot/SWISSPROT;Acc:Q9Y639]
ENSG00000170027	14-3-3 protein gamma (Protein kinase C inhibitor protein 1) (KCIP-1). [Source:Uniprot/SWISSPROT;Acc:P61981]
ENSG00000132535	Discs large homolog 4 (Postsynaptic density protein 95) (PSD-95) (Synapse-associated protein 90) (SAP90). [Source:Uniprot/SWISSPROT;Acc:P78352]

Figure 4.3 Portion of output from a BioMart example for identifying Ensembl genes with GO annotations associated with synaptic transmission.

database. We begin as in the previous example by choosing a "gene" focus with the Ensembl human database. This time, we filter on the GO ID code GO:0050804 for "regulation of synaptic transmission," which we obtained from the GO database via a link next to the GO-ID inputs on the MartView filter page. For attributes, we can choose the Ensembl gene ID and whatever other properties of the selected genes in which we are interested. Figure 4.3 shows a subset of the resulting output.

Identifying all SNPs located on a set of genes with specified gene IDs is also easy with BioMart. We again select MartView's gene focus, this time selecting Ensembl gene IDs and SNP IDs as output attributes. As in the first example, we can filter our output using a list of Ensembl gene IDs. If desired, we can further limit our output using the SNP filter option, so as to only select genes that have coding or non-synonymous SNPs.

For our final BioMart example, we will choose a slightly more complicated, but realistic, query. In this case, let us assume that we want to identify mouse homologs of human genes located on the X chromosome, and that we are particularly interested in identifying which homologs are on the mouse X chromosome and which are not. For this example, let us further assume that we want to restrict our search to genes that have been validated with manual curation.

Because we want manually curated genes, this time we begin our query by selecting the human VEGA gene set as our data focus. We then filter our query so it is restricted to genes on the (human) X chromosome. Next, we look for mouse homologs by clicking

Human Ens. Gene ID	Mouse Ens Gene ID	Mouse chromosome
ENSG00000205070	ENSMUSG00000069038	Y
ENSG00000102144	ENSMUSG00000066632	12
ENSG00000174028	ENSMUSG00000029672	6
ENSG00000123130	ENSMUSG00000047565	15

Figure 4.4 Portion of output from two-dataset BioMart example for identifying mouse homologs of human X chromosome genes that are not on the mouse X chromosome.

the second "Dataset" button (see the lower left part of Figure 4.2a) and selecting the mouse genome from the second-dataset pull-down menu. Finally, we select whatever human and mouse attributes we want to display, including the human gene IDs and the IDs and chromosomal locations of the mouse homologs identified. A portion of the resulting display is shown in Figure 4.4.

4.4 The UCSC Table Browser and Gene Sorter

UCSC provides two tools for interactive batch querying. For simple, gene-based queries there is the Gene Sorter, which has an interface that is similar in spirit to the gene focus component of Ensembl's BioMart. We previously encountered the Gene Sorter in Chapter 2, and we actually performed some batch queries with it (without calling them batch queries) when we asked the Gene Sorter to retrieve all genes that had, say, similar expression patterns to a query gene.

For more complex interactive batch queries, including ones that are not focused specifically on genes, UCSC has the Table Browser. In contrast to using the Gene Sorter or BioMart, using the Table Browser requires one to specify which database tables contain the data of interest. Moreover, the UCSC database actually consists of multiple related databases – one for each genome assembly as well as multiple auxiliary databases. Consequently, we will need to identify both the database and the table containing the required annotations. In addition, we will need to know how the table is structured, that is, what is the layout of the data in the table.

At first glance, locating the necessary database and table and determining the table's structure may seem daunting. UCSC uses multiple databases and each of these databases may have hundreds of tables, some with millions of entries. However, in most cases, it is not necessary to have a detailed understanding of the organization of the UCSC databases to use the Table Browser.[1] Descriptions of the most commonly used UCSC database table formats are given in Appendix 3. In addition, as we will see, we can usually identify the required table from the Table Browser interface itself, or else from the "Details" page of the associated track in the Genome Browser. We can then determine the table structure from the table description in the Table Browser.

---

[1] If you are interested in learning more about the architecture of the UCSC databases, one source is the (no longer maintained) UCSC table-definition documentation at http://genome. ucsc.edu/goldenPath/gbdDescriptionsOld.html.

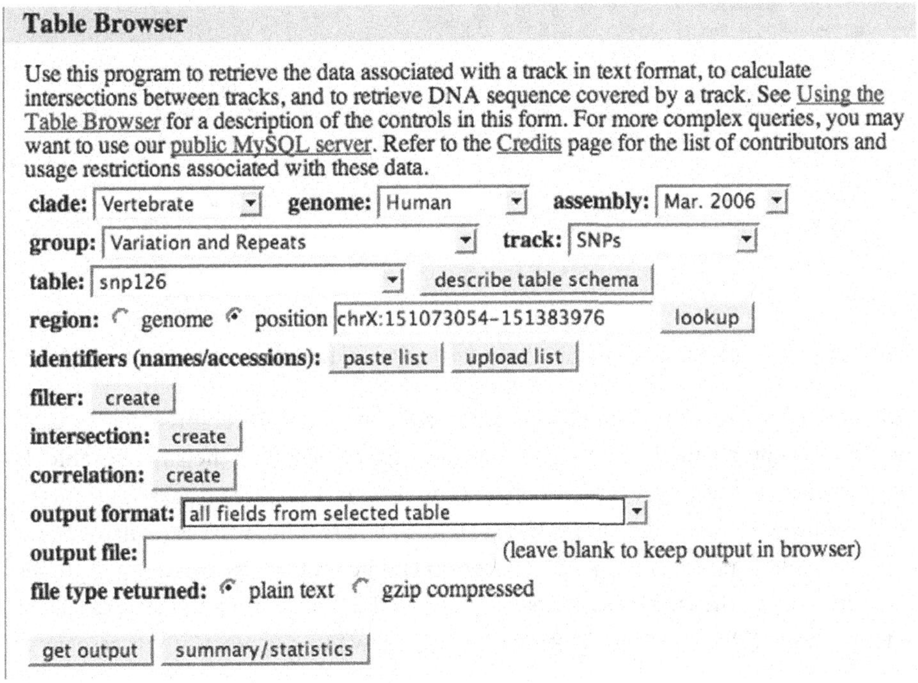

Table Browser

Use this program to retrieve the data associated with a track in text format, to calculate intersections between tracks, and to retrieve DNA sequence covered by a track. See Using the Table Browser for a description of the controls in this form. For more complex queries, you may want to use our public MySQL server. Refer to the Credits page for the list of contributors and usage restrictions associated with these data.

clade: | Vertebrate ▾ | genome: | Human ▾ | assembly: | Mar. 2006 ▾ |

group: | Variation and Repeats ▾ | track: | SNPs ▾ |

table: | snp126 ▾ | describe table schema |

region: ○ genome ● position | chrX:151073054-151383976 | lookup |

identifiers (names/accessions): paste list | upload list |

filter: create |

intersection: create |

correlation: create |

output format: | all fields from selected table ▾ |

output file: | | (leave blank to keep output in browser)

file type returned: ● plain text ○ gzip compressed

get output | summary/statistics |

Figure 4.5 Screenshot of Table Browser interface.

4.4.1 Using the Table Browser

As with BioMart, using the Table Browser consists of essentially three steps: choosing a genome and a focus, filtering the data, and selecting the desired output data and format. However, selecting a focus is more involved with the Table Browser because the "focus" can be any table in the entire database. Typically, one begins at the Table Browser input page at http://genome.ucsc.edu/cgi bin/hgTables. One can also reach this page from within the UCSC Genome Browser by clicking on the "Table" button at the top of the browser display. A screenshot of the Table Browser interface is shown in Figure 4.5.

4.4.1.1 Finding the right table

Often the appropriate table to query can be determined directly from the "track group" and "track" used in the associated genome browser query. For example, if one wants to check which genomic locations overlap known SNPs, one might start by entering "SNPs" into the "track" selector on the Table Browser input form, after having entered "Variations and Repeats" under the "group" selector. Once this is done, the table "snp126" is automatically selected (assuming that we are using the March 2006 human genome database, as illustrated in Figure 4.5).

Sometimes, there may still be more than one table that can be selected at this stage. In this case, one may need to look at the table descriptions of the possible tables (by clicking the "describe table schema" button shown in Figure 4.5) to determine

Schema for SNPs - Simple Nucleotide Polymorphisms (dbSNP build 126)

Database: hg18 **Primary Table:** snp126 **Row Count:** 12,351,941
Description: Polymorphism data from dbSnp database or genotyping arrays

field	example	SQL type	description
bin	585	smallint(5) unsigned	Indexing field to speed chromosome range queries.
chrom	chr10_random	varchar(15)	Reference sequence chromosome or scaffold
chromStart	490	int(10) unsigned	Start position in chrom
chromEnd	491	int(10) unsigned	End position in chrom
name	rs28694205	varchar(15)	Reference SNP identifier or Affy SNP name
score	0	smallint(5) unsigned	Not used
strand	+	enum('?','+','-')	Which DNA strand contains the observed alleles
refNCBI	A	blob	Reference genomic from dbSNP
refUCSC	A	blob	Reference genomic from nib lookup
observed	A/T	varchar(255)	The sequences of the observed alleles from rs-fasta files
molType	genomic	enum('unknown','genomic','cDNA')	Sample type from exemplar ss
class	single	enum('unknown','single','in-del','het','microsatellite','named','no var','mixed','mnp','insertion','deletion')	The class of variant (simple, insertion, deletion, range, etc.)
valid	unknown	set('unknown','by-cluster','by-frequency','by-submitter','by-2hit-2allele','by-hapmap')	The validation status of the SNP
avHet	0	float	The average heterozygosity from all observations
avHetSE	0	float	The Standard Error for the average heterozygosity
func	intron	set('unknown','locus','coding','coding-synon','coding-nonsynon','untranslated','intron','splice-site','cds-reference')	The functional category of the SNP (coding-synon, coding-nonsynon, intron, etc.)
locType	exact	enum('unknown','range','exact','between','rangeInsertion','rangeSubstitution','rangeDeletion')	How the variant affects the reference sequence
weight	1	int(10) unsigned	The quality of the alignment

Figure 4.6 Portion of a Table Description screenshot for the snp126 SNP table in the hg18 database.

which table contains the annotations of interest. For example, Figure 4.6 illustrates the information available from the Table Description page for the snp126 table. In contrast, for some database tables there may not be any associated tracks or groups on the Table Browser input page at all. In this case, there are other ways of identifying the required table. First, if the needed annotation can be found in the Genome Browser, one can examine the Details page associated with that annotation. In many cases, the Details page will have a direct link (labelled by "View table schema") to its associated table.

Occasionally, even using the Details pages will not indicate which table contains one's required data. In this case, you may need to select the "All tables" option under the Table Browser's group selection option and then scan through the entire (long) list of table names to identify the likely table containing the required fields. You can then verify whether this is indeed the table you need by clicking on the "Table description" button.

4.4.1.2 Filtering, intersecting, and correlating tables

Once one has identified one's table of interest, one usually wants to limit the number of records that are extracted from the table. A simple example occurs when the required part of the table can be specified as a single continuous region of a chromosome. In this case, one simply inserts the chromosome name and start and end positions into the genomic input position in the Table Browser input window.

The Table Browser input window also provides access to several other tools for limiting table output (see Figure 4.5). The Filter tool limits the records to be retrieved on the basis of data table values. Selecting the Filter tool presents a menu of data filtering options similar to those presented in BioMart, with the primary difference that in this case, filtering is on the values of specific fields in the database table rather than on the attributes of a data mart object. Figure 4.7 shows the Filter tool page for the snp126 table. After selecting our data filters, it is often a good idea to select the "Summary/statistics" option to determine how many records pass the filter to confirm that the number of output records seems appropriate (i.e., neither zero nor too large).

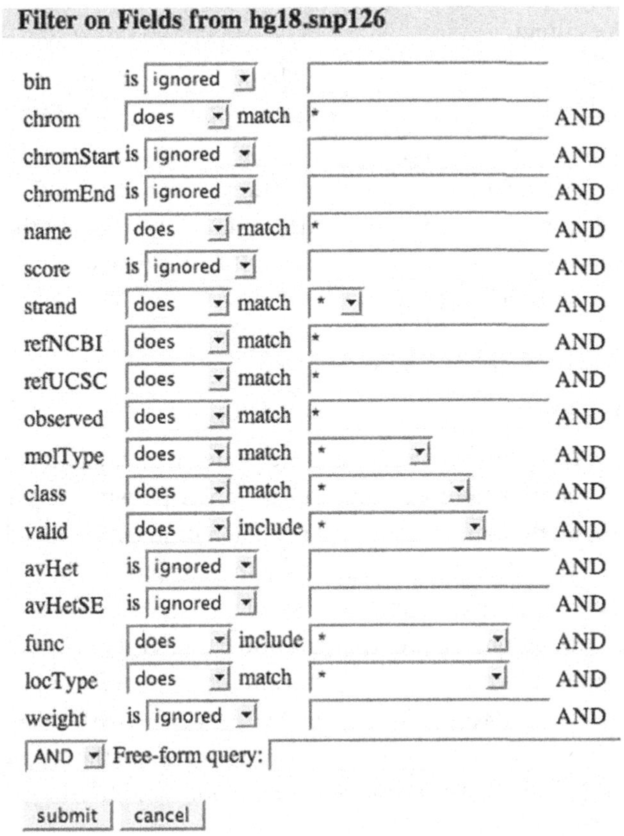

Filter on Fields from hg18.snp126

bin	is ignored		
chrom	does match	*	AND
chromStart	is ignored		AND
chromEnd	is ignored		AND
name	does match	*	AND
score	is ignored		AND
strand	does match	*	AND
refNCBI	does match	*	AND
refUCSC	does match	*	AND
observed	does match	*	AND
molType	does match	*	AND
class	does match	*	AND
valid	does include	*	AND
avHet	is ignored		AND
avHetSE	is ignored		AND
func	does include	*	AND
locType	does match	*	AND
weight	is ignored		AND

AND   Free-form query:

submit   cancel

Figure 4.7 Filter page for the snp126 SNP table. Note that one can filter on any field in the snp126 table (see Figure 4.6) as well as create a general SQL query using the "Free-form query" input box.

The Intersection tool provides a different method of restricting table output, namely by only extracting records that overlap regions from a second table. The Intersection tool can be quite useful in restricting the returned data to specific regions of interest in the genome. However, the Table Browser's Intersection tool also has some significant limitations (for example, only a limited number of output formats are allowed from Table Browser intersections). Consequently, for complex genome-intersection applications, it may be preferable to use Galaxy's "genomic interval" intersection tool, which we describe in the next chapter.

Finally, the Table Browser Correlation tool enables one to do nucleotide-by-nucleotide correlations of the values of two tables over a set of regions. For example, one can use the Correlation tool to investigate whether GENCODE exon locations are more highly correlated (i.e., have greater overlap) with RefSeq exons or with VEGA exons. Or one could look at whether there are genomic regions where conservation among genomes of multiple species (as measured by phastCons) is correlated with local GC%.

4.4.1.3 Specifying table output

Once the required records have been determined by specifying the input tables and possibly performing filtering or table intersection, the final step in using the Table Browser is to specify the record output format. The Table Browser offers a wide range of output format options, including obtaining all the fields or selected fields of the raw table data; BED files of genes, transcripts, exons, or introns; output in the form of custom tracks or hyperlinks to the genome browser; FASTA sequence data and MAF multiple alignment data, among others.

4.4.2 *Applications of the Table Browser and Gene Sorter*

4.4.2.1 Characterizing genes

Let us now apply the Table Browser and the Gene Sorter to some batch querying examples. For our first query, we look for human gene transcripts with a large number of exons, for example, more than twenty. With the Table Browser, performing such a search is easy. We first select a gene table, say, the ensGene table of Ensembl gene transcripts. We can identify the name of this table by selecting "Genes and gene predictions" from the track group pull-down menu and Ensembl genes from the track pull-down menu. We then select the genomic region in which we are interested. Next, we click the "Filter" button and set the filter to include only genes with an exonCount greater than 20. Finally, we choose "Selected fields from table" from the output format pull-down menu and specify which table fields we want, for example, the gene name and the exon count.

For our second example, we will revisit the conversion of Ensembl IDs to RefSeq IDs. Because this gene-based query is more easily carried out in the Gene Sorter, we select the Gene Sorter tool by clicking the "Gene Sorter" button at the top of the screen. Once we are at the Gene Sorter input page, we can first upload the Ensembl IDs by selecting the Gene Sorter Filter tool, and then configure the output to include both Ensembl IDs and RefSeq IDs. Clicking the Gene Sorter's "Go!" button results in a listing of Ensembl and RefSeq IDs.

4.4.2.2 Characterizing genomic regions

For our next examples, let us consider queries to characterize an arbitrary set of genomic regions. For example, we might have a BED list of novel, candidate disease-polymorphism locations and want to determine whether the polymorphisms include any known SNPs. To do this, we would first make a custom track out of the BED list and upload the custom track to the Table Browser. (The test data used in this example are located in the file roi.hg18.snp.track, which can be downloaded from the publisher's web site for the book.)

Note that the coordinates in our BED file must be from the same assembly that we are using with the Table Browser. If our coordinates are from a different assembly, we will need to convert them to the genome assembly we are using, or else change the

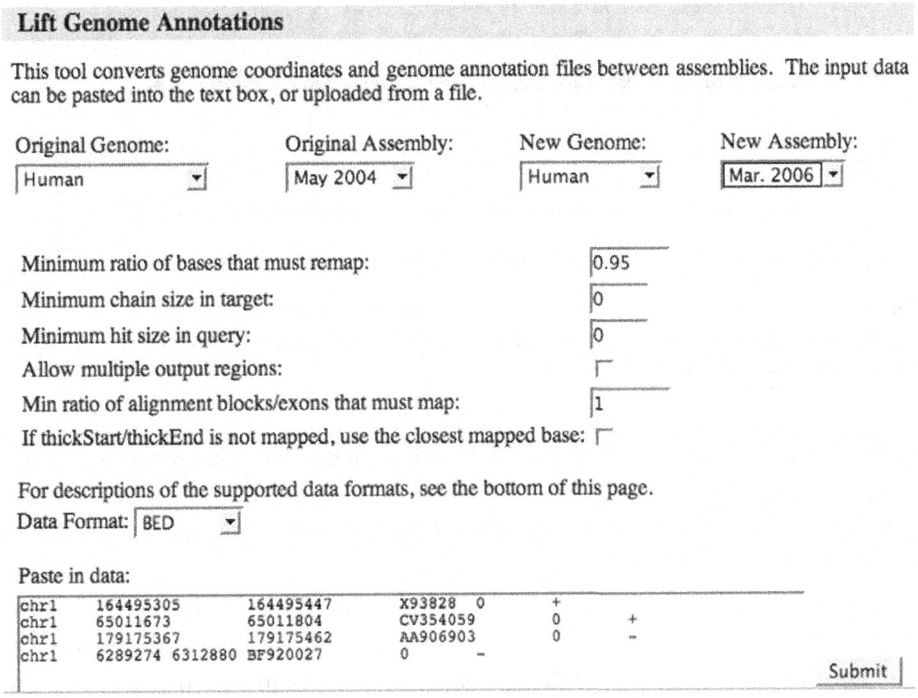

Lift Genome Annotations

This tool converts genome coordinates and genome annotation files between assemblies. The input data can be pasted into the text box, or uploaded from a file.

Original Genome:	Original Assembly:	New Genome:	New Assembly:
Human ▾	May 2004 ▾	Human ▾	Mar. 2006 ▾

Minimum ratio of bases that must remap: 0.95

Minimum chain size in target: 0

Minimum hit size in query: 0

Allow multiple output regions: ☐

Min ratio of alignment blocks/exons that must map: 1

If thickStart/thickEnd is not mapped, use the closest mapped base: ☐

For descriptions of the supported data formats, see the bottom of this page.

Data Format: BED ▾

Paste in data:

```
chr1    164495305       164495447       X93828  0       +
chr1    65011673        65011804        CV354059        0       +
chr1    179175367       179175462       AA906903        0       -
chr1    6289274 6312880 BF920027        0       -
```

Submit

Figure 4.8 Partial screenshot of UCSC coordinate LiftOver tool interface.

```
chr3    53563273        53563274        rs10707946      0       +
chr4    158503564       158503565       rs17850676      0       +
chr5    4170006         4170007         rs6894646       0       +
chr8    37060699        37060700        rs4739466       0       +
chr9    131899567       131899568       rs11794486      0       +
```

Figure 4.9 Result of a Table Browser search for all SNPs overlapping a list of regions of interest.

selected assembly in the Table Browser to match our BED file coordinates. To convert BED coordinates, we can use the UCSC LiftOver coordinate conversion tool, which is available via the "Utilities" button on the UCSC Browser home page. Figure 4.8 shows the data input page used by the LiftOver tool.

Having created our custom track, we next select the SNP table in the Table Browser and use the table-intersection tools to intersect the SNP table with the custom track to identify any previously known SNPs that overlap the candidate polymorphism dataset. The result of a BED format output is shown in Figure 4.9.

As another example, we might have a list of highly expressed human genome regions from a tiling-array experiment. We might want to know which transcripts overlap any known ESTs. As in the previous example, we begin by uploading our list of interesting genome regions as a custom track to the Table Browser. This time, however, we would intersect our custom track with either the intronEst table (if we are interested in human spliced ESTs) or the xeno_est table, if we are looking for ESTs

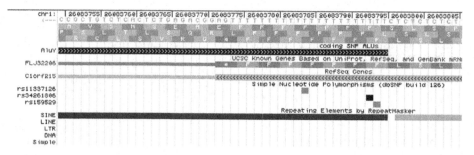

Figure 4.10 Example of an Alu that overlaps a coding region, as identified by a Table Browser batch query. The figure includes a custom track, called coding SNP Alus, of all the Alus that were extracted by the Table Browser query.

from other species. The result would be the subset of our list of regions that overlap at least one EST.

4.4.2.3 Alus and coding exons

As a final Table Browser example, we consider a hypothesis in protein evolution, namely, that repetitive sequences, such as Alus, play a role in the evolution of protein coding sequences. Alu elements (Alus) are short repetitive sequences that are common in primate genomes. In the presence of certain enzymes, known as transposases, Alus replicate and move to other locations within the genome. It has been hypothesized that such Alu duplication and movement can contribute to evolutionary change, including the creation of new exons. This might occur if an Alu were inserted into an existing intron, and subsequent mutations of the Alu created additional splice junctions within the intron (Lev-Maor et al., 2003).

A possible marker of such evolutionary activity would be the presence of Alus overlapping coding sequences, especially near intron-exon boundaries. To search for such evidence, we begin by creating a custom track including all coding exons. We create this track in the Table Browser by selecting all genes from, for example, the knownGene table and selecting the "One BED per exon" option, restricted to coding exons, as the output format. Then we select the repeat table in the Table Browser. This table (e.g., table rmsk in database hg18) can be identified by choosing "Variations and repeats" from the track group pull-down menu and "Repeatmasker" from the track pull-down menu. We next need to filter the rmsk table with the constraint repFamily matches Alu* (where "*" is the matching "wild card" operator). Finally, by intersecting the filtered repeat table with our custom track of coding exons, we can create a list or a custom track of Alu sequences that overlap coding exons, one of which is shown in Figure 4.10.

Using the Table Browser interface is more complex than using BioMart or the Gene Sorter. However, in return, we have seen some queries that were relatively easily addressed with the Table Browser but were difficult or impossible to answer with other tools. For example, one cannot currently query multiple genomic locations with BioMart (unless each of these regions has an associated gene ID) so that we could not look for SNPs or ESTs overlapping custom "regions of interest" as we did with the Table Browser. In addition, we would not be able to find all genes that have repeat sequences overlapping coding exons because repeat-sequence annotations are not available within BioMart. (Note that repeat-sequence annotations are stored in the Ensembl database; they can be retrieved one at a time using the Ensembl Genome Browser, or in batch mode, using SQL or the Ensembl API. However, because repeat sequences are not currently included in Ensembl's BioMart databases, they cannot be accessed using MartView.)

Finding genes with more than twenty exons is also not currently feasible with BioMart. On the other hand, finding mouse homologs of human genes on chromosome X, which we were able to do easily with Biomart, would be difficult or impossible to do with the Table Browser. Perhaps the most useful lesson to learn here is that if the sort of query you need to perform is difficult to do with the batch querying tool provided by one genome database, it may be worthwhile to check whether the data you need might be easier to obtain with the tools of another genome database.

4.5 Direct SQL querying

Essentially, all modern relational databases, including the genome databases, can be accessed directly via the SQL database language. In most cases, SQL querying is most conveniently accomplished indirectly, via a web-based user interface such as the Table Browser, Gene Sorter, or BioMart, or with library routines provided within an API for a conventional computer language such as Perl or C.

However, it is occasionally useful to directly query a genome database with SQL. This might be the case if a programmed or web-based batch query yields unexpected results or produces an error message. Direct SQL querying is also useful after one has just installed a mirror database and wants to confirm that everything has been installed and configured properly. Additionally, direct SQL querying is useful if one wants to educate oneself regarding the table structure of the database. (This is especially useful with Ensembl because Ensembl's batch-querying tool, BioMart, does not provide direct querying of the Ensembl database tables.) Finally, for the user who happens to be skilled and experienced with SQL, direct SQL database querying may simply be easier or faster than using a web interface or an API.

In any case, to execute SQL queries, you will need to install a local copy of the MySQL client program (freely available from http://www.mysql.com) and to connect to a "mirror" of the UCSC or Ensembl genome database. This database copy can be either a publicly available mirror or a privately installed one. Here, we restrict ourselves to the simpler approach, which is using a public mirror. If you do use a public mirror, you need to confirm that there are no firewall or other security systems in place at your computer location that block outgoing ports, thereby preventing you from accessing a remote database server.

As an initial example of direct SQL querying, we might just want to determine which databases are installed in the public mirror site. For the Ensembl site, we can execute the following command, from the Unix command line[2]:

```
$ mysql --user=anonymous --host=ensembldb.ensembl.org -A \
   -e "show databases;"
```

Once we have determined the names of the constituent databases, we can determine the tables in any one of them with commands like

```
$ mysql --user=anonymous --host=ensembldb.ensembl.org -A \
   -e "show tables;" homo_sapiens_core_41_36c
```

The commands with the UCSC mirror are entirely analogous except that the "user" and "host" would be "genome" and "genome-mysql.cse.ucsc.edu," respectively. It is also possible to connect directly to Ensembl's BioMart database at "martdb.ensembl. org," with "user" set to "anonymous," and "port" set to "3316."

With the UCSC system, it is possible and often easier to obtain database information via the Table Browser rather than with SQL. Strictly speaking, the data obtained via Table Browser and via direct SQL querying are not identical because the Table Browser queries the UCSC Genome Browser database itself, whereas SQL queries UCSC's public mirror of the genome browser database. However, because these two databases are resynchronized daily, any differences between them are likely to be minor. A more significant difference between Table Browser and SQL querying is that with SQL, one can only access the UCSC relational databases, whereas the Table Browser can also access some of UCSC's auxiliary data files, which contain sequence and multiple alignment data. In contrast, with Ensembl all local (i.e., non-DAS) data is stored in the relational databases and, hence, can be accessed via SQL. Ensembl data that has been transferred to Ensembl BioMart (which includes some Ensembl DAS data) can be accessed via MartView or via SQL querying of the Ensembl's martdb database.

Chapter summary

- Ensembl BioMart and the UCSC's Table Browser and Gene Sorter enable batch genome-database querying without the need of programming or using SQL.
- Using BioMart, the Table Browser, or the Gene Sorter involves essentially three steps:
  1. Choosing a species, assembly, and either a database table (Table Browser), a query-focus (BioMart), or a gene set (Gene Sorter)

---

[2] Note that we will indicate the Unix command-line prompt with the dollar-sign symbol "$" (i.e., the $ is not part of the command that you type). Also note the required backslash "\" which immediately precedes the carriage return if a command needs to be extended beyond a single line.

2. Filtering or limiting the records one wants to retrieve

3. Selecting the output fields or other output format that one wants to use

- Because Ensembl uses a data mart implementation, no knowledge of the database table structure is required for using BioMart.
- Using the Table Browser does require knowledge of the UCSC database table schema. However, as a result, nearly all data available in the UCSC databases can be accessed via the Table Browser.
- Both the Ensembl and UCSC genome databases also provide public mirrors for direct SQL access.

Exercises

1. Using BioMart, identify the human Ensembl genes that have the GO annotation of "leukocyte activation." Can you determine the human VEGA genes with GO annotation of "leukocyte activation" using BioMart? Can you find the Ensembl and VEGA genes with GO "leukocyte activation" annotation with the Gene Sorter, Table Browser, or other UCSC tools?

2. Create a list of Alu repeats that overlap coding exons in humans using the UCSC Table Browser as described in the text. How many of these regions also overlap known SNPs?

3. Identify all single-exon Ensembl genes in the "ENCODE" regions.

4. Identify genes on the human X chromosome that are highly expressed in the brain. You may want to use the Gene Sorter, the Table Browser, BioMart, or direct SQL querying.

5. Using direct SQL querying, determine how many databases are installed in the Ensembl public mirror. How many tables are in the Ensembl database homo_sapiens_core_42_36d? How many databases are installed in the UCSC public mirror? How many tables are in UCSC database hg18? (Note that you will need to have the MySQL client software installed for this exercise.)

5

Interactive Batch Post-Processing with Galaxy

In Chapter 4, we addressed various biological questions requiring sequence and annotation data from multiple genomic regions with batch querying techniques. In the present chapter, we move from simple batch querying to batch querying with batch post-processing. We will see how the range of realistic questions that we can answer is greatly expanded. We will begin with a brief overview of batch post-processing, including a sampling of the kinds of queries that can be addressed with batch post-processing. This is followed by a detailed description of Galaxy, a web-based interactive tool that enables biologists with no programming experience to perform quite sophisticated bioinformatics post-processing analyses.

5.1 Batch post-processing overview

We have already seen how BioMart, the Table Browser, the Gene Sorter, and SQL can be used to extract sets of related data from the Ensembl and UCSC genome databases, and in the process answer many types of biological queries. However, a much wider range of practical questions can be answered if we are able to perform additional manipulations on the retrieved data. We will refer to such additional analysis of data retrieved from a genome database as *batch post-processing*.

The types of data manipulations that may be required can be as basic as performing simple arithmetic computations or as complex as performing multiple data analyses with bioinformatics programs, such as BLAST, ClustalW, PAML, or one of the EMBOSS sequence analysis programs. The kinds of applications we will be able to address with these new tools – and will describe in this and the following chapters – range from the identification of gene transcripts that are candidates for nonsense mediated decay (NMD) and the determination of the length distributions of introns of genes with varying functional annotations, to the detection of candidate RNA-editing sites and the detection of highly conserved nucleotides in multiple genomic sequence alignments.

Batch post-processing can be performed either by writing computer programs or interactively, in which case no programming background is required. In addition,

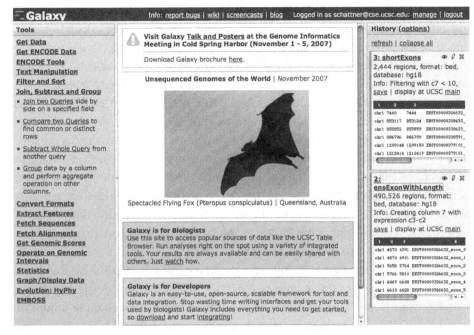

Figure 5.1 Screenshot of the main Galaxy data entry page. The display is partitioned vertically into three sections. The left section lists the available Galaxy tools. The middle section shows the details of the currently selected tool. The right section shows a summary of recently created datasets.

hybrid approaches exist in which it is possible to use an interactive interface to generate a computer script without actually needing to write the script itself. We will present examples of interactive batch post-processing using the Galaxy system in this chapter. We will describe programmed batch post-processing in the following chapters.

5.2 Introduction to Galaxy

To address the need for genomic data post-processing by nonprogrammers, the Center for Comparative Genomics and Bioinformatics at Pennsylvania State University has developed the "Galaxy" web site, located at http://g2.bx.psu.edu. Galaxy (Blankenberg et al., 2007) provides a uniform interface to sequence and data manipulation tools. A screenshot of the main data entry page is shown in Figure 5.1.

Galaxy is not the only integrated toolset that has been developed for biological sequence and annotation analysis. Others include NCBI's Genome Workbench (http://www.ncbi.nlm.nih.gov/projects/gbench), PlatCom (Choi et al., 2005), GenePattern (Reich et al., 2006), SRS (Zdobnov et al., 2002), the SDSC Molecular Biology Workbench (Subramaniam, 1998), and MiGenAS (Rampp et al., 2006). However, in contrast to other tools, Galaxy is specifically designed to handle genome-size data sets and to interface directly with the major genome databases.

Using Galaxy involves three steps: acquiring the data sets; filtering, manipulating, and reformatting the data sets; and carrying out post-processing calculations. As shown in the screenshot in Figure 5.1, the supported data manipulation tools are listed on the left side of the Galaxy page. Galaxy currently supports data import from multiple sources, including the UCSC genome databases (via Galaxy's Table Browser interface), from BioMart, from the Microbial Genome Project at NCBI, and from the NHGRI's ENCODE Project. In addition, almost any form of custom user data that can be converted to tabular form (e.g., a spreadsheet) can be uploaded to the Galaxy site.

In Galaxy, all nonsequence data are stored and manipulated in the form of tables, which are called "queries" in Galaxy. Because data manipulation in Galaxy is performed largely with tables, working with Galaxy may be somewhat more reminiscent of using the UCSC Table Browser than of using BioMart. That said, Galaxy can be used equally well with data from BioMart because once data has been extracted from BioMart, the data is also in tabular format.

Once input data sets have been uploaded to Galaxy, the next step is to combine and filter the data sets into a single processed dataset. Such input processing may range from filtering on any data-table field and performing table intersections or unions to complex multitable and genomic-interval manipulations. Galaxy's table editing and filtering tools provide essentially all of the data-filtering capability available in the Table Browser, Gene Sorter, or BioMart. However, Galaxy also contains data filtering and editing features that are significantly more powerful or easier to use than those in other tools.

Finally, after we have edited and filtered our desired data tables and sequence sets, we are ready to apply any of Galaxy's post-processing tools to each data record in the dataset. The post-processing may be as simple as performing column arithmetic, as in subtracting the value in a "transcript start" column from the value in a "transcript end" column to create a "transcript length" column. Or the manipulation on each record may be quite complex and may include applying functions from any of several standard bioinformatics data-processing packages that are incorporated in Galaxy, including the EMBOSS sequence manipulation package (Rice et al., 2000) and the HyPhy phylogenetics program suite (Pond et al., 2005).

5.3 Galaxy features

5.3.1 Table and interval manipulation

Galaxy's table and interval manipulation tools are among its most powerful features. However, some of these features may feel somewhat unintuitive initially, especially if you are not familiar with table manipulation in the relational database language SQL. Nevertheless, they are not difficult to master, and it is well worth the effort involved in learning them.

Two of Galaxy's most powerful table manipulation tools are the "join" and "compare" tools, with which one can extract related data from multiple tables. With "join,"

a)

Query1:

```
chr1 10 20 geneA
chr1 50 80 geneB
chr5 10 40 geneL
```

Query2:

```
geneA tumor-supressor
geneB Foxp2
geneC Gnas1
geneE INK4a
```

b)

Joining the 4th column of Query1 with the
1st column of Query2 will yield:

```
geneA chr1 10 20 tumor-supressor
geneB chr1 50 80 Foxp2
```

c)

Finding lines of the First query whose 4th column
matches the 1st column of the Second query:

```
chr1 10 20 geneA
chr1 50 80 geneB
```

Figure 5.2 Galaxy join and compare tools: (a) "join" and "compare" input; (b) "join" result; (c) "compare" result.

you combine records from two tables that are related by having identical values in some column. An example is shown in Figure 5.2. In this example, we have a list of "interesting" genes in one table and a list of gene descriptions (which are not necessarily the same genes) in a second table. Let us say that we want to identify the interesting genes that have descriptions, and to add those descriptions to the interesting-gene table. We accomplish this by using the "join" command (under "Join, subtract, and group" in the tool menu on the left side of the Galaxy screen). Specifically, we join on the common gene field in column 4 of table 1 and column 1 of table 2. The result, shown in Figure 5.2b, contains a record for each "interesting" gene that has a description, along with its description. In contrast, we would use the "compare" tool if we simply wanted to extract the records of table 1 for which a description exists (or the records for which a description does not exist) without actually modifying table 1 by adding the description. The result of comparing tables 1 and 2, using matching rows, is shown in Figure 5.2c.

Galaxy's "genomic interval tools" are similar in spirit to its table manipulation tools. However, in this case table records are compared on the basis of whether the

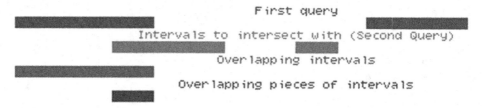

Figure 5.3 Example illustrating the two different types of interval intersection supported by Galaxy: "overlapping intervals" and "overlapping pieces of intervals."

genomic intervals they include overlap or are in close proximity. One useful genomic interval tool is Galaxy's Intersection tool, which is similar to, but more powerful than, the intersection tool in the Table Browser. This is because Galaxy's Intersection tool provides two types of genomic region intersections: "overlapping intervals" and "overlapping pieces of intervals" (see Figure 5.3). In contrast, the Table Browser Intersection tool only offers "overlapping intervals" table intersection, which is often not the type of interval intersection that one needs. A second advantage of using Galaxy's Intersection tool is that one can obtain the intersected table in multiple output formats, including any or all of the table's fields. In contrast, the Table Browser Intersection tool provides only a limited range of output formats (e.g., BED) for intersected tables.

5.3.2 Format conversion, attributes, and history files

Another useful feature of Galaxy is its data conversion capability. This is particularly useful when preparing data for input to standard bioinformatics applications programs, which often do not recognize the data formats produced by the genome databases. For example, aligned sequences are retrieved from the UCSC Table Browser in MAF or AXT formats; however, these are not suitable formats for most sequence and alignment analysis tools. To address this, Galaxy's conversion tool can convert AXT or MAF files to FASTA format so they can be used with standard sequence post-processing programs.

Each Galaxy dataset also has a set of "attributes" that describe the format of the data in the dataset and are used by Galaxy to determine what manipulations can be performed on the data. Typical attributes may be "table," "BED," or "interval." Tables with attributes such as "BED" or "interval" also need to have attributes indicating which columns contain the chromosome, coordinates, and (optionally) strand as well as information indicating from which genome assembly the coordinates derive.

In most cases, Galaxy automatically determines genomic-interval attributes when you load the data into Galaxy. However, in some cases – for example, if you load custom data from you own computer – you may need to explicitly add this information manually by using the Edit Attributes tool, which is indicated for each dataset via the pencil icon adjacent to the dataset name (see Figure 5.1). In addition, after you have performed some table manipulations (e.g., joins), interval attribute annotation may be lost, and you will need to re-enter them before genomic interval operations can be performed on the transformed table.

Another useful Galaxy feature is its History mechanism. With History, you can save all of your intermediate results and thereby more easily repeat the exact steps that you performed to reproduce some result. Moreover, one can create multiple histories – each with its own name – for different tasks, thereby keeping the datasets that were used for each task separate.

Galaxy has many other useful features that we have not described. We will see some of them in the examples in the next section. In addition, the reader is referred to the documentation at the Galaxy web site and, in particular, to the excellent series of videocasts (see http://g2.trac.bx.psu.edu/wiki/ScreenCasts) that describe Galaxy's features in detail.

5.4 Galaxy examples

We are now ready for some examples to see Galaxy at work. We will start with relatively easy examples, requiring only a few steps within Galaxy. Then we will move to more complex examples to illustrate the power of Galaxy at addressing realistic bioinformatics questions.

5.4.1 Finding short exons and short introns

For our first example, we will consider the question of how short exons can be. This question can be answered in Galaxy with just a few steps. First, we need to acquire a table of exon coordinates. We can accomplish this by selecting "Get Data" and then selecting, for example, the UCSC main Table Browser from the Galaxy "Get Data" menu. This will bring up a screen including the now-familiar Table Browser input page within the Galaxy window. We then can choose a dataset in the usual manner with the Table Browser, say, the ensGene table in hg18, and select an output format of BED with one record per exon. The output screen will now display an option for sending the result to Galaxy, which we will select.

Next, we need to create an additional column in our table containing the exon length. We create this column by using the Text Manipulation–Compute Expression tool to subtract the exon start value from the exon end value for each table record. Finally, we extract the short exons by filtering the table with the constraint that the value in the exon-length column is less than, say, ten (using the Filter tool under "Filter and sort"). By modifying our procedure slightly, we can also look for short introns (see Exercise 5.2).

As an aside, we note that many of the short "exons" and "introns," which we find in this manner, are likely to be artifacts caused by such factors as sequence polymorphisms and sequencing and assembly errors. Figures 5.4 and 5.5 illustrate the kind of situations you may find in your list of "short" exons and introns. Figure 5.4 shows three sets of mRNAs with very short gaps that, at first glance, look

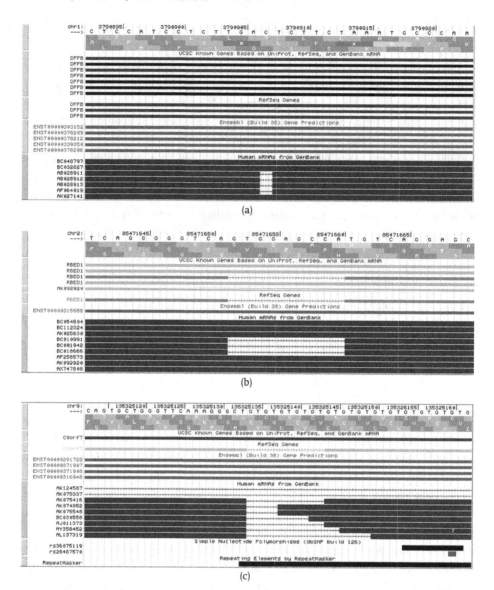

Figure 5.4 Browser display of three mRNA-genome alignments where short gaps probably result from indel polymorphisms. See text for details.

like introns. However, these gaps almost definitely do not represent real introns. For example, the single nucleotide gap of Figure 5.4a is almost certainly the result of a deletion polymorphism. Similarly, because the acceptor splice junction sequence in Figure 5.4b is CA rather than the canonical AG, and because the gap is in the untranslated regions (UTR) of the transcript (and hence where an indel polymorphism would not cause a frameshift mutation), the "intron" is more likely to be caused by a deletion polymorphism rather than by splicing. Finally, the short gaps overlapping

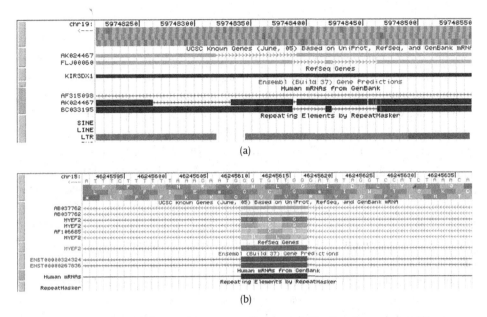

Figure 5.5 Browser display of two mRNA-genome alignments indicating the presence of short insertions. (a) Several short insertions probably resulting from indel-polymorphism or sequencing error. (b) 8 nt insertion with canonical splice junctions, which is probably an actual exon.

a known repetitive region in Figure 5.4c are also most likely the result of deletion polymorphisms. Note that in several – but not all – cases, the gaps shown in the mRNA tracks in Figure 5.4 have been removed in the UCSC, RefSeq, and Ensembl gene tracks.[1]

Figure 5.5 shows examples of short insertions and gaps in transcript alignments and gene annotations. The mRNA tracks of Figure 5.5a show a single nucleotide insertion and several short gaps that almost certainly result from an insertion polymorphism or sequencing error related to the overlapping repetitive element. Note that the single nt insertion has been removed from the RefSeq track but not out of one of the tracks in the UCSC Known Genes display. On the other hand, the 8 nt exon in Figure 5.5b is most likely real because the exon is flanked by the canonical donor (acceptor) splice-sequences GT (AG) and, in fact, the putative exon is preserved in the RefSeq, Ensembl, and UCSC gene tracks.

In contrast, if we want to find genes with short exons or introns that we can be confident are not artifacts, we could repeat this procedure using a manually curated gene set such as the CCDS or VEGA genes. But we will leave that as an exercise.

---

[1] The extent to which annotation tracks in the various browsers include such anomalous short introns and exons depends, in part, on the transcript alignment methods they use. For more information on the different alignment algorithms used by the browsers, see Appendix 4 and references therein.

5.4.2 Characterizing disease genes

For our second example, we will determine RefSeq or GenBank accession IDs, Swiss-Prot IDs, as well as gene structures, for all genes that have a specific SwissProt disease annotation, namely, myasthenia gravis.

With Galaxy, we can create the required dataset in a few steps. In this example, the most difficult step is the first step, which actually is not related to Galaxy at all. Rather, the difficulty involves identifying the UCSC database tables containing the data we need. We will start by going to the UCSC Human Genome Browser (we will use build hg17 for this example) and entering "myasthenia gravis" in the "Position or search term" input field. This results in several UCSC "Known Gene" hits. Next, we need to identify the UCSC tables that contain gene structure and disease annotations. Because our initial query resulted in UCSC Known Genes, we suspect that the knownGene table will have at least some of the data we need, and looking at the knownGene table in the Table Browser shows that it does contain SwissProt and GenBank/RefSeq IDs as well as gene structure information.

But examining the knownGene table description shows that it does not contain any disease annotations. Instead, below the table description, one finds a list of more than fifty other tables that are linked to the knownGene table. It is likely that one of these tables will contain the desired disease data – though this is not necessarily true because the disease table might only be linked to the knownGene table indirectly via yet a third table. In the present example, the list of tables linked to the knownGene table includes one with the promising table name of proteome.spDisease, and clicking on the link to the proteome.spDisease table shows that indeed this is the table where disease annotations are stored.

There is one more minor hurdle to overcome before we can locate the proteome.spDisease table. If we return to the main Table Browser input page, we will not find the proteome.spDisease table even if we select "All tables" from the group pull-down menu. The reason is that the table list in the Table Browser menu only includes the tables in the currently selected database, which, in the present case, is hg17. To get around this, we note that when we selected the "All tables" group, the Table Browser presented us with a new pull-down menu, not previously visible, called "Database" (see Figure 5.6). Here we can select any of the auxiliary databases that are linked to hg17. From the name of the table we need, we are led to select the proteome database, where we finally can find the proteome.spDisease table.

Once we have identified the tables we need, we simply download the hg17 known-Gene and proteome.spDisease tables into Galaxy using "Get Data" and UCSC main Table Browser from the Galaxy menu. Next, using Galaxy's Select tool (under the "Filtering and sorting" menu) to find occurrences of "myasthenia gravis" in the spDisease table, we create an intermediate table of those proteins whose descriptions refer to myasthenia gravis. Then we extract the first two columns of the intermediate table (using the Cut tool under "Text manipulation") corresponding to the SwissProt

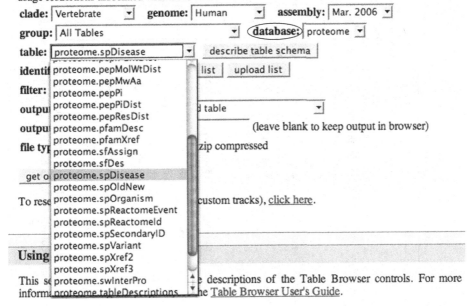

Figure 5.6 Screenshot of Table Browser Interface with "All Tables" selected from the group pull-down menu. Note that with "All Tables" selected, an additional pull-down menu, "Database," becomes available (highlighted in the figure with an ellipse), enabling one to select tables from the auxiliary UCSC databases, for example, the proteome database, which includes the SwissProt disease-annotation table.

accession numbers and display IDs. Finally, because we will need the gene structures of the genes corresponding to these proteins, we perform a "join" (under "Filter, sort, join, and compare") between our intermediate table with the knownGene table, linking them via their common SwissProt ID field. The resulting data is shown in Figure 5.7.

5.4.3 Finding regions with multiple overlapping ESTs

We recall that in the previous chapter, we were able to identify which regions from a list of genomic "regions of interest" (say, locations of high expression in a tiling micro-array experiment) overlapped at least one EST. However, we may also want to know which regions are covered by at least, say, three different ESTs. (The reason for this is that EST data is sometimes of limited quality and, consequently, it is preferable to restrict one's attention to cases with multiple EST coverage.)

To identify those regions that overlap three or more ESTs, we need to perform some data post-processing, which is why we were not able to perform this query

mRNA	Swiss	Description	chr	st	txStart	txEnd	cdsStart	cdsEnd	Ct	ExonStarts	ExonEnds
AF305908	P28329	CLAT_HUMAN	chr10	+	50492088	50543156	50494104	50543098	16	50492088,50494074, 50494553,50497776, 50498546,50500148, 50503524,50505659, 50524556,50526558, 50527559,50529935, 50533146,50533437, 50540696,50542828,	50492527,50494144, 50494654,50497968, 50498665,50500202, 50503705,50505837, 50524726,50526659, 50527688,50530058, 50533288,50533500, 50540834,50543156,
NM_000080	Q04844	ACHE_HUMAN	chr17	–	4741839	4747148	4742809	4747137	12	4741839,4743074, 4743271,4743541, 4744866,4745063, 4745598,4746005, 4746290,4746523, 4746694,4747091,	4742965,4743181, 4743458,4743656, 4744981,4745264, 4745699,4746161, 4746400,4746568, 4746837,4747148,
NM_005100	Q02952	AKA12_HUMAN	chr6	+	151653622	151770022	151653811	151766989	4	151653622,151718995, 151761959,151768975,	151653973,151719152, 151767001,151770022,
NM_020549	P28329	CLAT_HUMAN	chr10	+	50492061	50543156	50492241	50543098	15	50492061,50494553, 50497776,50498546, 50500148,50503524, 50505659,50524556, 50526558,50527559, 50529935,50533146, 50533437,50540696, 50542828,	50492527,50494654, 50497968,50498665, 50500202,50503705, 50505837,50524726, 50526659,50527688, 50530058,50533288, 50533500,50540834, 50543156,

Figure 5.7 Table of protein data from myasthenia gravis annotation example. This table was created by a "join" of the UCSC knownGene and proteome.spDisease tables and filtered for disease entries matching myasthenia gravis. Column headings have been added to the figure for clarity.

Galaxy Flowchart: Count ESTs Overlapping Regions of Interest (ROIs)

Acquire data
1. Upload ROIs from local computer to Table Browser as custom track
2. On Table Browser, intersect spliced EST table with ROI custom track, save result as "estROI"
3. Upload ROIs from local computer to Galaxy as BED file
4. Upload estROI from Table Browser to Galaxy as BED file

Initial data processing on Galaxy
5. Perform genomic-interval-join of ROIs and estROI data sets

Count ESTs for each ROI on Galaxy
6. Count # of records in joined data set with each ROI name
7. Filter to select ROIs with count > count minimum.

Figure 5.8 Flowchart of manipulations required for performing an overlapping-EST count example in Galaxy. See text for details.

in the Table Browser. However, with Galaxy, the procedure is easy. The method is outlined in Figure 5.8. We begin by uploading a BED file of our genomic regions-of-interest to UCSC as a custom track. Next, we intersect an appropriate UCSC EST table (e.g., the intronEst table for spliced same-species ESTs) with our custom track to find all overlapping ESTs, just as we did in the previous chapter. However, now we upload both the set of overlapping ESTs (from the Table Browser) as well as the original BED file of interesting regions (from our local computer) to Galaxy.

We can now use Galaxy to perform a "genomic join" between the tiling-array regions and the set of ESTs. We use Galaxy's genomic-interval operations, selecting the option to return only records that are actually joined. In this way, we have created a list of records that each include the ID of one of the tiling-array regions as well as the ID of one of the ESTs that overlaps the region. Now we can apply Galaxy's statistics Count tool to determine the number of records in the joined table corresponding to each tiling-array ID, which will equal the number of ESTs that overlap that region. Finally, by filtering the table of counts with the constraint that the count is greater than two (again using the Filter tool), we identify the regions with three or more overlapping ESTs.

It is worth noting that an alternative and superficially simpler approach would have been to just download the entire spliced EST table to Galaxy and then to perform all the manipulations, including the initial table intersection, on Galaxy. However, because the spliced EST table has approximately four million records, downloading the entire EST table to Galaxy would, at least, be very inefficient and might well not work at all.

5.4.4 Conservation at polymorphism sites

For our next example, we return to a query that we encountered in Chapter 2 in the context of genome browsing. There we were investigating a novel polymorphism and wanted to determine, among other things, whether the polymorphism site was conserved in related species and, for the ones that are not conserved, whether the new allele is found in any other species. With only one polymorphism site, this was easy to do with a genome browser.

But now let us assume we have a list of the coordinates of fifty candidate polymorphisms, and we want to carry out these tests for each of them. Using the genome browser to evaluate each site, one at a time, would be very cumbersome at best. However, performing this analysis on all fifty sites with Galaxy is not difficult. We simply upload a BED file of our polymorphism sites to Galaxy. Then we use the "Extract MAF blocks given a set of genomic intervals" tool under "Fetch alignments," using the list of polymorphism coordinates as our genomic intervals and selecting any available MAF file (such as multiz17way for build hg18) for the alignments. A small part of the result of such a query (representing just two sites) is shown in Figure 5.9. From these single-nucleotide alignments, we can tell by visual inspection in which cases the site is conserved and, if not, whether the new allele occurs in the homologous sequence in any other species. Note that the reason Galaxy is more useful here than the Table Browser is that the Table Browser Intersection tool can only retrieve the alignments for an entire aligning region and, in particular, cannot extract single-nucleotide alignments, as required in this application.

5.4.5 Screening for NMD candidates

The identification of mammalian mRNAs that are candidates for degradation via the NMD pathway is another application of batch querying and post-processing that can be performed with Galaxy. NMD is a biological process by which specific RNA molecules are degraded rather than being translated into protein (Green et al., 2003). NMD is widely used to protect the organism against aberrantly transcribed or spliced RNAs as well as against RNA viruses; in addition, there is evidence that NMD may serve other regulatory functions in the cell as well.

The precise sequence motifs that trigger the NMD response vary among species and are only partially understood. In mammals, the signal is generally the presence of a stop codon in an exon other than the final or next-to-last exon, or else the presence of a stop codon in the next-to-last exon that is more than fifty nucleotides from the 3′ end of the exon. Figure 5.10 is a schematic representation of typical mRNA structures that would be translated into protein as well as other mRNA structures that would be subject to NMD.

Let us now screen for genes that might be subject to NMD using Galaxy. We need to follow the steps shown in flowchart form in Figure 5.11. First, we need to acquire data from a gene or gene-prediction table at the Table Browser. We will use the ensGene table of Ensembl gene predictions for this example. We will need to transfer the data in this table to Galaxy in two distinct formats. First, we will use the "Download selected fields" output option to download the Name, ExonCount,

```
##maf version=1
a score=2312839.000000
s hg18.chr1                    31199407 1 + 247249719 G
s xenTro1.scaffold_532          223247 1 +    917834 T
s mm8.chr4                     24994864 1 - 155029701 G
s galGal2.chr23                 461960 1 +   5666127 G
s bosTau2.chr2                 75172077 1 +  86543008 G
s monDom4.chr4                  3926273 1 - 430141050 A
s panTro1.chr1                 31773307 1 + 229575298 G
s rheMac2.chr1                 33722232 1 + 228252215 G
s rn4.chr5                    149550170 1 + 173096209 G
s oryCun1.scaffold_196310         8631 1 +     84859 G
s canFam2.chr2                 15497952 1 -  88410189 G
s dasNov1.scaffold_27063          8207 1 -     43833 C
s loxAfr1.scaffold_6828          53222 1 +     95803 G
s echTel1.scaffold_302436        17060 1 +    150751 G
s tetNig1.chr21                1186878 1 +   5821691 C
s fr1.chrUn                   87114897 1 + 349519338 C
s danRer3.chr23               36902940 1 +  55418239 C


a score=746343.000000
s hg18.chr1                    33570571 1 + 247249719 C
s panTro1.chr1                 34205276 1 + 229575298 C
s rheMac2.chr1                 36116336 1 + 228252215 C
s rn4.chr5                     25124814 1 - 173096209 C
s mm8.chr4                     26782375 1 - 155029701 C
s oryCun1.scaffold_94308          9049 1 +     24239 C
s bosTau2.chr2                 71837066 1 +  86543008 C
s canFam2.chr2                 17417941 1 -  88410189 C
s dasNov1.scaffold_4980          19689 1 -    100180 C
s loxAfr1.scaffold_5522          22308 1 +     80412 C
s echTel1.scaffold_314821        34322 1 -     92458 C
s monDom4.chr4                  4825893 1 - 430141050 C
s xenTro1.scaffold_1087         158201 1 -    316202 G
s fr1.chrUn                  306507035 1 + 349519338 A
s tetNig1.chr11                3051437 1 -  11113812 A
s danRer3.chr23               30614371 1 +  55418239 C
```

Figure 5.9 Portion of output from a Galaxy example, described in the text, that shows the level of vertebrate sequence conservation at two single-nucleotide polymorphism sites in the human genome.

CdsStart, CdsEnd, and Strand fields of each Ensembl gene from the ensGene table. Then we will download the ensGene table a second time in the format of one BED per exon.

Before we can screen for NMD candidates, we need to perform some initial "clean-up" of the data. First, we need to create a table column containing the "index" of the exon, that is, a number that indicates whether the exon is the transcript's most 5′

a) Normal mRNA structures

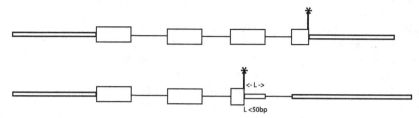

b) mRNA NMD candidates

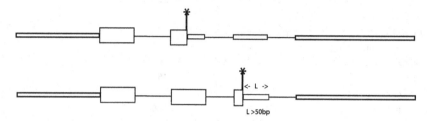

Figure 5.10 Cartoon showing intron-exon structure of mammalian mRNAs. Introns are indicated by thin lines. Portions of exons that include coding and untranslated sequences are indicated by thick and thin blocks, respectively; 5′ end of all sequences are at the left side of the figure. (a) Normal mRNAs, which are not candidates for NMD – stop codon is either in the final exon or within 50 nt of the 3′ end of the second-last exon. (b) NMD candidates. The stop codon occurs either before the second-last exon or more than 50 nt from the 3′ end of the second-last exon.

exon (exonIndex = 0), the second-most 5′ exon (exonIndex = 1), and so on. To create this index, we can take advantage of the way UCSC names exons when using one-BED-per-exon output format. Specifically, the exon index is embedded in the exon name where, for example, a downloaded exon with the name ENST00000319604_exon_2_0_chrX_9845063_f has an exon index equal to 2. Consequently, we can extract the exon index from the exon name by using Galaxy's "Convert underscores to tabs" tool from the "Convert delimiters" option of the "Text manipulation" menu. Then we extract the Name, ExonStart, ExonEnd, and ExonIndex columns from the exon-bed dataset with the Cut tool. (This is not strictly necessary but it makes the tables smaller and hence easier to read.) Next, we "join" the exon-bed and gene datasets so the resulting table will have all the required data for each location (i.e., Name, ExonStart, ExonEnd, ExonIndex, ExonCount, CdsStart, CdsEnd, and Strand) in a single line. Finally, we split the dataset into two, one with the positive-strand exons and the other with the negative-strand exons. This is necessary because the NMD calculations will be different in each case.

We now need to identify the exons that include stop codons because these exons determine which mRNAs are NMD candidates. For the positive-strand genes, this is straightforward: we extract those records for which the value of ExonStart is less than that of CdsEnd and the value of CdsEnd is less than that of ExonEnd. Then

Galaxy Flowchart: NMD Candidates

Acquire data from UCSC Table Browser
1. Acquire Gene dataset (Name, ExonCount, CdsStart,CdsEnd, Strand)
2. Acquire Exon dataset using one-bed-per-exon BED output format

Initial data processing
3. In Exon dataset, convert underscore to tab.
4. In Exon dataset, cut (Name, ExonStart, ExonEnd, ExonIndex)
5. Join Gene and Step-4-result on Name, creating data set with all needed fields:
Name, ExonCount, CdsStart,CdsEnd, Strand, ExonStart, ExonEnd, ExonIndex
6. Create separate datasets for positive and negative strand genes:
    6a. Filter Step-5-result with strand = ' + ' => positive-strand dataset
    6b. Filter Step-5-result with strand = ' − ' => negative-strand dataset

Find NMD candidates on positive strand.
7. Extract exons including stop codon
    7a. Filter Step-6a-result using ExonStart < CdsEnd
    7b. Filter resulting dataset using CdsEnd < ExonEnd
8. Extract exons with stop codon not in last or second-last exon:
Filter Step-7-result with ExonIndex < ExonCount − 2
9. Extract NMD candidates exons with stop codon in second-last exon:
    9a. Filter Step-7-result with ExonIndex = ExonCount − 2
    9b. Filter Step-9a-result with (ExonEnd − CdsEnd) > 50

Find NMD candidates on negative strand.
10. Extract exons including stop codon
    10a. Filter Step-6b-result dataset using ExonStart < CdsStart
    10b. Filter resulting dataset using CdsStart < ExonEnd
11. Extract exons with stop codon not in last or second-last exon:
Filter Step-10-result with ExonIndex > 1
12. Extract NMD candidates exons with stop codon in second-last exon:
    12a. Filter Step-10-result with ExonIndex = 1
    12b. Filter Step-12a-result with (CdsStart − ExonStart) > 50

Figure 5.11 Flowchart of manipulations required for performing NMD screen example in Galaxy. See text for details.

we can identify (positive-strand) NMD candidates by seeing whether the stop-codon containing exons satisfy either

```
exonIndex < ExonCount − 2,
```

or else

```
exonIndex = ExonCount − 2 and (ExonEnd − CdsEnd) > 50.
```

The situation for negative-strand genes is a bit more complicated because of the fact that UCSC stores gene data in *strand coordinates*. Strand coordinates are explained in detail in Appendix 1, but the essential point for this example is that for negative-strand genes, the position of the gene's stop codon will be located at CdsStart (rather

Figure 5.12 Gene structure of ENST00000243040, which was identified by Galaxy from the NMD screen. Note that the stop codon is in the third exon (as indicated by the conjunction of the thick and thin exon lines within the exon), whereas the entire transcript consists of five exons.

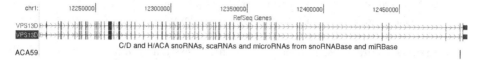

Figure 5.13 Relationship between snoRNA and its host gene. One notes that in this example, the intron that contains the snoRNA is at the extreme 3′ end of the gene and is shorter than the median (or average) intron in the host gene.

than at CdsEnd) and the terminal exon will have an ExonIndex equal to zero. Consequently, for negative-strand genes, the criteria for a stop codon being located within the last exon will be

```
ExonStart < CdsStart, CdsStart < ExonEnd, and ExonIndex = 0.
```

Carrying out these manipulations for both positive- and negative-strand genes yields the set of Ensembl gene predictions that are candidates for NMD. One example, ENST00000243040, is shown in Figure 5.12. We note from the figure that the predicted gene has five exons and that its stop codon is in exon three.

5.4.6 snoRNA host-gene characterization

For our final Galaxy example, we will investigate the length distribution of human introns that have a snoRNA embedded in them. snoRNAs are small non-protein-coding RNAs found in eukaryotes and archaea that play important roles in the development of the cellular machinery used to splice and translate messenger RNA molecules. In mammals, the genes that code for snoRNAs are located in the introns of protein coding genes, often in genes that code for ribosomal proteins.

Figure 5.13 shows the structural relationship between one snoRNA, ACA59, and part of the VPS13D gene, ACA59's *host gene* – that is, the protein-coding gene that contains the intron (the *host intron*) in which the snoRNA is embedded. In this case, the snoRNA is seen to be located in the intron at the extreme 3′ end of the gene. Also, the host intron is shorter than most of the other introns in VPS13D. Such observations led to hypotheses that perhaps snoRNA host introns are systematically shorter than other introns or are located at the 3′ ends of their host genes.

We now investigate the hypothesis that snoRNA host introns are shorter than other introns. For pedagogic purposes, we will address a slightly modified version of this question, namely, whether introns that contain snoRNA genes are systematically shorter (or longer) than other introns of snoRNA host genes. Modifying this example

Galaxy Flowchart: SnoRNA host-gene length distributions

Acquire data from UCSC Table Browser
1. Acquire snoRNA dataset as BED
2. Acquire Gene dataset using one-bed-per-Gene output format
3. Acquire Gene-intron dataset using one-intron-per-Gene output format

Initial data processing
4. Intersect Gene with snoRNA using overlapping intervals => hostGeneClusters dataset
5. Select longest transcript from each gene cluster => hostGene dataset
6. Convert underscores to tabs in Gene-intron data set
7. Compare Gene-intron with hostGenes => hostGeneIntrons
8. Intersect hostGeneIntrons with snoRNAs using overlapping intervals => hostIntrons
9. Subtract hostGeneIntrons from snoRNAs => hostGeneOtherIntrons

Calculate intron length distributions
10.Create intron-length column for hostIntrons by subtracting start from end
11.Calculate hostIntron median and mean lengths using summary statisitics tool
12.Create intron-length column for hostGeneOtherIntrons by subtracting start from end
13.Calculate hostGeneOtherIntrons median and mean lengths

Figure 5.14 Flowchart of manipulations required for performing intron-length distribution example in Galaxy. See text for details.

to other hypotheses – for example, that snoRNA host introns are longer or shorter than the average introns of *all* genes – is straightforward.

An outline of the algorithm is shown in Figure 5.14. To begin, we need to download the genomic coordinates of a set of snoRNAs to Galaxy. For our example, we will use the coordinates of the known human H/ACA snoRNAs from the Table Browser. We acquire this dataset by selecting the wgRna table of hg18 in the Table Browser. We filter the wgRna table output using the Table Browser Filter tool with the "Type" field matching "HAcaBox" and download the data to Galaxy in BED format. Next, we need to download a gene set, for example, the Ensembl genes ensGene table, from the Table Browser to Galaxy. We will need to download this data table to Galaxy twice, once selecting the whole-gene-per-BED output format and the second time using the one-intron-per-BED output format.

Now we need to edit and filter the data files. First, we need to isolate the records of those ensGene genes that "host" (that is, contain) snoRNA genes. We can do this by intersecting, with overlapping intervals, the whole-gene-per-BED ensGene table with our snoRNA table.

The next step is to identify the introns of the snoRNA host genes. However, before we can do this we need to address an issue that could inappropriately bias our results. The issue is that our host-gene table will often contain multiple transcripts for a single snoRNA. That is, the table will contain a record for every isoform of any gene that contains the snoRNA in one of its introns. This is not what we need for estimating host-gene intron-length distributions because genes with many isoforms would be weighted more heavily in our average than genes with only a single isoform. Ideally, we would want to average the intron lengths over all host-gene isoforms. As a simpler alternative, we will select a single representative transcript for each group of isoforms. Which isoform we select is somewhat arbitrary, but the selection should at least be performed systematically. We could select the isoform with, say, the highest mRNA expression, or simply the one with the most introns, or having the longest transcript length. With Galaxy, we are restricted to choose the longest (or shortest) isoform as the single representative transcript because these are currently the only ways of selecting a representative isoform. To do this, we select the Galaxy Cluster tool under "Operate on genomic intervals" and then select the "Find largest interval in each gene cluster," making sure to change the minimum cluster size to equal one. If we apply this tool to the set of host genes that overlap H/ACA snoRNAs, we will create a host-gene table with exactly one isoform for each host gene. We will call the resulting table the hostGenes table.

Now we are ready to identify the introns of the host genes. We accomplish this by first using the "underscore-to-tab" Conversion tool, as we did in the NMD example, so that the intron records have a field containing the same transcript ID as found in the gene records. Having done this, we can use the Galaxy Compare tool to select only those records in the intron-BED gene table that have the same ID as a record in the hostGenes table. We call this the hostGeneIntrons table. Next, we need to split the hostGeneIntrons table into two subtables – a hostIntrons table, which includes just those introns that actually include an embedded snoRNA, and a hostGeneOther-Introns table containing those host-gene introns that do not themselves overlap a snoRNA. We can obtain the hostIntrons table by intersecting with overlapping intervals the hostGenes table with the original snoRNA table. The hostGeneOtherIntrons table is created similarly by using the Subtract Intervals tool to find the rows in hostGeneIntrons that do not overlap intervals in the snoRNA table.

We can now perform the actual length computations. First, we apply the Edit–Compute Expression tool to create a new column of intron lengths in both the hostIntrons and hostGeneOtherIntrons tables by subtracting the intron start value from the intron end value for each table record. Finally, we can apply the Summary Statistics tool to calculate the mean and median values for the intron lengths of each of the two tables.

5.5 Extending Galaxy

Galaxy is a powerful tool. However, Galaxy also has limitations. First, there may be some necessary data manipulations for a specific application that have not been implemented in Galaxy. Second, the Galaxy web site is a shared, online resource. In principle, performance may be impacted if other users of the Galaxy site make large demands on the Galaxy system, although to date, usage of Galaxy has not been heavy enough for this to be an issue.

One way of addressing these potential limitations is by installing a local copy of Galaxy and adding one's own custom post-processing tools. However, although adding custom data-analysis tools to Galaxy is not difficult, developing such new tools and creating the necessary local Galaxy mirror do require computer and programming skills. We will not discuss this approach further here because it is well documented on the Galaxy web site. In particular, procedures for installing a Galaxy mirror and extending Galaxy functionality are described in the HowToInstall and the AddToolTutorial pages of the Galaxy online documentation (http://g2.trac. bx.psu.edu).

Chapter summary

- Practical batch-querying problems often require data post-processing in addition to database access.
- Batch post-processing can be performed either by using interactive tools or by writing a computer program.
- Galaxy provides batch-querying post-processing capabilities for the nonprogrammer.
- Galaxy can take input from Ensembl, UCSC, NHGRI, and other data sources as well as custom user-supplied data.
- Galaxy includes powerful tools for joining, comparing, and subtracting records from multiple datasets both on the basis of table values and genomic intervals as well as wrappers for several widely used bioinformatics data processing programs.

Exercises

1. Use Galaxy to determine which human Consensus CDS genes have the shortest exons. How many base pairs are the shortest annotated exons? Look at a few of them in your favorite browser.
2. Use Galaxy to determine which human VEGA genes have the shortest introns. How short are they? Look at a few of them in your favorite browser.
3. Carry out the example from the text to use Galaxy to identify which regions from a list are intersected by more than three ESTs. Repeat the exercise, this time identifying the regions that intersect more than three xeno (i.e., not same-species) ESTs.
4. Following the procedure described in the text, use Galaxy to determine how many mouse RefSeq transcripts are predicted to be candidates for NMD.
5. Using Galaxy, create a list of Alu repeats that overlap coding SNPs in humans. Compare this approach to that described in Chapter 5 for performing this task solely using the Table Browser.
6. Predictions of gene locations from the gene-prediction programs are sometimes considered to be "real" if there is an mRNA from GenBank that overlaps with the prediction. Simply having an EST (but no mRNA) overlap the prediction is often not considered sufficient evidence. Use Galaxy to identify human GENSCAN predictions that do not intersect any same-species GenBank mRNA but do intersect at least two ESTs.

6

Introduction to Programmed Querying

In the present chapter, we introduce programmed batch post-processing, which we also refer to as programmed querying. We first describe the types of situations for which programmed querying is advantageous. Then we describe SQL-based programming, which enables programmed querying of the Ensembl and UCSC genome databases without needing to use the Ensembl or UCSC APIs. Then we give an overview of the Taverna Toolkit, which provides an alternative approach to generating computer scripts for programmed querying. Finally, we briefly introduce the Ensembl and UCSC APIs, which will be described in detail in the following chapters.

6.1 Programmed batch post-processing

Even if one does not need to add any custom tools to an interactive system like Galaxy, there are still drawbacks to interactive data post-processing. These disadvantages relate to the difficulty of documenting and exactly reproducing the results of interactive analyses. With Galaxy, these problems are partially addressed by means of Galaxy's History mechanism, which stores all the intermediate data files of the interactive procedure. However, even with stored intermediate data files, it can be difficult to remember and reproduce the precise steps of a long interactive procedure. It is for this reason that we included detailed procedural descriptions (as shown in Figures 5.8, 5.11, and 5.14) for the more involved examples presented in the previous chapter.

In addition, repetitively carrying out multistep interactive manipulations is tedious and time-consuming, and simply performing many manual operations can lead to errors. These issues are particularly relevant when the required data analysis is relatively complex and needs to be carried out multiple times. For example, we might want to repeat our NMD analysis with other human gene sets (e.g., RefSeq, VEGA, N-SCAN) and compare the results, or we might want to repeat the analysis with mouse genes. We might have a new gene-prediction program, with adjustable parameters that we want to test by seeing how many (presumably false positive) NMD candidates it produces, and how the number of NMD candidates changes as we vary the adjustable parameters. We will not want to repeat all of the Galaxy table manipulations every time we modify a single parameter.

The Galaxy Development Team has plans that, in the future, Galaxy users will be able to write a "script" or "stored procedure" containing a list of Galaxy commands to be executed automatically with user-specified input files and parameters. However, currently without this capability, ensuring the reproducibility of a multistep data analysis on the Galaxy system is challenging. In contrast, with a programmed query, the computer code itself serves to document exactly what procedures were used and ensures that exactly the same steps are performed each time the program is executed.

6.2 SQL programming

The most straightforward approach to performing programmed post-processing is simply to write one's own data analysis program in whatever programming language one likes and then to embed appropriate SQL programming commands into one's code to access the needed data from the Ensembl or UCSC databases.[1]

This SQL-programming approach is illustrated in the "toy" program shown in Figure 6.1. The program code is available from the publisher's web site for this book and can be executed (after having been made executable with the Unix chmod utility and been placed in the $PATH program path) by executing the command:

```
$ ucscPerlDbiExample.pl
```

The program is a toy program because all it does is extract table names from the UCSC database and print them. However, for the reader experienced in SQL programming, it should be apparent how to extend this program skeleton so that it can extract other data from the UCSC database and do something useful with them.

The program works as follows. First, the "use" statement in line 5 specifies that the Perl DBI module will be accessed (and you will need to download and install this module from CPAN, the Perl module repository, at http:// www.cpan.org, if you do not already have it). In lines 7 through 9, the parameters required to access one of the mouse databases (mm6) of the UCSC database public mirror are specified.

In line 12, the DBI "connect" method is executed to make the actual database connection to the UCSC mirror. Lines 15 and 16 define and prepare the SQL data-retrieval command we want to send to the database. For our demo program, we will simply retrieve a list of the names of the tables in the mm6 database. The command is executed in line 17 and, in line 20, the results for each retrieved record are read into an array we call @ary. In general, each field of the record would be a separate

---

[1] Currently, NCBI does not provide programmed access to the MapViewer database. NCBI does offer bioinformatics programming tools via the NCBI C++ Toolkit, which offer some of the functionality of the Ensembl and UCSC APIs. However, we will not describe these tools here and, instead, refer the interested reader to the online documentation at http://www.ncbi.nlm.nih.gov/books/bv.fcgi?call=&rid=toolkit.TOC.

```
1   #! /usr/bin/perl -w
2   # get data from UCSC DB in perl
3
4   use strict;
5   use DBI;
6
7   my $host = "genome-mysql.cse.ucsc.edu";
8   my $db = "mm6";
9   my $user = "genome";
10
11  # connect to database
12  my $dbh = DBI->connect ("DBI:mysql:$db:$host", $user)
13          or die ("Cannot connect to database");
14  # issue query
15  my $command = "SHOW TABLES";
16  my $sth = $dbh->prepare ($command);
17  $sth->execute ()
18          or die ("Cannot execute query");
19  # read results of query
20  while (my @ary = $sth->fetchrow_array ())
21  {
22          print join ("\t", @ary), "\n";
23  }
24
25  $dbh->disconnect ()
26          or die ("Cannot disconnect from database");
27  exit (0);
```

Figure 6.1 Perl program, using standard SQL, to list the table names in UCSC's mm6 mouse database. Note that line numbers have been added to the code for clarity.

component of the array, though in the present case, there is only one field per record and, hence, only one component in the array. Finally, in line 22, we print out the data, that is, the name of each mouse database table.

For readers with Perl and MySQL programming experience, it should be apparent how one could easily modify the data retrieval and data processing command in the "while" loop so that different kinds of genomic data could be extracted from the database and manipulated in various ways. Readers without such Perl and MySQL experience are referred to any of several excellent introductions to this subject, such as chapters 5 through 7 of DuBois (2005).

We will not describe the direct SQL method of programmed database access further because we will soon describe the Ensembl and UCSC APIs, which usually are easier and more powerful ways of accessing the Ensembl and UCSC databases. That said, it is worth mentioning that sometimes the basic SQL-programming approach can be advantageous. First, if one only occasionally needs programmatic access to one of the genome databases, one can accomplish this without installing the Ensembl or UCSC code or learning the Ensembl or UCSC APIs (which both have nontrivial learning curves). Second, you are not restricted in what programming language you use – in particular, you do not need to use Perl to analyze Ensembl data or C to manipulate UCSC data. And last, if you have a commercial application and you need data from the UCSC databases, you can access the UCSC data without using UCSC's database code, which requires a license for commercial use.

6.3 The Taverna Toolkit

The Taverna Toolkit provides a very different approach to batch post-processing from the other approaches we have seen so far. Taverna is a computer scripting language designed specifically for creating bioinformatics data processing pipelines. In addition, Taverna includes a graphical user interface (GUI) that enables one to build one's own Taverna analysis pipeline without having to do any explicit computer programming. Instead, with the Taverna interface, one creates a script that retrieves data from different databases and then calls bioinformatics programs to analyze the data. The only requirement is that the data and analysis programs be either on the user's local system or else be accessible over the Internet via web service protocols (Zimmerman et al., 2005). In particular, such standard tools as BLAST, Repeatmasker, Clustalw, and the EMBOSS sequence analysis programs are all currently available over the Internet via web service protocols.

For example, in one of its original applications (Stevens et al., 2004), Taverna was used to search for genes in a medically important region of the human genome, the Williams-Bueren Syndrome (WBS) region, so called because patients who suffer from WBS often have chromosomal deletions in this region. Searching for WBS candidate disease genes is complicated by the fact that the region has significant amounts of repetitive sequences. Consequently, at the time when this analysis was performed, the region still had large gaps in its assembly.

For this application, a Taverna pipeline was developed that took newly acquired sequence data and automatically passed the data to the Repeatmasker program (to mask repetitive-sequence regions, which limit the effectiveness of BLAST) and then to the BLAST program to determine if the sequence was located in the WBS region. Finally, if the new sequence was in the WBS region, the sequence was passed to the NIX gene-finding program to determine if any novel putative genes might be located in the new sequence.

Because in this application, the sequence data to be analyzed had only just been sequenced, the data was not yet in the UCSC or Ensembl databases. Consequently, the methods that we have presented so far – as well as those we will present in the following chapters – would be difficult or impossible to apply. As a result, using

Taverna was an effective approach for this batch data analysis application. In contrast, for applications involving data already deposited in the UCSC or Ensembl databases, the advantages of the Taverna approach are less compelling.

There are also disadvantages to approaches using web-based data analysis. For example, the pipeline will fail if any one of the required web services is not available. In addition, the tedious but critical task of reformatting data from the output of each stage of the pipeline so that it is properly formatted to be input for the next analysis program must generally be performed by a custom-written program. In contrast, in more integrated analysis systems (e.g., Galaxy), this data format conversion is typically carried out by the data analysis system itself.

That said, numerous important bioinformatics analysis programs, which can be accessed via web services and hence by Taverna, are not included in the Galaxy Toolkit. Consequently, if one is faced with an application involving processing newly sequenced data or needing access to web-based bioinformatics tools, it is worth considering whether Taverna may be an appropriate tool for the task. Additional information on Taverna is available at the Taverna web site (http://taverna.sourceforge.net) as well as in the Taverna papers (Hull et al., 2006; Oinn et al., 2004).

6.4 Genome-database APIs

If one does need to write a computer program to perform genomic data processing, it is of course possible to write the program "from scratch" in any computer language of one's choice. For example, we illustrated this approach with the Perl SQL program in Section 6.2. However, for many applications, and especially for larger projects, this is an inefficient approach. One reason is that genome databases are generally implemented as relational databases with data stored in tables. In contrast, most modern computer programs store related data as a single data "object" (in C++, Java, or object-oriented Perl) or data "structure" (in C). Consequently, one generally needs software that can convert data back and forth between the tabular form required by the relational database organization and the hierarchical, nontabular data structures that are more natural for application programs.

Writing one's own code to convert data between database format and program data structures is a nontrivial task. Moreover, both the Ensembl and UCSC teams have already written extensive software to perform these tasks. Consequently, it is often preferable to use the extensive API libraries of data extraction and manipulation tools that have already been developed for these purposes. In this way, one has immediate access to functions that perform essentially all the data manipulations that we are accustomed to performing interactively on the genome browsers. In the following chapters, we will introduce the Ensembl and UCSC APIs for programmed querying of genome databases and illustrate how they can be applied to a variety of practical biological problems.

Chapter summary

- Genome-database batch post-processing programs can be written in any computer language that provides an SQL interface to the database.
- Taverna provides an interface for generating computer programs to access bioinformatics data processing pipelines using web services.
- Writing batch post-processing programs for use with the Ensembl and UCSC databases is greatly facilitated by using the APIs available for accessing these databases.

Exercise

1. Modify the SQL Perl program in the text so that it displays the gene names of all the mouse mm6 Ensembl genes between specified genomic coordinates on a specified mouse chromosome. You will need to know some SQL and Perl for this example. (We will learn easier ways of performing this sort of programmed query in the following chapters.)

7

_____

Using the Ensembl API

In the present chapter, we begin our detailed introduction to the Ensembl Perl API. We start with an overview of the API followed by a description of the required software and necessary setup procedures. Next, we present a brief review of the BioPerl software package, on which much of the Ensembl code was originally based. Next, we look at examples of typical Ensembl API syntax and the way that it is similar to, yet also differs from, BioPerl usage. We also address the question of locating specific Ensembl functions within the large Ensembl code base. This is followed by an examination of two complete Perl programs that illustrate how to incorporate the API modules into one's own programs. Finally, we present some comments on Ensembl's Java API. In the following chapter, we continue our introduction to programmed querying of Ensembl data by describing more advanced programming techniques, as well as by presenting an overview of the local-mirror installation procedure for an Ensembl database.

7.1 Overview of the Ensembl API

Facilitating programmed querying of the Ensembl databases has long been a design priority of the Ensembl development team. To achieve this goal, the Ensembl team has developed a reasonably "formal" API for accessing the Ensembl databases within the Perl programming language.[1] By a formal API, we mean essentially a "contract" between the Ensembl designers and the application programmer saying that, "If you (i.e., the applications programmer) access our software exclusively using the tools we have provided for this purpose, then we promise that the system will operate as we claim it will and, moreover, we promise that it will continue to operate in the advertised manner even if we change the underlying implementation of the tools."

_____

[1] Historically, Ensembl has offered two APIs, one in Perl and one in Java. However, as we briefly discuss at the end of this chapter, the Java API is becoming obsolete. Hence, unless otherwise noted, the term "Ensembl API" will refer to the Ensembl Perl API.

Using the Ensembl API provides several advantages. The API includes thoroughly tested software with functionality to perform a wide range of genome data manipulations. The Ensembl API software is open source, and its use is completely free for all users. In addition, the software is well documented and is actively supported by Ensembl for general database querying by researchers outside the Ensembl Browser Development Team. The API is standardized and API versions are synchronized with Ensembl database releases. Consequently, code written with the API can be applied equally well to every genome included in the associated Ensembl database release. Moreover, by using the API, one is assured that one's code will continue to function with the published interface in future releases, if at all possible, and that if interface changes are unavoidable, the changes will be well documented and tools will be provided to enable programmers to convert their code to the modified interface.

Another advantage of using the Ensembl API is that Ensembl emphasizes the use of community-standard tools, protocols, and data formats, such as BioPerl, DAS, GFF, and XML, and generally makes its code interoperable with software from other genome database projects, such as those of the single-genome databases and GMOD. Finally, Ensembl has the useful feature that any programmed query that works with a local Ensembl database can be executed equally well against the public Ensembl mirror. This feature is useful for software testing, as well as for eliminating the need to install a local mirror database at all, if one's programmed-querying needs are not extensive.

To accomplish these goals, Ensembl uses object-oriented programming techniques. As a result of Ensembl's object-oriented design, the Ensembl API user does not need to be concerned with the details of how the "object methods" – that is, the subroutines that one calls – are implemented by the system. In particular, when using the objects of the Ensembl API, one does not need to understand the table structure of the Ensembl database. The API automatically performs all database-to-application-program data conversions. Consequently, the task of applications programming becomes much easier and the resulting programs are immune to database schema changes.

7.2 Software and programming requirements

To take advantage of the Ensembl API, one needs to have a basic understanding of the Perl programming language, including an understanding of how to use Perl references, modules, objects, and methods, and a basic knowledge of Perl is assumed in the remainder of this chapter as well as in the next chapter. Note that advanced programming skills, such as object-oriented programming, though helpful, are not required to use the API. In particular, the skills learned in a standard introductory Perl programming course should be sufficient for understanding the examples presented. For those readers who do not have such skills, there are many excellent texts available, including ones that are appropriate for self-study. For example, I have found *Perl Core*

Language (Holzner, 1999), *Beginning Perl for Bioinformatics* (Tisdall, 2001), and *Mastering Perl for Bioinformatics* (Tisdall, 2003) to be good books for learning Perl.

Various standard software programs need to be installed on one's system before one can use the Ensembl API. You will need to have a local copy of the Perl interpreter. You will also need a locally installed version of the MySQL client program. (Note that installation of the MySQL server program is not necessary.) The MySQL client program is freely available for download at http://www.mysql.com. Other software that is very useful to have – though not strictly required – include software to download data from the Internet, such as NcFTP (http://www.ncftp.com), the grep text searching utility (http://www.gnu.org/software/grep), the CVS source-code archiving utility (http://www.nongnu.org/cvs), and the rsync data transfer and synchronization program (http://samba.anu.edu.au/rsync).

If you are using a Unix-based system, some of these tools will already be installed on your system. If they are not already installed, downloading them is free and relatively straightforward, and the required procedures are described in the documentation that accompanies the programs.

7.2.1 *Installing BioPerl and Ensembl code*

To run the Ensembl API, you will need to install BioPerl as well as the Ensembl API. Specifically, Ensembl requires installation of the core component of BioPerl (version 1.2.3 or later), called bioperl-live, as well as the Perl modules upon which BioPerl depends. Instructions for obtaining these files are included in the Ensembl API documentation. You can download a copy of the API itself from http://www.ensembl.org/info/using/api.

Installation of BioPerl and the Ensembl API is not difficult and basically follows the standard three-step Perl installation commands of "perl Makefile.PL," "make," and optionally, if you have system-administrator privileges, "make install." If you do not have system privileges, or you do not want to install BioPerl or Ensembl in the system Perl libraries, you will need to include command-line arguments or Perl "lib" pragmas in each of your programs to enable the Perl interpreter to locate the BioPerl and Ensembl libraries. The Ensembl and BioPerl installation README files, which are included with the download distributions, describe these straightforward procedures in more detail.

The Ensembl API installation instructions (see http://www.ensembl.org/info/using/api) specify installing the most recent version of the Ensembl API software. Although this is likely to be what you ultimately want to do, as an initial step, you may want to install version 42 because this is the version of Ensembl that I used to write and test the Ensembl programs described in this book. Downloading version 42 simply involves replacing the current Ensembl version number with "42" in the download commands, for example, for downloading the Ensembl "core" API:

```
$ cvs -d :pserver:cvsuser@cvs.sanger.ac.uk:/cvsroot/ensembl \
  checkout -r branch-ensembl-42 ensembl
```

Once you have version 42 running properly, moving to the current Ensembl API release is straightforward.

You will definitely need to download the Ensembl core API. Depending on your applications, you may want to download additional Ensembl APIs, such as the comparative genomics API, ensembl-compara (which we will need for the multiple-sequence alignment example in Chapter 8), or the functional genomics API, ensembl-functgenomics. In addition, you will need to download the Ensembl Browser software, called ensembl-draw, to perform the "grep" examples described below. Downloading these other components of the API simply requires modifying the final argument in the previous CVS command; for example, the command to download the ensembl-draw API is:

```
$ cvs -d :pserver:cvsuser@cvs.sanger.ac.uk:/cvsroot/ensembl \
  checkout -r branch-ensembl-42 ensembl-draw
```

7.3 Database access

In addition to installing the Ensembl code, you will need to establish some form of database access to a mirror of the Ensembl genome databases. This requirement exists because the genome-browser databases themselves do not have the capacity to handle programmatic querying. In general, there are three ways to create database access: using a public mirror database, downloading individual database tables and files, and creating one's own private mirror.

In the present chapter, we will only describe using the Ensembl public mirror database at ensembldb.ensembl.org. For occasional programmatic database querying, this is the easiest approach. Since ensembldb.ensembl.org is an exact mirror of the Ensembl Browser database, any code using the Ensembl API should run properly on the Ensembl public mirror. You will need to confirm that there are no firewall or other security systems where your computer is located that prevent you from accessing a remote database over the Internet. In the next chapter, we will describe database access via setting up a local mirror of an Ensembl database. The approach of downloading individual database tables and files is more applicable to the UCSC databases than to Ensembl, and hence will be described later in the context of the UCSC system.

7.4 Using non-Unix systems

I will assume that you are working on a Unix-like system, such as linux or Mac OS X. The Ensembl API and databases have been successfully installed and thoroughly tested on a wide variety of Unix systems and the necessary procedures are well documented. Moreover, if database installation problems are encountered on Unix systems, a query addressed to the Ensembl mailing list is likely to find someone who has already encountered the same problem and identified a solution to it.

In contrast, little documentation is available regarding the use of the genome-database APIs on computers running the Microsoft Windows operating systems. To be sure, Perl code is relatively portable between computer operating systems. MySQL software for Windows systems is also available. That said, porting a system as large as a genome database API to a different type of operating system is a complex task, and one should expect that subtle potential incompatibilities that need to be addressed will arise. Anecdotal reports from users who have attempted such Windows ports indicate that this is indeed the case.

In general, if one needs to perform programmed querying and one absolutely must use a Windows-based system, the best route is probably via the Cygwin emulation program. Using Cygwin, available from http://www.cygwin.com, a Microsoft Windows user is able to run Unix programs essentially as if they were using a Unix machine. If one is only running the Ensembl API, as opposed to installing a local mirror database, running under Cygwin should work. An alternative approach for accessing the Ensembl API under Windows is to install the Windows Perl environment provided by Active State and available at http://www.activestate.com. That said, using the Ensembl (or UCSC) APIs under Microsoft Windows is not recommended and will not be discussed further in this book.

7.5 Perl and BioPerl

In addition to a basic understanding of Perl, some familiarity with the BioPerl package is important for using the Ensembl API. BioPerl is a collection of Perl modules that facilitate the development of Perl scripts for bioinformatics applications. BioPerl provides modules for sequence manipulation, for accessing of databases that use a range of data formats, and for executing and parsing the results of various molecular biology programs, such as BLAST. In this section, we will present a brief overview of the BioPerl package, emphasizing the components that are most relevant to the Ensembl API. Readers who are not already at least somewhat familiar with BioPerl are referred to the detailed BioPerl tutorial, bptutorial, which is available at http://www.bioperl.org/Core/Latest/bptutorial.html.

BioPerl consists of a "core" software distribution, called bioperl-live, as well as several auxiliary distributions with names such as bioperl-run, bioperl-db, and bioperl-ext. The auxiliary distributions include more specialized software modules, which typically require somewhat more involved installation procedures. Use of the Ensembl API does not depend on any of the auxiliary BioPerl modules, and they will not be discussed further here.

The BioPerl package is implemented via a collection of Perl objects, each representing an important bioinformatics concept. The software includes objects for biological sequences, for sequence alignments, for sequence features (e.g., genes and transcripts), and for sequence variations (e.g., SNPs). In addition, there are objects for implementing many common general bioinformatics tasks such as parsing database records or BLAST analysis reports and for reading and writing data from files and databases in various standard formats.

The central BioPerl object is the sequence object, Seq, which is used for performing most sequence manipulations on DNA, RNA, or protein sequences. (We note that there are other sequence objects in BioPerl as well, for more specialized use. See the BioPerl documentation for more details.) When using BioPerl, Seq objects are typically created for you automatically when you read in a file containing sequence data using the BioPerl sequence input-output object, SeqIO. For example, reading a set of GenBank records into BioPerl using SeqIO is as simple as the following two lines of Perl code:

```
use Bio::SeqIO;
$inSeqs = Bio::SeqIO->new(-file => "inputfilename", -format => genbank);
```

Once the sequences have been read in as Seq sequence objects, they can easily manipulated by calling Seq's various methods, for example:

```
while ( $seqobj = $inSeqs->next_seq() )
    {
    # obtain the sequence as a string
    $seqString = $seqobj->seq();
    # retrieve part of the sequence as a string
    $partSeq = $seqobj->subseq(5,10);
    # a description of the sequence
    $description = $seqobj->description();
    # truncate nucleotides 5 to 10 as new Seq object
    $truncSeq = $seqobj->trunc(5,10);
    # reverse complements sequence
    $revcomSeq = $seqobj->revcom;
    # translation of the sequence
    $transSeq = $seqobj->translate;
    # etc ...
    }
```

If the sequences have annotations associated with them, as will generally be the case if they are extracted from the Ensembl database, the annotations will be stored as sequence feature objects (called SeqFeature objects) that can be retrieved with code with this format:

```
# to retrieve all of the "top level" sequence features
# (e.g. all gene annotations)
        $seqobj->get_SeqFeatures
# to retrieve all sequence features, including sub-sequence features
# (e.g. all genes with their associated transcript and exon annnotations
        $seqobj->get_all_SeqFeatures;
```

In addition, using BioPerl's Bio::Tools::GFF format parser, SeqFeature objects can also be created directly from annotations stored in a flat file containing annotation data in GFF format for GFF data through version 2.5. Accessing feature data from files written in GFF3 is slightly more complicated but can be accomplished using BioPerl's Bio::DB::SeqFeature::Store::GFF3Loader object. Appendix 2 includes an introduction to the GFF and GFF3 data formats.

Typical syntax for using Bio::Tools::GFF would be

```
$gffio = Bio::Tools::GFF->new(file => "gffFileName", -gff_version => 2);
while ($feature = $gffio->next_feature()) {
  # do something with feature
}
```

In a similar manner to the object representation of sequences and sequence annotations, pairwise and multiple-sequence alignments are stored as alignment objects in BioPerl, called SimpleAlign objects. SimpleAlign objects are read and written to files with the AlignIO object in a manner completely equivalent to the SeqIO object for sequences, for example:

```
use Bio::AlignIO;
$io = Bio::AlignIO->new(-file => "alignmentFileName",-format => "clustalw" );
$threshold_percent = 60;
while ($aln = $io->next_align()) {
        $consensus = $aln->consensus_string($threshold_percent);
        # etc
}
```

Once read into memory, alignment objects have their own associated methods for computing percent identity and a consensus sequence, extracting individual sequences from the alignment, and so on. For more examples of BioPerl functionality and usage, the reader is referred to the BioPerl documentation.

7.5.1 Finding BioPerl objects and methods

Often the most difficult parts of writing a Perl program using BioPerl are identifying the BioPerl object that you need, determining what methods are available for that object, and finding the proper syntax for using those methods. For common functions, a simple approach is to look at the sample code in the BioPerl tutorial and modify it to fit your needs. In addition to the BioPerl tutorial, many examples of code usage can be found in the program scripts located in the BioPerl examples and scripts subdirectories. Yet more examples of code usage can be found in the BioPerl "t" subdirectory, which contains test code for nearly all of the methods that can be called by a BioPerl object.

An alternative, and more general, approach for locating methods associated with BioPerl objects, including methods that may not be used in the BioPerl tutorials or example scripts, is to use the automatically generated documentation called bioperl pdoc, located at http://doc.bioperl.org (see Figure 7.1). With pdoc, you can find BioPerl

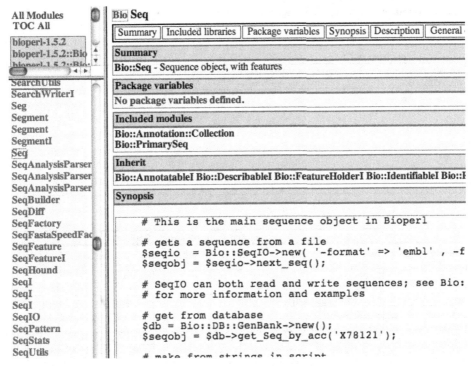

Figure 7.1 Screenshot of a portion of the BioPerl pdoc page. The upper left side of the display contains a selectable list of BioPerl object directories, whereas the lower left side is a list of objects contained in the selected directory. The right side of the display shows documentation for the selected object, in this case, the Bio::Seq object, including the name of the object's "parent" objects, examples of proper syntax for using the object, and a list of methods defined by the object.

objects and methods by first scrolling through the list of objects on the left side of the screen display. Once one has selected the object of interest, the right side of the display presents a description of the methods associated with that object and sample syntax illustrating how they are properly used, as well as the object code itself.

Although this procedure often works well, there is an important detail that needs to be considered. That is, even after you have located the object that performs a certain data manipulation in pdoc, it is sometimes still not clear where to find the documentation for all of the object's methods. This is because pdoc only displays documentation for methods that are *explicitly* defined for that object. In contrast to explicit definitions, Perl objects can also "inherit" functionality (i.e., methods) from other Perl objects, essentially by being "special cases" of more general objects. Although such software inheritance is often useful, it can make finding documentation with pdoc more difficult.

For example, say you want to find documentation on the parse method of the BioPerl Genscan object. You would not find this documentation in the pdoc entry for the Genscan object. In fact, the pdoc documentation of the Genscan object

Search **class names** by string or Perl regex (examples: Bio::SeqIO, seq, fasta$)

| Genscan | Submit Query |

OR select a class from the list:

| Bio::DB::GFF::Aggregator::ucsc_genscan | UCSC genscan aggregator |
| Bio::Tools::Genscan | Results of one Genscan run |

sort by method ▾

methods for **Bio::Tools::Genscan**			
next_feature	Bio::Tools::Genscan	A Bio::Tools::Prediction::Gene object.	while($gene = $genscan->next_feature())
next_prediction	Bio::Tools::Genscan	A Bio::Tools::Prediction::Gene object.	while($gene = $genscan->next_prediction
noclose	Bio::Root::IO	value of noclose (a scalar)	$obj->noclose($newval)
otherwise	Bio::Root::Root	not documented	not documented
parse	Bio::Tools::AnalysisResult	not documented	not documented
qualify	Bio::Tools::Genscan	not documented	not documented
qualify_to_ref	Bio::Tools::Genscan	not documented	not documented
rmtree	Bio::Root::IO	number of files successfully deleted	Bio::Root::IO->rmtree($dirname);
stack_trace	Bio::Root::RootI	array containing a reference of arrays	@stack_array_ref = $self->stack_trace

Figure 7.2 Partial screenshot of the Deobfuscator user interface. The top of the display lists modules that matched a keyword input, in this case, "Genscan." The bottom of the display lists the methods available for the selected object by inheritance, including those methods defined implicitly.

does not even show that a parse method exists. Rather, the pdoc documentation for the Genscan parse method is found in the listing for the AnalysisResult object from which the Genscan object inherits the parse method. Although the documentation for the Genscan object does list the (four) other objects from which it inherits methods, it does not identify which methods are inherited from where. In cases where an object inherits methods from many different sources, tracking down the object containing the actual method definition (and hence its documentation) can be tedious.

To assist one in finding one's way through the tree of inherited methods, the BioPerl distribution (as of BioPerl version 1.5) includes a useful, though somewhat obscurely named, tool – the Deobfuscator. (The Deobfuscator is also available on the web at http://bioperl.org/cgi-bin/deob_interface.cgi.) The user interface to the Deobfuscator is shown in Figure 7.2. For our Genscan example, we can enter "Genscan" or "Bio::Tools::Genscan" into the upper part of the display and after clicking "Submit Query," the lower part of the display lists all of the methods available to Genscan objects, including those that are inherited. The display also indicates from which objects implicitly defined methods are inherited. In our case, the display shows that Bio::Tools::Genscan's parse method is defined in the AnalysisResult object. Clicking on the associated link retrieves the code and any documentation for the parse method from the AnalysisResult documentation.

7.6 BioPerl and the Ensembl API

Once one is comfortable with BioPerl objects, the transition to Ensembl objects is straightforward. All of the basic BioPerl sequence, annotation, and alignment tools exist in the same or very similar implementations in Ensembl. This is not surprising because the Ensembl software was originally developed using BioPerl as a source-code base.

From the user's perspective, the main new element in Ensembl Perl is that most of one's input data will be from a connection to an Ensembl database rather than from flat files on one's own computer. Accessing this Ensembl data using the API is essentially a three-step process. First, one identifies the object that is used by Ensembl to store the required data type in computer memory. Next, one determines the database in which this data is stored and the database-adaptor object required for extracting this data from the database. Finally, one uses the methods associated with the data object to perform the necessary data manipulations.

An example should clarify this process. Let us assume we want to perform some manipulations on the coordinates of a set of mRNA transcripts – say, to compute distances between their stop codons and their exon boundaries to determine whether the transcripts are predicted to be degraded by NMD. The first step is to identify the class of Ensembl objects that permit computations of distances between stop codon and exon boundary coordinates. In this case, it is pretty trivial to guess the name of the required object: specifically, an Ensembl "transcript object." (We will shortly describe general methods for identifying the Ensembl object associated with a specific type of genome annotation.)

Next, we need to identify the appropriate database and database-adaptor object. The required database depends on the species of interest and the Ensembl release number. Note that the Ensembl database release number needs to agree with the version number of the Ensembl API code that you are using. Specifically, the first number of the Ensembl database release name must be the same as the number of the Ensembl API release. For example, if you are using version 42 of the Ensembl API and you want to query the Ensembl Human Genome Database, you will need to use the Ensembl homo_sapiens_core_42_36d database. If you do not know the precise Ensembl database name for a species, it can be determined directly with MySQL using the command

```
$ mysql --user=anonymous --host=ensembldb.ensembl.org -A \
    -e "show databases;"
```

Having identified the proper database, we next need to create a database-adaptor object. For annotations that are associated with a region of a chromosome (called a "slice" in Ensembl parlance), for example a gene, exon, or transcript, this object is a "slice-adaptor" object. By looking up the syntax for slice-adaptor objects in the API documentation, we find that the required code would look like the following:

```
$db = new Bio::EnsEMBL::DBSQL::DBAdaptor(-host => 'ensembldb.ensembl.org',
                -user => 'anonymous',-dbname => 'homo_sapiens_core_42_36d');
$slice_adaptor = $db->get_SliceAdaptor();
$slice = $slice_adaptor->fetch_by_region('chromosome','X', 5.29e5,5.3e5);
$genes = $slice->get_all_Genes();
foreach $gene (@$genes) {
        $transcripts = $gene->get_all_Transcripts();
        foreach $transcript (@$transcripts) {
        # perform calculation on transcript ...
        }
}
```

Note that the @$genes syntax is required because the Ensembl data-adaptor object returns a Perl reference (i.e., a pointer) to a list of data objects rather than the list of objects itself.

7.6.1 Finding Ensembl objects

As was the case with using BioPerl, often the most challenging part of using the Ensembl API is just identifying the appropriate Ensembl object that implements the required data manipulations. Sometimes the name of the Ensembl object will be suggested by the type of annotation it describes, as in the case of Ensembl Gene, Transcript, or Exon objects. However, this is not always the case. This may be true even though it may be clear that *some* Ensembl object with the required properties must be available because the needed data is available via the Ensembl Genome Browser.[2]

One approach for locating Ensembl objects is to use a text search program, such as the Unix grep utility, to scan the entire Ensembl code base or the ensembl-draw subdirectory, which contains the routines that are used to display the Ensembl Genome Browser. For example, say we are interested in Ensembl's histone modification annotations. We know that the Ensembl database contains histone modification annotations because they can be displayed in the Ensembl Browser and are not a DAS track. However, if we want to write a program that accesses multiple histone modification annotations, we will need to identify the object and database-adaptor object associated with these annotations.

If we execute the following grep command in the ensembl-draw directory,

```
$ grep -irn histone.modification  .
```

---

[2] The statement that every data manipulation performed in the Ensembl Browser must be implemented in the Ensembl API is not strictly correct. In particular, the Ensembl DAS tracks are not created by Ensembl and, consequently, the code used in their creation will in general not be included in the Ensembl code base.

we find the entry:

```
./modules/Bio/EnsEMBL/GlyphSet/histone_modifications.pm:1:package
Bio::EnsEMBL::GlyphSet::histone_modifications;
```

(Note that the options i, r, and n tell grep to ignore capitalization differences, search recursively in subdirectories, and report back the name of the files in which the text is found, respectively. Also note that the final dot "." in the command, telling grep to search the current working directory, is required.)

If we now look at the histone_modifications.pm module, we find the following lines of code, which implement the retrieval of the histone modification data from the Ensembl database[3]:

```
my $pf_adaptor = $db->get_PredictedFeatureAdaptor();
my $features = $pf_adaptor->fetch_all_by_Slice($self->{'container'});
```

Performing a second grep, this time on "get_PredictedFeatureAdaptor," we learn that get_PredictedFeatureAdaptor is defined in the PredictedFeatureAdaptor.pm module in the ensembl-functgenomics/modules/Bio/EnsEMBL/Funcgen/DBSQL subdirectory, and that the object the adaptor object returns is a "PredictedFeature" object. Consequently, we now know that histone modification data is stored in the Ensembl API as a PredictedFeature object.

With this information, we can finally look up the PredictedFeature object in the Ensembl pdoc documentation, which is similar in format to the BioPerl pdoc documentation described previously. In fact, there are multiple Ensembl pdocs, one for each of the major subcomponents of the Ensembl software. For example, the pdoc for the Ensembl core modules is located at http://www.ensembl.org/info/using/api/Pdoc. In the present case, we need the pdoc for the Ensembl functional genomics component, which is located at http://www.ensembl.org/info/using/api/Pdoc/ensembl-functgenomics/index.html.

As with the BioPerl pdoc, the various Ensembl pdocs only document methods when they are explicitly defined, and do not document methods that are inherited from parent objects. Moreover, Ensembl does not currently provide a program that is equivalent to BioPerl's Deobfuscator to identify inherited methods. Consequently, one needs to explicitly traverse the entire inheritance tree of an Ensembl object to identify all of the methods associated with it. However, in most cases, Ensembl object inheritance trees are not very large and, in practice, finding the methods that one needs for any given object is generally not too onerous.

---

[3] Note that experienced Perl programmers should not be disturbed seeing Perl syntax like

```
$self->{'container'}
```

Although using direct hash references to a Perl object is not good application programming practice, here we are looking within the internal Ensembl code where such constructions are both appropriate and useful.

7.7 Programming examples using Ensembl and BioPerl

To further illustrate the Ensembl API, we next examine two complete programs. The first is a short demonstration program that simply prints a list of gene, transcript, and exon annotations for a genomic region. For our second example, we revisit the comparison of intron lengths between mammalian introns that contain embedded snoRNAs and those that do not.

7.7.1 Ensembl example 1: Retrieving gene data

Program ensemblTest1 displays gene annotations for a specific chromosomal region. The code for the program is shown in Figure 7.3. The code for this program (and all the other examples in the book) is available from the publisher's web site for the book.

Once the program is installed on your system, it can be made executable by using the Unix chmod utility and placing it in one of the directories that the operating

```perl
1   #! /usr/bin/perl -w
2   # test program
3   use strict;
4   use lib "$ENV{HOME}/programs/ensembl42/ensembl/modules/";
5   use lib "$ENV{HOME}/programs/ensembl/bioperl-live/";
6   use Getopt::Std;
7   use Bio::EnsEMBL::DBSQL::DBAdaptor;
8   use Bio::EnsEMBL::DBSQL::SliceAdaptor;
9   use vars qw( %option);
10
11  sub feature2string {
12  # Convert data from feature object to character string
13  my $f = shift;
14  my $stable_id = $f->stable_id();
15  my $display_id = $f->display_id();
16  my $seq_region = $f->slice->seq_region_name();
17  my $start = $f->start();
18  my $end = $f->end();
19  my $strand = $f->strand();
20  my $chromStart = $f->seq_region_start();
21  my $chromEnd = $f->seq_region_end();
22  my $chromStrand = $f->seq_region_strand();
23  my $sliceString = "$stable_id : $seq_region:$start-$end ($strand)\n";
24  my $chromString =
25          "$display_id : $seq_region:$chromStart-$chromEnd ($chromStrand)\n";
26  return ($sliceString, $chromString);
27  }
28
```

Figure 7.3 Source code for the gene-data retrieval example (program ensemblTest1.pl).

```
29  ####################
30  sub ensemblTest1 {
31  # retrieve and print gene information from database
32  my ($db) = @_;
33  my $slice_adaptor = $db->get_SliceAdaptor();
34  my $slice = $slice_adaptor->fetch_by_region('chromosome','IV',
35                                      5.29e5,5.3e5);
36  print "**Starting gene loop\n";
37    foreach my $gene (@{$slice->get_all_Genes()}) {
38        my ($gstring, $gChromString) = feature2string($gene);
39        print "$gstring$gChromString";
40        print "\t**Starting transcript loop\n";
41        foreach my $trans (@{$gene->get_all_Transcripts()}) {
42            my ($tstring, $tChromString) = feature2string($trans);
43            print "\t$tstring\t$tChromString";
44            my @exons = @{$trans->get_all_Exons()};
45            print "\t\t**Starting exon loop with ", scalar(@exons), " exons\n";
46            foreach my $exon (@exons) {
47                my ($estring, $eChromString) = feature2string($exon);
48                print "\t\t$estring\t\t$eChromString";
49                print "\t\t sequence: " , $exon->seq->seq , "\n";
50            }
51        }
52    }
53  }
54
55  ####################
56
57  my $USAGE =<<END_OF_USAGE;
58
59  Usage: ensemblTest1 [options]
60        Options:        -d <dbName>: default = saccharomyces_cerevisiae core 42 1e
61                        -h <hostName>: default = ensembldb.ensembl.org
62                        -u <userName>: default = anonymous
63  END_OF_USAGE
64
65  getopts('d:u:h:', \%option) || die("$USAGE");
66  my $host = $option{'h'} || 'ensembldb.ensembl.org';
67  my $user = $option{'u'} || 'anonymous';
68  my $dbname = $option{'d'} || 'saccharomyces_cerevisiae_core_42_1e';
69  #Now we can make a database connection:
70  my $db = new Bio::EnsEMBL::DBSQL::DBAdaptor(-host => $host,
71                          -user => $user, -dbname => $dbname);
72  if ($db != 0) {print "Made DB connection OK\n";}
73  else {print "No DB connection\n";}
74  ensemblTest1($db);
75  __END__
76
```

Figure 7.3 (continued)

system checks for executable programs (i.e., the directories in $PATH). We can now execute our program with the command

```
$ ensemblTest1.pl
```

By default, the program accesses the release 42 version of the *S. cerevisiae* database on the public Ensembl mirror, but these choices can be modified using the options -d, -h, and -u to specify other databases and host and user names. As the program is written, these options are not especially useful because the chromosomal coordinates of the features to be retrieved are "hard-coded" in the demo program. However, modifying the program so the user can select a different genomic region by adding additional options (or program arguments) is straightforward (and is left as an exercise).

7.7.2 ensemblTest1 – program implementation

The program is composed of four components: a list of the needed libraries and modules and their locations; the "main" program; the central subroutine, ensmblTest1, which retrieves the annotations and prints the results; and an auxiliary subroutine, feature2string, which formats the annotation data for printing.

The program begins with a list of "use" statements (lines 3 to 9) that specify libraries and modules needed by the program. The "use Bio::EnsEMBL::DBSQL::DBAdaptor;" and "use Bio::EnsEMBL::DBSQL::SliceAdaptor;" statements are needed to specify where within the Ensembl code tree the Perl interpreter will be able to find the two explicit invocations of Ensembl code. Similarly, the statement

```
use lib "$ENV{HOME}/programs/ensembl/ensembl/modules/";
```

is needed to direct the Perl interpreter to the location of the root directory of the entire Ensembl code tree. (If you run this program, the lib statement needs to point to the location of the main Ensembl directory in your system. Alternatively, if the Ensembl modules have been installed in the main Perl library tree structure, the lib statement can be removed altogether.)

The main program first processes any options using the standard Perl options module Getopt::Std (lines 65 through 68). The program then connects to the Ensembl database in lines 70 and 71 and reports whether the connection was successful in lines 72 and 73. If the connection was successful, the subroutine ensemblTest1 is called.

Subroutine ensemblTest1 (in lines 34 and 35) issues the command to the database to retrieve all features for the specified region whose coordinates are hard-coded in the arguments to the fetch_by_region method. The routine then cycles through all of the gene annotations for the region (line 37), all the transcript annotations for each gene (line 41), and all the exon annotations for each transcript (line 46) in the manner we illustrated in Section 7.6. For each gene, transcript, or intron, the routine calls the subroutine feature2string to convert annotation data stored within

```
Made DB connection OK
**Starting gene loop
YDR038C : IV:-1581-1694 (-1)
YDR038C : IV:527418-530693 (-1)
      **Starting transcript loop
      YDR038C : IV:-1581-1694 (-1)
      YDR038C : IV:527418-530693 (-1)
            **Starting exon loop with 1 exons
            YDR038C.1 : IV:-1581-1694 (-1)
            YDR038C.1 : IV:527418-530693 (-1)
            sequence: ATGAGCGAGGGAACTGTCAAAGAA...
```

Figure 7.4 Part of the output for the gene-data retrieval example.

the Ensembl Feature object corresponding to the gene, transcript, or exon into a formatted, printable-character string.

Finally, subroutine feature2string converts the Feature object data to character-string data (in lines 14 through 22) using various methods of the Ensembl Feature object (recall that we can find these methods by looking up Feature.pm in the Ensembl pdoc documentation). Once the desired annotations have been extracted, they are reformatted for printing in two character strings in lines 23 to 25.

Running the program should produce output such as that shown in Figure 7.4. Note that for each feature, the program displays its coordinates both in chromosomal and in "slice" coordinates (Ensembl's coordinate system conventions are discussed further in Appendix 1).

7.7.3 Ensembl example 2: Intron length comparisons

For our second example, we revisit the snoRNA host-intron length calculation, which we have already investigated with Galaxy in Section 5.4.6 of Chapter 5. We recall that we want to compare of the lengths of introns that have snoRNAs embedded in them (so-called snoRNA host introns) with the lengths of introns that do not contain snoRNAs.

The program is called ensemblIntronLengths.pl and is run with a command like

```
$ ensemblIntronLengths.pl hacaWgRna.hg18.bed
```

where hacaWgRna.hg18.bed is a data file containing a list of BED coordinates of snoRNA genes (in this case, mammalian H/ACA snoRNA genes, as acquired from the UCSC Table Browser for the wgRna table for UCSC hg18 database). The program provides options for selecting a different Ensembl host or database, as well as an option for printing some warning and debugging messages. Note that the user is responsible for ensuring that the coordinates of the snoRNA genes are from the same genome assembly as those of the database selected by the -d option. There is no problem with using coordinates extracted from the UCSC hg18 database because the program accesses (by default) release 42 of the Ensembl human database, and both UCSC hg18 and Ensembl release 42 use the same genome assembly (NCBI assembly 36).

7.7.3.1 Outline of program implementation

The program listing is shown in Figure 7.5 and a flowchart describing the algorithm is shown in Figure 7.6. From the flowchart, we see that there are two main phases to the program. In the preprocessing phase, the program builds a hash, called $hostHash, that associates one transcript of each snoRNA host gene with a list of all the snoRNAs embedded in its introns. In the second phase, the host-gene hash is used to calculate the required intron length distributions.

```
1   #! /usr/bin/perl -w
2
3   use strict;
4   use lib "$ENV{HOME}/programs/ensembl142/ensembl/modules/";
5   use lib "$ENV{HOME}/programs/ensembl/bioperl-live/";
6   use Getopt::Std;
7   use FileHandle;
8   use Bio::EnsEMBL::DBSQL::DBAdaptor;
9   use Bio::EnsEMBL::DBSQL::SliceAdaptor;
10  use Bio::Range;
11  use vars qw(%option);
12
13  ####################
14  sub transcript_length {
15  my ($transcript) = @_;
16  return 0 unless $transcript;
17  return ($transcript->end_Exon->end() - $transcript->start_Exon->start() + 1);
18  }
19
20  ####################
21  sub transcript_intron_count {
22  my ($transcript) = @_;
23  return 0 unless $transcript;
24  my @introns = @{$transcript->get_all_Introns()};
25  my $count = scalar @introns;
26  return $count;
27  }
28
29  ####################
30  sub sortGeneTranscriptsByIntronCtAndLength {
31  my ($transcriptList)  = @_;
32  return 0 unless $transcriptList;
```

Figure 7.5 Source code for the ensemblIntronLengths.pl intron-lengths distribution program.

```
33  my @sortedTranscriptList =
34       sort {
35             transcript_intron_count($b) <=> transcript_intron_count($a) ||
36             transcript_length($b) <=> transcript_length($a)
37       } @$transcriptList;
38  return \@sortedTranscriptList;
39  }
40
41  ####################
42  sub get_transcript_with_most_introns {
43  my ($slice) = @_;
44  my $maxIntronT = 0;
45  my $maxIntronCount = 0;
46  foreach my $gene (@{$slice->get_all_Genes()})
47       {
48       my $t = $gene->get_all_Transcripts();
49       my $sortedTranscripts =
50             sortGeneTranscriptsByIntronCtAndLength($t);
51       my $maxCurrentT = $sortedTranscripts->[0];
52       if (transcript_intron_count($maxCurrentT) > $maxIntronCount)
53             {
54             $maxIntronT = $maxCurrentT;
55             $maxIntronCount = transcript_intron_count($maxCurrentT);
56             }
57       }
58  return $maxIntronT;
59  }
60
61  ###############
62  sub parseUcscBedLine{
63  # parse bedline in ucsc format and convert data to Ensembl coordinate format
64  my ($bedLine) = @_;
65  chomp ($bedLine);
66  my ($chrom, $start, $end, $name, $score, $strand) = split " ", $bedLine;
67  $start++; # NEED to offset coordinates by 1
68  $chrom =~ s/chr//;
69  $strand = $strand eq '+' ? 1 : -1 ;
70  return ($chrom, $start, $end, $name, $score, $strand) ;
71  }
72  ####################
73  sub get_Median {
74  my ($list) = @_;
75  my $listLength = scalar(@$list);
76  my $median;
77  @$list = sort { $a <=> $b } @$list;
```

Figure 7.5 (continued)

```
78 if ($listLength % 2 == 0)
79         {
80         my $upperMiddleIndex = $listLength/2;
81         my $lowerMiddleIndex = $listLength/2 - 1;
82         $median = .5 *($list->[$upperMiddleIndex] + $list->[$lowerMiddleIndex]);
83         } else {
84         my $middleIndex = ($listLength - 1)/ 2 ;
85         $median = $list->[$middleIndex];
86         }
87 return $median;
88 }
89
90 #####################
91 sub intronOverlaps{
92 # Converts Intron object into a Range object
93 # and checks overlap of intron with input range
94 my ($intron, $intronStrand, $range) = @_;
95 my $intronRange = Bio::Range->new(-start => $intron->seq_region_start(),
96         -end=> $intron->seq_region_end(), -strand => $intronStrand);
97 my $overlapResult = $intronRange->overlaps($range);
98 return $overlapResult;
99 }
100
101 #####################
102 sub intronOverlapsAnyBed{
103 my ($bedRangeList, $intron, $intronStrand) = @_;
104 foreach my $bedRange (@$bedRangeList)
105         {
106         if (intronOverlaps($intron, $intronStrand, $bedRange))
107                 {return 1;}
108         }
109 return 0;
110 }
111
112 #####################
113 sub intronLengthsForOneHostGene{
114 my ($transcript, $bedRangeList, $overlapList, $otherList) = @_;
115 my @introns = @{$transcript->get_all_Introns()};
116 my $transId = $transcript->display_id();
117 my $intronStrand = $transcript->strand();
118 foreach my $intron (@introns)
119         {
120         if (intronOverlapsAnyBed($bedRangeList, $intron, $intronStrand))
121                 {push @$overlapList, $intron->length();}
122         else
123                 {push @$otherList, $intron->length();}
124         }
125 }
126
```

Figure 7.5 (continued)

```
127 ####################
128 sub checkForOverlappingIntrons {
129 # determine how many introns overlap region (should = 1)
130 my ($transcript, $transId, $chrom, $bedRange, $name, $anomalousHostCount) = @_;
131 my @introns = @{$transcript->get_all_Introns()};
132 print " Transcript: " , $transcript->display_id() ,
133        " intron Ct = " , scalar(@introns) , "\n" if $option{'w'};
134 my $intronStrand = $transcript->strand();
135 my $foundOverlaps = 0;
136 foreach my $intron (@introns)
137        {
138        $foundOverlaps++ if(intronOverlaps($intron, $intronStrand, $bedRange));
139        }
140 if ($foundOverlaps == 0)
141        {
142        warn "##No overlapping intron in $transId for $name $chrom: ",
143              $bedRange->start, "-", $bedRange->end, "\n" if $option{'w'};
144        $anomalousHostCount->{'noHostCount'} += 1;
145        }
146 if ($foundOverlaps > 1)
147        {
148        warn "##More than 1 overlapping introns in $transId for $name $chrom: ",
149              $bedRange->start, "-", $bedRange->end, "\n" if $option{'w'};
150        $anomalousHostCount->{'multipleIntronCount'} += 1;
151        }
152 return $foundOverlaps;
153 }
154
155 ####################
156 sub checkTranscript {
157 # Confirm that BEDfile region is included in exactly one "host gene" intron
158 my ($transcript, $chrom, $bedRange, $name, $anomalousHostCount) = @ ;
159 if (!$transcript)
160        {
161        warn "#####Could not find transcript with intron for $name $chrom:",
162              $bedRange->start, "-", $bedRange->end, "\n" if $option{'w'};
163        $anomalousHostCount->{'noHostCount'} += 1;
164        return 0;
165        }
166 my $transId = $transcript->display_id() ;
167 my $foundOverlaps =
168        checkForOverlappingIntrons($transcript, $transId,
169              $chrom, $bedRange, $name, $anomalousHostCount);
170 #region should overlap exactly 1 intron
171 return 0 if ($foundOverlaps != 1);
172 return 1;
173 }
174
```

Figure 7.5 (continued)

```
175 #####################
176 sub buildHostHash {
177 # Build two-level hash associating transcript name with Transcript
178 # object and list of embedded regions
179 my ($bedFile, $slice_adaptor, $anomalousHostCount) = @_;
180 my $hostHash = {};
181 my $infh = new FileHandle "<$bedFile";
182 while (my $bedLine = <$infh> )
183         {
184         next if $bedLine =~ /^\s*#/ ;
185         my ($chrom, $start, $end, $name, $score, $strand) =
186         parseUcscBedLine($bedLine);
187         # get genes that overlap BED range
188         my $slice = $slice_adaptor->fetch_by_region('chromosome',$chrom,
189                                 $start,$end, $strand);
190         die "Could not get slice for $chrom:$start-$end\n" if (!$slice);
191         my $bedRange = Bio::Range->new(-start => $start,
192                                 -end=> $end, -strand => $strand);
193         my $transcript = get_transcript_with_most_introns($slice);
194         my $statusOk = checkTranscript($transcript, $chrom,
195                         $bedRange, $name, $anomalousHostCount);
196         next if !$statusOk;
197         my $transId = $transcript->display_id() ;
198         $hostHash->{$transId} = { "transcript" => $transcript,"bedRanges" => [ ]}
199             if (!exists $hostHash->{$transId});
200         my $currentHostGene = $hostHash->{$transId};
201         my $currentHostBeds = $currentHostGene->{"bedRanges"};
202         push @$currentHostBeds, $bedRange;
203         }
204 close $infh;
205 return $hostHash;
206 }
207
208
209 ####################
210 sub process_bedFile {
211 my ($bedFile, $db) = @_;
212 my $overlapList = [ ]; # reference to list of lengths of overlapping introns
213 my $otherList = [ ]; # reference to list of lengths of 'other' introns
214 # create hash to keep track of 'anomalies': Snos with no intron-containing
215 # host gene or Host genes with snos overlapping > 1 intron
216 my $anomalousHostCount = {'noHostCount'=> 0, 'multipleIntronCount' => 0};
217 my $startTime = time();
218 my $slice_adaptor = $db->get_SliceAdaptor();
219 my $hostHash = buildHostHash($bedFile, $slice_adaptor, $anomalousHostCount);
```

Figure 7.5 (continued)

```
220  foreach my $transcriptID (keys %$hostHash)
221          {
222          my $hostHashValue = $hostHash->{$transcriptID};
223          my $transcriptObject = $hostHashValue->{"transcript"};
224          my $bedRangeList = $hostHashValue->{"bedRanges"};
225          intronLengthsForOneHostGene($transcriptObject, $bedRangeList,
226                  $overlapList, $otherList);
227          }
228  print "Snos with no intron-containing host gene = ",
229          $anomalousHostCount->{'noHostCount'}, "\n";
230  print "Host genes with snos overlapping > 1 intron = ",
231          $anomalousHostCount->{'multipleIntronCount'}, "\n";
232  print "Host genes found = " , scalar (keys %$hostHash) , "\n";
233  print "Median value of lengths of ", scalar (@$overlapList) ,
234          " overlapping introns = " , get_Median($overlapList) , "\n";
235  print "Median value of lengths of ", scalar (@$otherList) ,
236          " other introns = " , get_Median($otherList) , "\n";
237  print "Elapsed Time = " , (time() - $startTime) , " secs.\n";
238  }
239  ####################
240
241  my $USAGE =<<END_OF_USAGE;
242
243  Usage: ensemblIntronLengths [options] myBedFile
244          where myBedFile is a bed file of genomic ranges
245          Options:        -d <dbName>: default = homo_sapiens_core_42_36d
246                          -h <hostName>: default = ensembldb.ensembl.org
247                          -u <userName>: default = anonymous
248                          -w display warnings
249  END_OF_USAGE
250
251  getopts('d:u:h:w', \%option) || die("$USAGE");
252  my ($bedFile) = @ARGV;
253  $bedFile || die("$USAGE");
254  my $host = $option{'h'} || 'ensembldb.ensembl.org';
255  my $user = $option{'u'} || 'anonymous';
256  my $dbname = $option{'d'} || 'homo_sapiens_core_42_36d';
257  my $db = new Bio::EnsEMBL::DBSQL::DBAdaptor(-host => $host,
258                  -user => $user, -dbname => $dbname);
259  process_bedFile($bedFile, $db);
```

Figure 7.5 (continued)

$hostHash is a two-level hash whose structure is illustrated in Figure 7.7. At the
top level, the keys of the hash are the IDs of the embedding transcripts. The values
of the hash are two-element anonymous hashes, with the two keys "transcript" and
"bedRanges." The value of the "transcript" hash element is a Perl reference that points
to the transcript's Ensembl Transcript object, whereas the value of the "bedRanges"
element points to a Perl array of BioPerl Range objects, one for each snoRNA that is

Build hash of transcripts of host genes

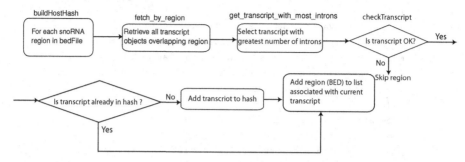

Calculate lengths of all introns in host genes

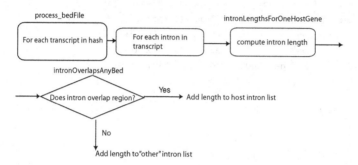

Figure 7.6 Flowchart for the ensemblIntronLengths.pl intron-lengths distribution program. The top part of the figure shows the construction of the hash that associates host genes with their embedded snoRNAs. The bottom half of the figure outlines the steps involved in extracting all the introns and computing the intron lengths. The principal steps in the algorithm are indicated by rounded rectangles, with the program subroutine or Ensembl API method used to implement each step noted outside of the corresponding rectangle.

```
$hostHash =
  {
  $transcriptID_1->
    {
     'transcript'-> $referenceToTranscriptObect_1,
     'bedRanges'->
        $referenceToArrayOfRangeObjectsForSnoRnasInTranscript_1
    },
  $transcriptID_2->
    {
     'transcript'-> $referenceToTranscriptObect_2,
     'bedRanges'->
        $referenceToArrayOfRangeObjectsForSnoRnasInTranscript_2
    },
  ...
  }
```

Figure 7.7 Schematic illustrating the two-level "hostGene" hash structure used to store transcript and embedded snoRNA coordinates in the program ensemblIntronLengths.pl. A top-level "key" specifies the host gene transcript ID. The values of the hash are two-element anonymous hashes, with the two keys "transcript" and "bedRanges."

embedded in an intron that has been spliced out of the transcript. (Note that multiple snoRNAs are sometimes embedded in the introns of a single host gene.)

As described previously in the Galaxy implementation of this problem (see Section 5.4.6), each host gene will typically have multiple transcripts corresponding to the multiple isoforms of the gene. To avoid biasing the subsequent counting statistics, the program needs to select a single representative isoform for each host gene. For the present example, we will choose the isoform with the largest number of exons as the representative transcript, although other choices could be made equally well.

In the second part of the program, shown in the lower half of the flowchart in Figure 7.6, the program loops through all of the host-gene transcripts in $hostHash and, for each transcript, retrieves a list of Ensembl Intron objects corresponding to each of the transcript's introns. The program then computes the length of each intron and stores that length in one of two lists, depending on whether the current intron overlaps any of the regions in the BED file or not. Finally, after all the host-gene transcripts have been processed, the program computes the median length of all the introns that overlapped regions in the input list (e.g., those overlapping a snoRNA) and the median length of all the introns that did not overlap the corresponding feature.

7.7.3.2 Program implementation details

Now let us look at the program implementation in more detail, focusing especially on the program components involving the Ensembl API. The main program (Figure 7.5, lines 251 through 259) is found at the end of the code, after all the subroutines. The main program simply reads in the program argument and options (lines 251 through 256), creates a database adaptor (lines 257 through 258), and passes a reference to the adaptor along with the name of the BED file to the subroutine process_bedFile (line 259).

The process_bedFile subroutine (lines 210 through 238) is the principal subroutine of the program. process_bedFile first creates a Slice Adaptor object to retrieve gene information from the database (line 218) and then passes a reference to the Slice Adaptor object to the routine buildHostHash. Subroutine buildHostHash then creates the $hostHash hash structure, which associates each host-gene transcript with its embedded snoRNAs (line 219). After the $hostHash data structure has been built, process_bedFile cycles through each of the host genes in the hash (lines 220 through 227) and for each one, calls the intronLengthsForOneHostGene subroutine (lines 225 through 226) to calculate the intron lengths for the host gene. Finally, after all the host genes have been processed, the subroutine calls the get_Median routines (lines 234 and 236) to calculate the medians of the lists of intron lengths and displays the results.

The subroutine buildHostHash (lines 176 through 206) builds the hash structure that associates snoRNAs with host genes. The subroutine is implemented by looping through the input BED file and for each BED record, first parsing the record using the parseUcscBedLine subroutine (lines 62 through 71) to perform simple data transformations on the chromosome and strand fields to convert them from UCSC's BED format to Ensembl format. In particular, note that the offset of 1 executed in line 67 is a result of UCSC and Ensembl's different coordinate conventions and is described in more detail in Appendix 1. After reading in the list of regions in the BED file, buildHostHash constructs a Slice object to retrieve the annotations associated with each region (lines 188 through 189). buildHostHash then calls the subroutine get_transcript_with_most_introns (line 193) to select the Transcript object associated with the overlapping transcript with the most introns.

In turn, subroutine get_transcript_with_most_introns (lines 42 through 59) takes a Slice object as an argument and starts by retrieving all the Transcript objects associated with all the Gene objects that overlap the region specified in the Slice object (lines 46 through 48). get_transcript_with_most_intron then sorts the list of transcripts for each overlapping Gene object and identifies the transcript that has the largest number of introns in the transcript, or is the longest transcript, if two or more transcripts have the same (maximum) number of introns (lines 49 through 51). Finally, subroutine get_transcript_with_most_introns compares the number of introns of the currently selected transcript with the maximum-intron count from the previously tested genes and saves the intron count and a reference to the current Transcript object if a new maximum has been identified (lines 52 through 56). The actual sorting of the Transcript objects is performed by the subroutine sortGeneTranscriptsByIntronCtAndLength (lines 30 through 39).

Once get_transcript_with_most_introns has returned a host-gene transcript, the buildHostHash subroutine then calls an error-checking routine checkTranscript (lines 194 through 195) to check for anomalous cases, such as finding that the BED region overlaps a host-gene exon or that the BED region overlaps more than one of the host-gene transcript's introns. These occurrences should not happen, but the input BED file may have incorrect data, and the data in genome databases may have errors as well. It is usually wise to include such "sanity checks" just to be safe. Finally, if the transcript does not have any errors, buildHostHash adds the host gene to $hostHash if it is not yet present in the hash (lines 198 through 199), and adds the BED coordinate string of the snoRNA to a Perl array associated with the hash element for the host gene (lines 200 through 202).

The actual intron length computations are carried out in the intronLengthsForOneHostGene subroutine (lines 113 through 125). This subroutine applies the get_all_Introns method to the current Transcript object (line 115), retrieves the lengths of each intron using the Intron object's length() method (lines 121 and 123), and appends the lengths to either the list of host introns or the list of non-host introns, depending on whether the intron overlapped any region in the BED file or not (lines 120 through 124).

The last subroutine we will look at is intronOverlaps (lines 91 through 98). This subroutine determines whether the coordinates of an intron, as represented by an

Ensembl Intron object, overlap those of a snoRNA gene, as represented by a BioPerl Range object. To determine whether the regions overlap, intronOverlaps calls the overlaps() method of the BioPerl Range object (line 97). However, before intronOverlaps can call this method, the subroutine needs to convert the Ensembl Intron object into a BioPerl Range object. The reason for this change is that the default coordinates of an Ensembl Intron object – as specified by the Intron object's start() and end() methods – are Ensembl slice coordinates, whereas the coordinates of a Range object are chromosomal coordinates. The intronOverlaps subroutine resolves this incompatibility by building a Range object for the intron using chromosomal coordinates by using the seq_region_start() and seq_region_end() methods (lines 95 through 96). Once both the intron and snoRNA locations are in chromosomal coordinates, the Range overlaps() method correctly determines whether they overlap or not.

7.7.3.3 Executing the program

After making the program executable and placing it in a $PATH directory, we now run the program with the command:

```
$ ensemblIntronLengths.pl hacaWgRna.hg18.bed
```

We obtain the following output:

Snos with no intron-containing host gene = 10
Host genes with snos overlapping > 1 intron = 1
Host genes found = 65
Median value of lengths of 86 overlapping introns = 1075.5
Median value of lengths of 856 other introns = 932
Elapsed Time = 219 secs

From the result, we see that there is a 15% difference between the median lengths of introns that host snoRNAs and the other introns in the host genes (i.e., 1,075 versus 932). However, if we do a statistical analysis of this difference – for example, by performing a t test on the two length distributions that we have found – we find that this difference is not statistically significant ($p = 0.3$). There are a couple of other interesting things that we can learn from the results of our program. First, we notice that there is no annotated Ensembl host gene for ten of the snoRNAs, indicating that these snoRNAs are intergenic, or (more likely) that the Ensembl gene-prediction algorithm did not detect the host genes. Second, and more surprising, the program has also identified one case where a snoRNA apparently overlaps more than one intron. To further understand this unexpected result, we can re-run the program, this time adding the -w flag to display more diagnostic information. From the resulting output, we learn that it is transcript ENST00000357861 that has been flagged as having two introns that overlap the snoRNA. Further inspection of this transcript in a genome browser shows that the overlapped "exon" is only two nucleotides in length and is almost definitely an annotation artifact (see Figure 7.8). It is precisely because

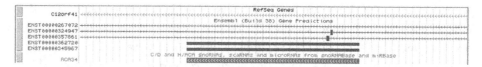

Figure 7.8 Display of a portion of the ENST00000357861 intron-exon structure. The gene structure prediction includes an anomalous two-nucleotide exon, as a result of which snoRNA ACA34 appears to overlap two introns.

misannotations are not unusual that including test code for unexpected annotations is important.

In one sense, our results are disappointing in that we have not found a statistically significant difference in the lengths of human H/ACA snoRNA host introns and other host-gene introns. However, the point of the example has been to demonstrate that by writing a relatively simple program, we were able to answer a nontrivial biological question. In contrast, to have written such a program without the use of the Ensembl API (or the UCSC API) – for example, by writing a general SQL program as in Section 6.2 – would have required substantially more effort. (The reader who is not convinced of this statement is encouraged to write such a program.)

Compared to performing this analysis with Galaxy, as we described in Chapter 5, writing a program using the Ensembl API has both advantages and disadvantages. Clearly, performing a single interactive analysis is faster than writing and debugging a computer program. However, if we need to run the analysis repeatedly – say, we want to look at introns embedding C/D snoRNAs rather than H/ACA snoRNAs, or we want to compare the results obtained in the mouse or rat genome with that obtained in the human genome – using a computer program becomes more attractive. In addition, if we wanted to see the effect of using a different criterion for selecting the representative host gene, we would just have to change one subroutine (see Exercise 7.4), whereas with Galaxy, we would need to have a local Galaxy implementation to which we would need to write and add a custom tool.

7.8 Ensembl Java API

Historically, Ensembl has supported two distinct APIs – one for Perl language programmers, the other for Java programmers. Currently, there are several reasons why using the Ensembl Perl API is recommended and why we exclusively describe the Perl API in this book. First, the Ensembl Genome Browser itself is implemented with the Ensembl Perl API, implying that code for any data manipulations found in the Ensembl Browser will be available somewhere within the Ensembl Perl software. In contrast, although many types of data manipulation found on the Ensembl web site can also be carried out using the Java API, there is no guarantee that this will always be the case. Second, the range of bioinformatics software available in the Perl language – mostly via the BioPerl project – is significantly larger and has a larger user base than the comparable code available via Java and the BioJava project (e.g., a search for "BioPerl" on Google yields about four times as many hits as a "BioJava" query). Finally, and most importantly, as of December 2006 (Ensembl release 42) the Ensembl developers are no longer formally supporting the Java API.

In principle, the Java programmer who wants programmed access to the Ensembl databases can still use the archival Ensembl Java API. Documentation for the most recently supported version of the Java API can be found in the Ensembl archives at http://oct2006.archive.ensembl.org/info/software/java/index.html. In addition, the BioJava project (http://biojava.org) still provides some support for Java querying of Ensembl databases. However, many of the features of the Ensembl Browser and database are already incompatible with the archival Java API, and over time, the Ensembl Java API is likely to grow increasingly out of date and difficult to use.

Chapter summary

- The Ensembl API is patterned to a large degree on objects and methods found in the BioPerl software package.
- Identifying the proper software objects to store different types of genomic data and database-adaptor objects is often the crucial step in using the Ensembl API.
- Ensembl pdocs documentation contains definitions and sample usage of Ensembl objects. Ensembl tutorials, scripts, and "t" files contain complete programs illustrating the usage of those objects.
- Any programs written with the Ensembl API run equally well on the Ensembl public database as on a local, private mirror.
- All Ensembl software is open source, and software use does not require licensing for any application.

Exercises

Note that you will need to install BioPerl to carry out Exercises 1, 3, 4, and 5, and you will need to install the Ensembl API to complete Exercises 3, 4, and 5.

1. Write a Perl script, using BioPerl, to read in a file of FASTA sequences, and for each sequence:
   a. Print the name and description of the sequence.
   b. Print the reverse complement of the first thirty nucleotides of the sequence.
   c. Print the amino acid translation of the reverse complement of the first thirty nucleotides of the sequence.
2. Find the code for the revcom method used in the BioPerl Bio::Seq module using BioPerl pdoc or the Deobfuscator. In what module is the revcom method actually defined?
3. Modify the program ensemblTest1.pl so that the program accepts arguments specifying the chromosome and start- and end-coordinates of the region from which to access the gene annotations.

4. Modify the intron lengths program so that it selects the longest gene transcript as the representative gene rather than the transcript with the largest number of introns. How do the results for the relative sizes of snoRNA hosting introns and host-gene introns that do not contain embedded snoRNAs change?

5. Write a Perl program using the Ensembl API to search for NMD candidates. Run the program against the human Ensembl gene set as well as against the human VEGA gene set. Compare the percentage of genes that are NMD candidates in the two data sets. Are the results similar in mouse?

8

Programmed Querying with Ensembl, Continued

In Chapter 7, we introduced the Ensembl Perl API and presented two programs illustrating how to use the API to extract and manipulate genomic data from an Ensembl database. In the present chapter, we continue our discussion of programmatic querying of Ensembl data with three more advanced topics: accessing multiple-sequence alignment (MSA) data from Ensembl's comparative-genomics "compara" database, accessing data found on Ensembl DAS tracks, and installing and maintaining a local mirror of an Ensembl database.

8.1 Using ensembl-compara

The Ensembl API software consists of several components, including the Ensembl core API, ensembl-functgenomics for functional genomics (e.g., micro-array expression) data, ensembl-variation for variation (e.g., SNP) data, and ensembl-compara for comparative genomics, for example, multiple-sequence alignment (MSA) data. The examples presented in the last chapter only used code from the Ensembl core API. Because accessing MSAs is necessary for several of the biological examples we consider in this book (e.g., assessing sequence conservation or identifying if a polymorphism in one species is the dominant variant at the homologous site in another species), we now describe the ensembl-compara component of the Ensembl API.

In Chapter 7, we created an explicit database-adaptor object for each Ensembl database that we needed to access. This approach is fine if one only needs to connect to one or two Ensembl databases. However, this method has disadvantages if one wants to access an MSA. This is because with Ensembl, accessing MSAs requires connecting to the genome database of every species in the alignment. Consequently, using explicitly defined database adaptors would require changing the database adaptors each time the set of species in the alignment changed. In addition, whenever Ensembl released a database update, the database adaptors would need to be modified, or else they would not point to the most current versions of the database.

To address these issues, the Ensembl API includes a "Registry file" (named reg.conf) that contains a listing of all of the database IDs for each Ensembl release. Using the

Registry file, one can write programs that load all of the database adaptors required for MSAs without explicitly listing all of the required database names. However, writing the code to access an MSA via the Registry file is still a bit tricky because each of the sequences in the MSA is stored in a separate database. For example, to access a human-mouse-rat alignment, the software needs to extract the human sequence from Ensembl's Human Genome Database, the mouse sequence from the Ensembl mouse database, and the rat sequence from the rat database.[1] In addition, the ensembl-compara API software needs to access yet another database (the ensembl-compara database itself) to assemble the various sequences into the actual alignment.

Retrieving alignments via the Ensembl Registry file involves five steps. First, the program needs to "load" the Registry file. Second, the program accesses the Registry file to create an Ensembl database-adapter object (called a GenomicAlignBlockAdaptor object) that can extract blocks of genomic MSAs. Third, the program must create a MethodLinkSpeciesSet object, which links together the name of the desired alignment method (e.g., PECAN or BLASTZ) with the set of species for which the alignment is needed. Fourth, the program selects one species in the alignment to call the "query" sequence and creates an Ensembl Slice object specifying the alignment region (i.e., chromosome and coordinates) in the query-species genome. Finally, the program retrieves the alignment by passing Perl references for the Slice object and the MethodLinkSpeciesSet object to the fetch_all_by_MethodLinkSpeciesSet_Slice of the GenomicAlignBlockAdaptor object.

The tricky part is the third step because there is no method or subroutine built into the Ensembl API for creating MethodLinkSpeciesSet objects. However, in the next section, we will present an example of such a subroutine that is taken from the DumpMultiAlign.pl program in the Ensembl comparative genomics tutorial (called ComparaTutorial and available as part of the Ensembl software distribution). If you need a subroutine with these capabilities, you can just copy this subroutine directly into your application program (just as I did for the example below).

8.1.1 Ensembl-compara example

Let us now use the ensembl-compara API to display MSAs extracted from the Ensembl databases. The objective of our example program, ensemblComparaExample.pl, is similar to that for our conservation-at-polymorphism-site example with Galaxy (see Section 5.4.4). Specifically, we want to be able to pass the program a list of coordinates (in a BED file) to indicate the regions for which we need alignments. The program should then display all of the alignments that overlap each region.

The program will be called with the command:

```
$ ensemblComparaExample.pl [options] ensemblCompara.test.bed
```

---

[1] As we will see in Chapter 10, this is quite different from the way that UCSC stores MSA data.

```
Homo sapiens:    NCBI36
Mus musculus:    NCBIM36
Rattus norvegicus:     RGSC3.4
Bos taurus:    Btau_2.0
Canis familiaris:      BROADD2
Pan troglodytes:       CHIMP2.1
Macaca mulatta: MMUL_1
Monodelphis domestica: BROADO3
Gallus gallus: WASHUC2


CLUSTAL W(1.81) multiple sequence alignment
HsX/100162142-100162165 TGCAGTCCATCTTGCATCCTCCAC
PtX/100571604-100571627 TGCAGTCCATCTTGCATCCTCCAC
MmX/99725832-99725855   TGCAGTCCATCTTGCATCCTCCAC
MmX/129586439-129586462 TGCAGTCCACCTGGCATCCTCTAC
RnX/121709625-121709648 TGCACTCCACCTGGCATCCTCTAC
CfX/78022478-78022501   TGCAGTCCACCTGGCATCCTCTAC
Gg4/2085509-2085532     TGCATTCCACCTGGCATCCTCTAT
                        **** **** ** ******** *
Elapsed Time = 42 secs.
```

Figure 8.1 Output generated by the ensemblComparaExample.pl program using the test data file shown in the text. Note that difference in the start positions in the human genome between the input file (100162141) and the output in the figure (100162142) is the result of UCSC and Ensembl's different coordinate numbering conventions.

where ensemblCompara.test.bed contains the list of genomic regions in BED format. The program has numerous options, including ones for specifying what kind of alignment to use, what species to include, and which species' coordinates are used in the BED file.

For example, if we apply our program to a very simple test BED data file consisting of the single line:

```
chrX   100162141   100162165   cxorf34   0   -
```

we obtain the result shown in Figure 8.1. Note that we need to ensure that the coordinates in the input BED file correspond to the species and assembly of the selected query sequence.

As we noted in Chapter 5, when we described the Galaxy implementation of this task, simply displaying an alignment is typically not all we would want to do. In a more realistic application, we would want to perform additional data processing – perhaps determining the alignment consensus sequence, or counting what fraction of the sequences have the consensus nucleotide at a given alignment position. However, once we have access to the alignment in computer memory (e.g., as a BioPerl SimpleAlign object), it is not difficult to perform these additional tasks by taking advantage of methods available for a SimpleAlign object, which are detailed in the BioPerl pdoc documentation for SimpleAlign.pm.

8.1.1.1 Program implementation

The ensemblComparaExample.pl program source listing is shown in Figure 8.2 and a flowchart of its algorithm is shown in Figure 8.3. As depicted in the flowchart, the execution of the program has two phases. In the first phase (outlined in Figure 8.3a and implemented in lines 194 through 199 of Figure 8.2), the program follows the

```
1  #! /usr/bin/perl -w
2
3  use strict;
4  use lib "$ENV{HOME}/programs/ensembl42/ensembl/modules/";
5  use lib "$ENV{HOME}/programs/ensembl42/ensembl-compara/modules/";
6  use lib "$ENV{HOME}/programs/ensembl/bioperl-live/";
7  use Bio::EnsEMBL::Registry;
8  use Bio::EnsEMBL::Utils::Exception qw(throw);
9  use Bio::SimpleAlign;
10 use Bio::AlignIO;
11 use Bio::LocatableSeq;
12 use FileHandle;
13 use Getopt::Long;
14
15 my $usage = qq{
16 Usage: ensemblComparaExample [options] myBedFile
17    where myBedFile is a bed file of genomic ranges
18  Getting help:
19    [--help]
20  General configuration:
21    [--db compara_db_name]
22       the name of compara DB in the registry_configuration_file or any
23       of its aliases. Uses "compara" by default.
24  For the query slice:
25    [--species species]
26       Query species. Default is "human"
27    [--noRestrictBlocks]
28     Display entire overlapping alignment, not just specified region
29  For the alignments:
30    [--alignment_type method_link_name]
31       The type of alignment. Default is "PECAN"
32    [--set_of_species species1:species2:species3:...]
33       The list of species used to get those alignments. Default is
34       "human:mouse:rat:cow:dog". The names should correspond to the name of
35       core database in the registry_configuration_file or any of its
36       aliases
```

Figure 8.2 Source code of the ensemblComparaExample.pl program for displaying a multiple-sequence alignment using the Ensembl API.

```
37    Output:
38        [--output_format clustalw|fasta|...]
39              The type of output you want. "clustalw" is the default.
40        [--output_file filename]
41               The name of the output file. By default the output is the
42               standard output
43    };
44
45    my $db = "compara";
46    my $species = "human";
47    my $alignment_type = "PECAN";
48    my $set_of_species = "human:mouse:rat:cow:dog:chimp:rhesus:opossum: chicken";
49    my $output_file = undef;
50.   my $output_format = "clustalw";
51    my $help;
52    my $noRestrictBlocks = 0;
53
54    GetOptions(
55        "help" => \$help,
56        "noRestrictBlocks" => \$noRestrictBlocks,
57        "db=s" => \$db,
58        "species=s" => \$species,
59        "alignment_type=s" => \$alignment_type,
60        "set_of_species=s" => \$set_of_species,
61        "output_format=s" => \$output_format,
62        "output_file=s" => \$output_file,
63      );
64
65    ####################
66    sub get_species_set {
67    # Construct Ensembl MethodLinkSpeciesSet Object that
68    # contains database adaptors for all the species
69    # in the sequence alignment
70    my ($db, $set_of_species, $alignment_type) = @_;
71    my $genome_dbs;
72    my $genome_db_adaptor =
73        Bio::EnsEMBL::Registry->get_adaptor($db, 'compara','GenomeDB');
74    throw("Registry configuration file has no data for connecting to <$db>")
75        if (!$genome_db_adaptor);
76    foreach my $this_species (split(":", $set_of_species))
77        {
78        my $this_meta_container_adaptor =
79            Bio::EnsEMBL::Registry->
80                    get_adaptor($this_species,'core','MetaContainer');
81        throw("Registry configuration file has no data for <$this_species>")
82            if (!$this_meta_container_adaptor);
```

Figure 8.2 (continued)

```
83   my $this_binomial_id = $this_meta_container_adaptor->get_Species-> binomial;
84       # Fetch Bio::EnsEMBL::Compara::GenomeDB object
85       my $genome_db =
86          $genome_db_adaptor->fetch_by_name_assembly($this_binomial_id);
87       # Display assembly info
88       print $genome_db->name, ":\t", $genome_db->assembly, "\n";
89       # Add Bio::EnsEMBL::Compara::GenomeDB object to the list
90       push(@$genome_dbs, $genome_db);
91       }
92   # Getting Bio::EnsEMBL::Compara::MethodLinkSpeciesSet object
93   my $method_link_species_set_adaptor =
94       Bio::EnsEMBL::Registry->
95          get_adaptor($db,'compara','MethodLinkSpeciesSet');
96   my $method_link_species_set =
97       $method_link_species_set_adaptor->
98          fetch_by_method_link_type_GenomeDBs($alignment_type, $genome_dbs);
99   throw("The database do not contain $alignment_type data for $set_of_species!")
100      if (!$method_link_species_set);
101  return $method_link_species_set;
102  }
103
104  ####################
105  sub slice_from_Registry {
106  my ($species, $chrom, $start, $end) = @_;
107  my $slice_adaptor =
108      Bio::EnsEMBL::Registry->get_adaptor($species, 'core', 'Slice');
109  throw("Registry configuration file has no data for connecting to <$species>")
110      if (!$slice_adaptor);
111  my $query_slice =
112      $slice_adaptor->fetch_by_region('toplevel', $chrom, $start, $end);
113  throw("No Slice can be created with coordinates $chrom:$start-$end")
114      if (!$query_slice);
115  return $query_slice;
116  }
117
118  ####################
119  sub GAB_to_SimpleAlign {
120  my ($GAB) = @_;
121  my $simple_align = Bio::SimpleAlign->new();
122  my $all_aligns = $GAB->get_all_GenomicAligns;
123  # Create a Bio::LocatableSeq object from every GenomicAlign
124  foreach my $this_align (@$all_aligns)
125      {
126      my $seq_name = $this_align->dnafrag->genome_db->name;
127      $seq_name =~ s/(.)\w* (.)\w*/$1$2/;
```

Figure 8.2 (continued)

```
128    $seq_name .= $this_align->dnafrag->name;
129    my $aligned_sequence = $this_align->aligned_sequence;
130    my $seq = Bio::LocatableSeq->new(
131            -SEQ    => $aligned_sequence,
132            -START  => $this_align->dnafrag_start,
133            -END    => $this_align->dnafrag_end,
134            -ID     => $seq_name,
135            -STRAND => $this_align->dnafrag_strand
136        );
137    # Add this Bio::LocatableSeq to the Bio::SimpleAlign
138    $simple_align->add_seq($seq);
139    }
140  return $simple_align;
141  }
142
143  ################
144  sub parseUcscBedLine{
145  # parse bedline in ucsc format and convert data to Ensembl coordinate format
146  my ($bedLine) = @_;
147  chomp ($bedLine);
148  my ($chrom, $start, $end, $name, $score, $strand) = split " ", $bedLine;
149  $start++; # NEED to offset coordinates by 1
150  $chrom =~ s/chr//;
151  $strand = $strand eq '+' ? 1 : -1 ;
152  return ($chrom, $start, $end, $name, $score, $strand) ;
153  }
154
155  ################
156  sub alignments_for_one_BED {
157  my ($species, $chrom, $start, $end,
158    $method_link_species_set, $GAB_adaptor, $alignIO) = @_;
159  # Fetching the query Slice:
160  my $q_slice = slice_from_Registry($species, $chrom, $start, $end);
161  # Fetching all the GenomicAlignBlock corresponding to this Slice:
162  my $genomic_align_blocks = $GAB_adaptor->
163    fetch_all_by_MethodLinkSpeciesSet_Slice($method_link_species_set,$q_slice);
164  my $all_aligns;
165  # Create a Bio::SimpleAlign object from every GenomicAlignBlock
166  foreach my $this_GAB (@$genomic_align_blocks)
167    {
168    $this_GAB = $this_GAB->restrict_between_reference_positions($start, $end)
169    if (!$noRestrictBlocks);
170    my $simple_align = GAB_to_SimpleAlign($this_GAB);
171  #  my $simple_align = $this_GAB->get_SimpleAlign;
172    push(@$all_aligns, $simple_align);
173    }
```

Figure 8.2 (continued)

```
174 foreach my $this_align (@$all_aligns)
175    {
176    print $alignIO $this_align;
177    }
178 }
179
180 ####################
181 # main program starts here
182 my $startTime = time();
183 # Print Help and exit
184 if ($help) {
185   print $usage;
186   exit(0);
187 }
188 if ($output_file) {
189    open(STDOUT, ">$output_file") or die("Cannot open $output_file");
190 }
191 my ($bedFile) = @ARGV;
192 $bedFile || die("$usage");
193
194 Bio::EnsEMBL::Registry->load_registry_from_db(-host => "ensembldb.ensembl.org",
195             -user => "anonymous");
196 my $GAB_adaptor =
197    Bio::EnsEMBL::Registry->get_adaptor($db, 'compara', 'GenomicAlignBlock');
198 my $method_link_species_set =
199    get_species_set($db, $set_of_species, $alignment_type);
200
201 # construct a Bio::AlignIO object for printing the genomic alignments
202 my $alignIO = Bio::AlignIO->newFh(
203             -interleaved => 0, -fh => \*STDOUT,
204             -format => $output_format,  -idlength => 10);
205 # read BED file of regions of interest and convert to Ensembl format
206 my $infh = new FileHandle "<$bedFile";
207 while (my $bedLine = <$infh> ) {
208    next if $bedLine =~ /^\s*#/ ;
209    my ($chrom, $start, $end, $name, $score, $strand)
210       = parseUcscBedLine($bedLine);
211 # extract and print alignments overlappping current region
212    alignments_for_one_BED($species, $chrom, $start, $end,
213       $method_link_species_set, $GAB_adaptor, $alignIO);
214    }
215 print "Elapsed Time = " , (time() - $startTime) , " secs.\n";
216 exit;
217 __END__
218
219
```

Figure 8.2 (continued)

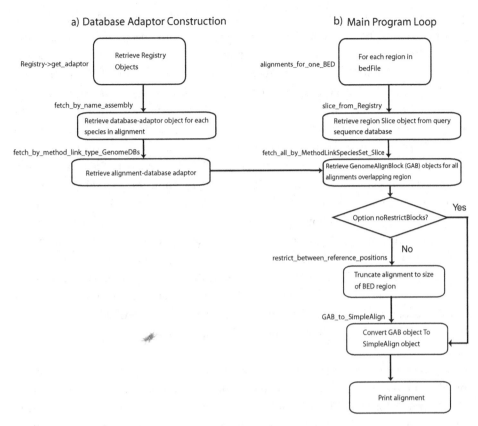

Figure 8.3 Flowchart for the ensemblComparaExample.pl program. The program subroutine or Ensembl API method used in each step is noted outside of the corresponding box. (a) Construction of the database adaptors needed to access the sequence-alignment data. (b) Steps involved in extracting the actual alignment for each region in the BED file using the database adaptors created in part (a).

general procedure described in Section 8.1 for constructing Ensembl database-adaptor objects. Specifically, first the main program loads the Ensembl Registry file (lines 194 and 195) and constructs a GenomeAlignmentBlockAdaptor (GAB) object (lines 196 and 197). A MethodLinkSpeciesSet object is then created by the get_species_set subroutine (called in lines 198 and 199). Once the database-adaptor objects have been constructed, the main program loop (outlined in Figure 8.3b) can be executed. For each BED region, the main program (lines 212 and 213) calls the subroutine alignments_for_one_BED, which creates a query slice for the region, retrieves the alignments that overlap the region, truncates the alignments to the length of the BED, and finally, prints the overlapping alignments. Let us look now at some of the subroutines in more detail.

The MethodLinkSpeciesSet object is created with the get_species_set subroutine, shown in lines 66 through 101. The subroutine is modified from the DumpMulti-Align.pl code in the ensembl-compara tutorial. Subroutine get_species_set first creates a database adaptor for the ensembl-compara database (lines 72 and 73). The subroutine then cycles through each species in the alignment (line 76). For each species, the subroutine first retrieves the species' "binomial ID," which is simply the standard two-word ID specifying the genus and species (e.g., "Homo sapiens" for human, or "Mus musculus" for mouse). In the Ensembl system, the binomial ID is contained in a BioPerl Species object, which is stored in an Ensembl Bio::EnsEMBL::DBSQL::MetaContainer object (lines 78 through 83). Next, in lines 85 and 86, subroutine get_species_set calls the fetch_by_name_assembly method with the species' binomial ID to retrieve the species' GenomeDb object (which specifies the required genomic assembly). The get_species_set subroutine then prints the assembly's ID (line 88) and adds the species' GenomeDb object to a list of all the required GenomeDb objects (line 90). Finally, in lines 93 through 99, the subroutine constructs the MethodLinkSpeciesSet object that links together the required alignment type with the list of the GenomeDb objects.

The get_species_set subroutine is a bit complicated because the subroutine needs to perform multiple steps to establish all the required database connections via the Ensembl Registry file. However, as noted previously, a detailed understanding of the steps involved in constructing the MethodLinkSpeciesSet object is not really necessary. Instead, one can simply copy the get_species_set subroutine in any program in which one needs to extract paired- or multiple-sequence alignment data from Ensembl.

Once we have created the MethodLinkSpeciesSet object, retrieving and displaying alignments using the alignments_for_one_BED subroutine (lines 156 through 178) is straightforward. The subroutine alignments_for_one_BED first calls the subroutine, slice_from_Registry (at line 160) to create a Slice object for the specified genomic region. Using this Slice object, along with the previously created GenomicAlignBlockAdaptor and MethodLinkSpeciesSet objects, the subroutine calls the fetch_all_by_MethodLinkSpeciesSet_Slice method to retrieve a list of Genomic-AlignBlock objects for all the overlapping alignments (lines 162 and 163). Next, alignments_for_one_BED truncates each alignment to the size of the region specified by the BED file coordinates.[2] This step is necessary because the fetch_all_by_MethodLinkSpeciesSet_Slice method returns objects corresponding to the *entire* length of an alignment, even if only a subregion of the alignment overlaps the query region. The truncation is performed by the restrict_between_reference_positions method of the GenomicAlignBlock object in line 168.

Finally, subroutine alignments_for_one_BED needs to convert the GenomicAlignBlock object into a BioPerl SimpleAlign object so that the alignment can be displayed in an easily readable format. The Ensembl API provides a method, called get_SimpleAlign, for carrying out such a conversion (see commented line 171). However, using get_SimpleAlign causes the species' ID information to be lost in the format conversion. (You can confirm this by modifying and executing ensemblCompara-Example.pl with line 170 commented out instead of line 171.) Because we want to

---

[2] This is true unless the "noRestrictBlocks" option was selected when the program was called.

display the species identification in the alignments, we will instead use a custom GenomicAlignBlock to SimpleAlign conversion routine, called GAB_to_SimpleAlign (lines 119 through 140). The key lines in GAB_to_SimpleAlign are lines 126 and 127, where the species' ID is parsed from the sequence name in the GenomicAlign object, and line 128, in which the species' ID is then concatenated with the chromosome name. Finally, the concatenated name is used as the sequence ID in the SimpleAlign object in line 134.

8.1.2 Finding Ensembl objects, revisited

We noted that we identified some of the Ensembl objects we needed (e.g., Genomic-AlignBlock or MethodLinkSpeciesSet) in the DumpMultiAlign.pl demo program. If we had not been aware of the DumpMultiAlign.pl program – and we had not used these Ensembl objects before – we would have probably had to locate them in the Ensembl code using grep and pdoc as described in the previous chapter. (Note that the point here is not that the Ensembl documentation is poor – it is not – but rather that just finding the documentation you need is sometimes difficult.)

For example, knowing that the Ensembl Browser creates MSAs using the PECAN algorithm, we can find the appropriate Ensembl object for MSAs by executing the command:

```
$ grep -rin pecan   .
```

from the main Ensembl directory. We would find that most of the retrieved results are in the ensembl-compara/modules/Bio/EnsEMBL/Compara/Production/Genomic-AlignBlock/ subdirectory, suggesting that we look for a GenomicAlignBlock object in the ensembl-compara pdoc. Doing so, we would have found most of the code we needed. In addition, if we had then run the command:

```
$ grep -rin GenomicAlignBlock   .
```

in the ensembl-compara/scripts subdirectory, we would have found numerous complete examples of the usage of GenomicAlignBlock and MethodLinkSpeciesSet objects as well.

8.2 Accessing Ensembl DAS data

As we have mentioned previously, programmed querying to access data from Ensembl's DAS tracks via the Ensembl API is not possible because most DAS annotations are not stored in the Ensembl databases. However, DAS annotations can be accessed directly from the individual DAS servers. Such access can be performed using the Bio::DAS Perl API. The syntax for using the Bio::DAS API is very similar to that used with the BioPerl and Ensembl APIs. For example, sequence feature annotations from the WormBase DAS server can accessed with code such as (code adapted from the Bio::DAS documentation):

```
use Bio::Das;

# contact a DAS server using the "elegans" data source
 $das     = Bio::Das->new('http://www.wormbase.org/db/das' => 'elegans');

# fetch a segment
 $segment = $das->segment(-ref=>'CHROMOSOME_I',
        -start=>10_000, -stop=>20_000);

# get features from segment
  for $feature ($segment->features) {
      $id = $feature->id;
      $type = $feature->type;
      $refseq = $feature->refseq;
      $reference = $feature->reference;
      $start = $feature->start;
      $stop = $feature->stop;
      @subs = $feature->sub_seqFeature;
}
```

Because all DAS servers transmit their data in the same (i.e., DAS) format, one only needs to learn a single data format and API to access data from any DAS server. However, one does need to know the host address of each DAS server that one needs to access. This information can generally be retrieved from the DAS Registry located at http://www.dasregistry.org.

Actually, since Ensembl (and UCSC) databases include DAS server capability as well, in principle one could directly access the Ensembl and UCSC databases using the Bio::DAS interface as well without needing to use the Ensembl or UCSC APIs. However, in most cases, using Bio::DAS to access the Ensembl or UCSC databases does not offer any advantages compared to using the Ensembl or UCSC APIs.

An alternative method for obtaining programmatic access to some of Ensembl's DAS annotations is via the Ensembl martdb database. This is possible because Ensembl mirrors some DAS annotations so that they can be accessed via MartView. The DAS annotations that are available this way include those that are listed as "external references" on the MartView Attributes input page. Examples of DAS annotations that are available via BioMart include RefSeq, SwissProt, and Unigene IDs. These annotations can then be accessed from Ensembl's martdb database (martdb.ensembl.org) either directly via SQL or with the BioMart API – which uses syntax quite similar to that of the Ensembl API. Because Ensembl's martdb database uses a non-default port number (3316), one needs to include the port number in the database connection command. For example, to determine the names of all the databases in martdb.ensembl.org using SQL, we would execute:

```
$ mysql --user=anonymous --host=martdb.ensembl.org \
    --port=3316 -A -e "show databases;"
```

We will not describe the DAS or BioMart APIs further here. The interested reader is referred to Appendix 2 for an introduction to the DAS format and to the Bio::DAS

documentation for more details on using the Bio::DAS API (http://search.cpan.org/~
lds/Bio-Das-1.06/Das.pm). The BioMart API is described in http://www.biomart.org/
user-docs.pdf.

8.3 Installing and maintaining an Ensembl mirror database

So far, we have executed all of our sample programs against the public Ensembl mirror
database at ensembldb.ensembl.org. In fact, an attractive feature of the Ensembl API
is that any program that can be run against a private Ensembl mirror can be run
equally well against the public mirror.[3] The ability to run one's programs against
a public database can be useful even if one also has installed a private mirror. This
is particularly helpful while debugging a newly developed program, so one can
determine whether unexpected program results are caused by program bugs or by
the configuration of the local mirror.

In fact, as any Ensembl API-based program can be executed equally well with the
public mirror, one might well ask why bother to install a private Ensembl mirror
at all. Indeed, for occasional use, exclusively accessing the public mirror can be an
effective way to programmatically query the Ensembl databases. For one thing, the
public databases will be kept up to date by the Ensembl Development Team. In addition, queries involving MSA data, as in our previous example, require downloading
multiple Ensembl databases (eight, in our example) even if our primary interest
is only in one of those species. Moreover, creating a local mirror of an Ensembl
database requires a considerable amount of free disk space and the human and computer resources to maintain and administer a MySQL database system. As of October
2007, installing a mirror of all of the Ensembl databases required about 500 to 600
gigabytes of disk space. In addition, the Ensembl database is continually increasing in
size; for current disk space requirements, one can check http://www.ensembl.org/info/
webcode.

Despite the additional effort required in installing a private Ensembl mirror, there
are situations where one will find it advantageous to mirror one or more Ensembl
databases. Because the public databases are shared resources, accessing databases
locally may improve performance compared to using the public mirror. Similarly,
one's own usage of the public databases needs to be restrained so as not to monopolize
the shared resources. In addition, it is of course impossible to customize a public
database. Lastly, some computer security systems prevent the accessing of remote
databases.

8.3.1 Installation preliminaries

Installing an Ensembl mirror database is well documented (http://www.ensembl.
org/info/webcode) and reasonably straightforward. As an illustrative example, we

---

[3] As noted in Chapter 10, this feature is not shared by the UCSC system.

outline the steps for installing a small Ensembl database, the core Ensembl database for yeast, *Saccharomyces cerevisiae*. In fact, if one wants to install any Ensembl database, it is wise to first install a small one (e.g., a yeast database) as a test to ensure that there are no problems with the download and install procedures on your system. Installation of the yeast genome database follows very similar procedures as that for larger genomes but will be much faster. Consequently, if there are going to be installation problems, you will generally be confronted with them much more quickly and will be able to test fixes to them more quickly as well. We will consider some of the issues specific to the installation of large databases in Chapter 10 in the context of installing the UCSC databases.

Before installing any of the Ensembl databases, one needs to have installed the Ensembl API code, as well as all of the necessary prerequisite software (e.g., BioPerl) as described in Chapter 7. As noted previously, the version of the Ensembl API you have must agree with the release number of the databases you want to install. You will also need to have installed a MySQL server and to have the necessary MySQL privileges to load a new database and create a new user on this database. In addition, you will need to have a web server program, such as Apache (http://www.apache.org) installed, if you want to mirror the Ensembl Browser in addition to the genome databases. For some of the steps, you also will need "superuser" access to the local host computer system. If you do not have such access, you will need to get assistance from your local system administrator.

We will assume in this chapter – and will continue to assume for the remainder of this book – that any mirror databases will be implemented with MySQL and any mirror browser is implemented using the Apache web server. Whereas, in principle, it is possible to implement the Ensembl (or UCSC) systems with other database management systems (DBMS) or web servers, in practice this is likely to be a very challenging endeavor, as there are places where references to the MySQL and Apache systems are hard-coded in the API software. Moreover, since MySQL and Apache are freely available, there are a few cases where there is a reason not to use these tools. One exception might be if one wishes to integrate Ensembl data with local data that is stored using a different DBMS, such as Oracle or Postgres. In this case, rather than directly use the Ensembl database structure, one may want to perform the data integration within a generic genome-database development environment, such as that provided by the GMOD development tools discussed in Chapter 11. In particular, the GMOD tools support DBMS other than MySQL.

8.3.2 Installing a mirror of an Ensembl database

Assuming that the necessary software has been installed, one can download and install an Ensembl database with the following – Unix Bash and MySQL – commands (where "/diskLocationWithAvailableSpace" is replaced by the actual disk directory location to be used, and the yeast database needs to be changed appropriately if you are installing a release other than 42_1e):

```
# First define a local hard-drive location with available space
$ export ENSEMBLROOT="/diskLocationWithAvailableSpace/ensemblData"
$ mkdir -p $ENSEMBLROOT/saccharomyces_cerevisiae_core_42_1e/

# Next perform the data download.
$ cd $ENSEMBLROOT/saccharomyces_cerevisiae_core_42_1e/
$ ncftp ftp://ftp.ensembl.org/pub/saccharomyces_cerevisiae_core_42_1e/data/mysql/
> get -R *
> quit
$ gunzip *gz

# Next create database and tables
$ mysql -uroot -p
password:
mysql> create database saccharomyces_cerevisiae_42_1e;
$ mysql -u root -p saccharomyces_cerevisiae_42_1e \
    < saccharomyces_cerevisiae_core_42_1e.sql
password:

# Load database tables
$ mysqlimport -u root -p saccharomyces_cerevisiae_42_1e -L *.txt.table

# grant database privileges to ensembl user
$ mysql -u root -p
password:
mysql> grant select on saccharomyces_cerevisiae_42_1e.* to \
    ensemblDbUser@localhost;
```

At this point, the data should be properly installed in the local database and accessible by the Ensembl API. To check that the installation has been completed successfully, one can perform a few basic tests. First, one can confirm that direct SQL access of the database is functional by executing a command such as:

```
$ mysql --user ensemblDbUser --host=localhost -A \
    -e "show tables;" saccharomyces_cerevisiae_42_1e
```

If SQL access is working properly, one can then run a simple program such as ensemblTest1.pl, described in the previous chapter, using the host name and user name of your new local mirror. If this program successfully retrieves and prints a list of gene data, then one can be confident that the local database installation has been successful.

Note that we have only described installing a mirror of the Ensembl core database for S. cerevisiae. Certain applications will require the installation of additional Ensembl databases. For example, to access sequence alignments locally using a program like ensemblComparaExample.pl, we would need to mirror the ensembl-compara database as well as the Ensembl databases of each of the other species in the desired alignments. In addition, we would need to configure a local version of the Ensembl Registry file. Similarly, if we need to mirror Ensembl's BioMart, we would need to install the Ensembl BioMart database. We will not describe these tasks further here, instead referring the reader to the installation and registry sections of the Ensembl software documentation site (http://www.ensembl.org/info/using/api) and the BioMart installation procedures (http://www.biomart.org/user-docs.pdf), where they are described in detail.

8.3.3 Keeping one's database up to date

We recall that Ensembl updates the data in its databases when it performs a new release – which occurs approximately every two months. Depending on one's applications, it may be important to use as current a version of the Ensembl databases as possible. At the present time, if one wants to update one's local Ensembl mirror after a new Ensembl release, one needs to perform a complete re-download and re-installation of the databases that are to be updated. No support is provided for incremental database updating. Consequently, programmed-querying applications for which it is important to use the most current version of the Ensembl data are usually best implemented by accessing the Ensembl public mirror databases.

In addition, the version of API software that one is using must be compatible with the build of the associated database. In Ensembl, tracking API-database version compatibility is implemented via a linked numbering system. For example, to access the homo_sapiens_core_42_36d database, one should use version 42 of the Ensembl API, where the initial number in the full Ensembl database name indicates the required API release number. If another API version (either an earlier or a later one) is used, a warning will be generated. The API will still attempt to execute the program but there is a good chance that the program will crash. Consequently, a new version of the Ensembl API should be downloaded and installed each time one wants to use a new release of the Ensembl databases.

Chapter summary

- Some Ensembl data, such as micro-array expression and multiple-sequence alignment data, are stored in specialized Ensembl databases rather than the "core" Ensembl databases.
- The Ensembl Registry file facilitates simultaneously accessing multiple Ensembl databases, which is necessary for retrieving Ensembl multiple-sequence alignment data.
- Although not directly accessible via the main Ensembl databases, Ensembl DAS data can be accessed directly from their DAS sources via the Bio::DAS interface or, in some cases, via Ensembl's martdb database.
- For reasons of perfomance, data security, and customization, one may want to install private mirrors of one or more Ensembl databases, which can be carried out in a straightforward manner.

Exercises

Note that you will need to install BioPerl and the Ensembl API to complete these exercises. You will also need to install the MySQL server to complete Exercise 2.

1. Run the ensemblComparaExample.pl program against the mouse genome using the mouse sequences identified when running the program against the human genome (i.e., using the mouse coordinates shown in Figure 8.1).

2. Follow the procedure outlined in the text to install a local copy of the Ensembl yeast database. Test your installation by directly querying the database with SQL and by executing the ensemblLocalTest1.pl program.

9

Introduction to the UCSC API

In the present chapter, we turn to UCSC's implementation of programmed genome-database querying. The UCSC software and API, written in C, were originally developed by Jim Kent and are typically referred to as the "kent source tree." Even if you never access the UCSC databases, if you ever write bioinformatics code in C, you are likely to save yourself hours of coding and debugging time by becoming familiar with the kent source tree. This is because of the large number of sequence manipulation and bioinformatics utility routines it provides. The functionality of this code is comparable to – and in some cases, more comprehensive than – the analogous code provided by other well-known bioinformatics projects, such as BioPerl, BioJava, BioPython, and the AJAX libraries of the EMBOSS project. In addition, use of the kent tree "core" code (i.e., that portion of the code neither related to the database interface nor BLAT, and not located in the jkOwnLib subdirectory) is free for all uses, including commercial applications.

We begin with a general overview of the UCSC API design. Then we describe programming requirements, procedures for obtaining, installing, and compiling the UCSC software, and methods for establishing UCSC database access. Next, we look at the principal components of the UCSC source code – including the general and database-specific code libraries, utility programs, database-construction programs, and browser-display programs – and describe the functionality each provides. Next, we consider the basic steps involved in constructing a program to access the UCSC databases via the UCSC API. We then describe methods for "navigating" around the code to find the specific library functions or subroutines that one needs. This is followed by illustrative code examples that apply these tools. In the following chapter, we continue our description of the UCSC API with more advanced examples as well as a description of the issues involved in installing and maintaining a private mirror of all or part of the UCSC databases.

9.1 Overview of the UCSC API

In many ways, the motivation and design philosophy behind the UCSC API are similar to those of the Ensembl API. Both are designed to ease the task of writing application code to access and manipulate data in their underlying databases in a way that does not require detailed knowledge of the database implementation and that will continue to function even if the details of the database implementation change over time.

However, there are significant differences between the UCSC and Ensembl APIs, beyond the obvious ones that they are written in different programming languages and access different genome databases. Some of the differences stem from differences in the objectives of the two groups. Whereas one of Ensembl's stated goals is to develop a programmer interface that can be used relatively easily by non-EBI researchers, the UCSC software has been designed primarily for efficient, internal software development. In contrast to the Ensembl API, the UCSC API is not formally described and, in fact, the UCSC software documentation does not even use the term "API." In addition, although the UCSC code is generally well documented, the documentation is somewhat dispersed and often located within the code itself. In particular, there is little in the way of software-usage tutorials and documentation for the programmer who is new to the UCSC API.

Using the UCSC API requires some knowledge of the table schema of the underlying database. Also, in the UCSC API, data is extracted in the form of C structures rather than as software objects with formally associated methods. To be sure, certain functions are associated with each C structure, as indicated by being defined in the same ".h include" file as the definition of the C structure. However, in contrast to a purely object-oriented approach, it is not unusual to manipulate data in a UCSC C structure with functions that are not explicitly associated with the structure.

Also, UCSC often uses custom-designed data formats to maximize system performance rather than rely on ones that already exist in the bioinformatics community. To be sure, because several of these internally developed data formats,[1] such as BED, PSL, and MAF, are not only useful for fast data transfer but have other desirable properties, they have become *de facto* standards as well. Moreover, the UCSC API does typically include some support for more traditional data formats and protocols. That said, currently there is generally more support for outside developers from Ensembl than from UCSC.

Although the UCSC code is generally backwardly compatible, in the sense that application programs will continue to work even after changes to the UCSC systems software and databases, there is no formal commitment to this effect, nor is there a commitment to provide advanced announcements of changes to library code that may affect the functionality of application programs. Similarly, application programs developed using the UCSC API for use with one species database will usually be portable for use with other UCSC species databases, but this is not always true. For example, database table names sometimes change between UCSC species databases.

---

[1] The principal UCSC data formats are described in Appendices 2 and 3. The reader will need to be familiar with this material to understand parts of this chapter.

As a result, using the UCSC API imposes somewhat more work on the programmer, especially in terms of identifying the appropriate tables and data manipulation functions to use. However, in some situations, the flexibility afforded by such explicit selection of tables and functions can be useful. Moreover, in practice, accessing the necessary database tables for use with specific C structures is simplified because there are utility programs available for converting between database tables and C structures and back again. In fact, in most cases, the necessary code to convert between database table (SQL) format and code (C) format is generated automatically by Jim Kent's autoSQL program, which is described in Chapter 11.

9.2 Software prerequisites

Software requirements for using the UCSC API are similar to those for using Ensembl's API (aside from the obvious differences relating to the fact that the UCSC API is in C rather than Perl or Java). Consequently, the reader is referred to the section on required software in Chapter 7. Also, C programming exerience is clearly necessary for using the UCSC API. For readers lacking such background, many introductory textbooks are available; I have found *Pointers in C* (Reek, 1998) particularly helpful.

As in Chapters 7 and 8, we will assume a system configuration based on Unix and MySQL, and all the caveats described in Chapter 7 for using other systems are equally relevant here. We will also assume that you have a recent version of the C compiler gcc. If your system does not currently have a recent gcc compiler, then you should download and install a copy of gcc, which is freely available from http://gcc.gnu. org.

9.3 Obtaining and maintaining the UCSC code tree

The UCSC API source code can be obtained using the CVS utility as described at http://genome.ucsc.edu/admin/cvs.html. Alternatively, the code can be downloaded from http://hgdownload.cse.ucsc.edu/admin/jksrc.zip. For CVS access, the basic commands (in Bash) are

```
$ export CVSROOT=:pserver:anonymous@genome-test.cse.ucsc.edu:/cbse
$ cvs login
CVS password: genome
$ cvs co -rbeta kent
```

Using CVS downloading has the advantage that keeping your source tree up to date will be easier. It is true that the library functions and utilities, which are the parts of the code you will be primarily using, are quite stable and not updated frequently. Nevertheless, if you use the UCSC API regularly, it is advisable to update your copy of the kent source tree periodically. Under CVS, updating your copy of the UCSC API just involves logging in to the UCSC pserver site and executing the CVS update command (see http://genome.ucsc.edu/admin/cvs.html for more details) :

```
$ cvs update -A -d -P -rbeta
```

In contrast, at this time if you acquire the UCSC source code tree by downloading the zip file, the only way you will be able to update your tree is by downloading another entire copy of the zip file (the zip file is updated approximately every two weeks).

9.4 Installing the UCSC code

Once the UCSC code is downloaded, it needs to be unzipped (if downloaded as a zip file) and compiled. Detailed API installation instructions can be found in the README. building.source file located in the src/product directory of the code tree.

If your system uses the same compiler and compiler version as UCSC – which is likely to be the case if your operating system is linux – then compilation should proceed without any problems. With linux, the only modifications to the README installation procedure that I recommend are to make the code readable by a symbolic debugger such as gdb. The necessary changes are as follows:

1) Make sure that all compilations use the gcc compiler debugging flag. This can be accomplished by editing the compiler-options line in the common.mk file (in the kent/src/inc subdirectory) to be

```
COPT=-g -O
```

(Alternately, one can explicitly include the debugging flag every time one runs a "make" command.)

2) Define an environmental variable called "STRIP" (ideally, in one's login file). With Unix Bash, this definition would be

```
export STRIP=true
```

(or "setenv STRIP true," in the C shell). This somewhat unintuitive definition specifies to the makefile used by the UCSC API to *not* strip debugging information from the object file. Note that the reason for compiling with debug options enabled is not because the UCSC code is buggy (it is not) but rather because it makes it easier to track problems in your own code later.

If your computer uses a different version of the gcc compiler than that used by UCSC, you may need to make a few additional modifications to the standard API compilation procedure. For example, using version 3.3 of the gcc compiler, which is included in some versions of Mac OS X, may produce some warning messages when compiling the UCSC code. In particular, you may get warning messages, such as

warning: ISO C requires whitespace after the macro name
warning: -Wuninitialized is not supported without -O

These warnings are harmless and can be safely ignored. However, there may also be warnings such as

warning: 'rcsid' defined but not used
warning: use of 'long double' type; its size may change in a future release

which may lead to a fatal compiler error. If this fatal error occurs, it can generally be converted to a harmless warning by issuing the Bash command:

```
$ export OSTYPE
```

(or else add the command "export OSTYPE" to your login command file). An alternative approach would be to edit the file common.mk in the kent/src/inc subdirectory to remove the compiler option -Werror from the definition of ${HG_WARN}.

The next step is to define an environmental variable (ideally, in your login file) to tell the system where the top source level of the kent code tree is located. For example, in Bash, the command would be

```
export KENTSRC='/pathToKentSourceTree/kent/src/'
```

where pathToKentSourceTree would, of course, be replaced by the actual path to the code in your system.

You can then compile the basic UCSC library functions (using the "make libs" command in the top directory of the source tree, as described in the README.building. source file). Finally, you may want to compile and test one or two of the utility programs in the $KENTSRC/utils subdirectory just to confirm that everything has installed properly. You may need to create a new subdirectory in your home directory – $HOME/bin/$MACHTYPE, where the environmental variable $MACHTYPE should be defined by your operating system, for example, "ppc" on my Power Macintosh. If this variable is not defined by your system, you will either need to define it in your startup file or else edit the variable $BINDIR in common.mk in $KENTSRC/inc, which specifies where executable programs should be stored. In either case, make sure to include your $BINDIR location in $PATH, the list of directories to be searched for executables.

With these preliminaries taken care of, you should be able to compile and link, say, the faSplit program in $KENTSRC/utils, and try it on a sample FASTA file. To find out just what this program (or any other UCSC code program) does, as well as what command-line syntax and options it expects, you can run the program without any arguments. The program will respond with a short usage message describing its functionality and syntax.

9.5 Database access

If you only want to use the kent library and utility programs, such as the faSplit program mentioned in the last section, you do not need to have any database connection to the UCSC genome databases. However, to take full advantage of the UCSC API, you need to be able to access data from the UCSC databases. In the present chapter, we will describe the easiest method for obtaining such database access – using the UCSC public mirror database at genome-mysql.cse.ucsc.edu. As with remote Ensembl access, you need to check that no local firewalls prevent you from accessing a remote database. In Chapter 10, we will describe alternative methods for accessing data in the UCSC databases that can be advantageous in certain situations. Note that accessing the UCSC databases with the UCSC API for commercial applications requires a license from UCSC.

9.6 Major components of the UCSC code source tree

The UCSC software is organized in a conventional hierarchical directory structure. Although the main kent/src directory (which you should have located at $KENTSRC) has some thirty or more subdirectories, many of them contain code that is not directly needed for genome-database querying. Rather, these directories contain code used by other UCSC or Jim Kent software projects such as alignment programs (e.g., BLAT and intronerator) or programs for running massively parallel computer systems (parasol).

In fact, most of the definitions of the structures and functions of interest to genome-database programming are located in the "include" files in the inc and hg/inc subdirectories of $KENTSRC. Implementations of these functions are found in the lib and jkOwnLib subdirectories, for functions defined in inc, and in the hg/lib subdirectory for functions in hg/inc. In most cases, structures and functions for general applications are defined in inc, whereas those that are useful primarily in the context of the UCSC Genome Browser and databases are defined in hg/inc.

9.6.1 Library functions

The inc subdirectory contains definitions of library functions that provide a wide range of sequence and data manipulation capabilities. Library subroutines are available for everything from managing C data structures such as linked lists, balanced trees, hashes, and directed graphs to developing routines for HTML or CGI code. Additional library functions are available for biological sequence and data manipulation tasks such as reverse complementation, codon and amino acid lookup, and sequence translation.

Many of the most commonly used library functions are defined in the file inc/common.h, which contains routines for memory management, error handling, linked-list management, basic string manipulation, and so on. Other files in the inc subdirectory contain more specialized functions. For example, biological sequence manipulation routines (e.g., for nucleotide translation or reverse complementation) are found in the dnautils.h and dnaSeq.h files, whereas string manipulations for strings whose length can vary during program execution (a capability that is awkward to implement in standard C) are found in dyString.h.

Structure and function definitions directly related to the UCSC genome databases are located in hg/inc. In this subdirectory, one can find structures and functions associated with essentially every type of database table found in the UCSC databases. Standard functions included in these files include ones for creating, loading, and freeing C structures associated with each database table. Additional functions implement data manipulations related to specific data types (analogous to methods associated with data objects). For example, hg/inc contains bed.h, genePred.h, and maf.h files, as well as more specialized files such as snp.h, tfbsCons.h, and mapSts.h for handling SNP and transcription-factor binding site and STS mapping data, respectively.

9.6.2 Utility programs

Many of the programs in the kent source tree are specific for the construction of the UCSC genome databases and displaying data on the UCSC Genome Browser, and are of limited interest to bioinformatics programmers outside the UCSC Development Team. However, the code also includes programs that implement common bioinformatics tasks and that are of more general interest. These programs are mostly found in the utils subdirectory. The source code for each utility program, along with its "make" file, is located within a separate subdirectory of the utils directory. Once the lib library functions have been compiled, any of these utilities can be built by executing its associated make file, after which the utility can be directly executed from the command line. The included utilities include programs for sorting, splitting, or merging FASTA sequences; record parsing and data conversion using GenBank, FASTA, nib, and BLAST data formats; sequence alignment; motif searching; hidden Markov model development; and much more.

Other useful programs include those used to build the UCSC databases, particularly for loading database tables (e.g., hgLoadBed and hgLoadPsl) and for parsing different types of data formats (e.g., gbToFaRa for parsing GenBank records). These programs are generally found in subdirectories of the hg directory. Again, each program or set of closely related programs is usually stored in a separate subdirectory and can be built by running the make file in that subdirectory. The principal overall documentation for these programs is found in the docs subdirectory of the hg/makeDb directory, which also contains the programs for building the UCSC databases.

In addition to being useful as stand-alone applications, the programs in the utils and hg subdirectories provide examples of correct syntax for calling the library functions in inc and hg/inc. Moreover, when one requires functionality not available in one of the libraries, the necessary code can often be found in one of the UCSC application programs and can be copied from there.

9.7 Using the UCSC code library functions

Developing code to access and manipulate data in the UCSC genome databases using the UCSC API involves a three-step process, similar to that we saw in Chapters 7 and 8 when using the Ensembl API. First, we need to write code to connect to the UCSC databases containing the needed data. Second, we need to identify the tables in the database containing the needed data and write code to load that data from the database into our program memory. Finally, we need to find the appropriate library data manipulation functions to perform the data post-processing part of the program.

UCSC database connection can often be accomplished with only a few lines of code. For example, connecting to the hg18 database on UCSC's public mirror may be performed with these two lines of code[2]:

```
hSetDbConnect("genome-mysql.cse.ucsc.edu", "hg18", "genomep", "password");
conn = sqlConnectRemote("genome-mysql.cse.ucsc.edu", "genomep", "password", "hg18");
```

The call to library function hSetDbConnect sets the database parameters within the UCSC API, whereas the call to sqlConnectRemote establishes the database connection itself. There are also numerous variants of the sqlConnRemote() function – such as sqlConnect(), hgAllocConn(),– that are sometimes useful and are defined in the hg/inc/jksql.h and hdb.h include files.

Once a database connection has been established, the API provides functions for accessing the database tables. For some commonly used data types, for example, PSLs and genePreds, one can load all or part of a table into a linked list of C structures with a single line of code such as

```
psl = pslReaderLoadRangeQuery(conn, table, chrom, start, end, NULL);
```

or

```
gp = genePredReaderLoadRangeQuery(conn, table, chrom, start, end, NULL);
```

psl and genePred C structures are defined in $KENTSRC/inc/psl.h and $KENTSRC/hg/inc/genePred.h, respectively, and are shown in Figure 9.1. Fields in these structures are very similar to those in the corresponding file and table formats, which are described in Appendices 2 and 3.

---

[2]: The careful reader will notice that with the UCSC API, the user name is "genomep" and the password "password" is required. In contrast, for command-line MySQL querying and SQL programmed querying, as illustrated in Chapters 4 and 6, respectively, one can use the user name "genome" without a password. The host and database names are always the same whether or not one is using the API.

a)
```
struct psl
/* Summary info about a patSpace alignment */
    {
    struct psl *next;  /* Next in singly linked list. */
    unsigned match;      /* Number of bases that match that aren't repeats */
    unsigned misMatch;  /* Number of bases that don't match */
    unsigned repMatch;  /* Number of bases that match but are part of repeats */
    unsigned nCount;     /* Number of 'N' bases */
    unsigned qNumInsert;      /* Number of inserts in query */
    int qBaseInsert;     /* Number of bases inserted in query */
    unsigned tNumInsert;      /* Number of inserts in target */
    int tBaseInsert;    /* Number of bases inserted in target */
    char strand[3];    /* + or - for strand */
    char *qName;  /* Query sequence name */
    unsigned qSize;     /* Query sequence size */
    int qStart;   /* Alignment start position in query */
    int qEnd;       /* Alignment end position in query */
    char *tName;  /* Target sequence name */
    unsigned tSize;     /* Target sequence size */
    int tStart;   /* Alignment start position in target */
    int tEnd;       /* Alignment end position in target */
    unsigned blockCount;      /* Number of blocks in alignment */
    unsigned *blockSizes;     /* Size of each block */
    unsigned *qStarts;   /* Start of each block in query. */
    unsigned *tStarts;   /* Start of each block in target. */
    char **qSequence;  /* query sequence for each block */
    char **tSequence;  /* target sequence for each block */
    };
```

b)
```
struct genePred
/* A gene prediction, with optional fields. */
{
    struct genePred *next;  /* Next in singly linked list. */
    char *name;   /* Name of loci, transcript, mRNA, etc */
    char *chrom;  /* Chromosome name */
    char strand[2];      /* + or - for strand */
    unsigned txStart;   /* Transcription start position */
    unsigned txEnd;     /* Transcription end position */
    unsigned cdsStart;  /* Coding region start */
    unsigned cdsEnd;    /* Coding region end */
    unsigned exonCount; /* Number of exons */
    unsigned *exonStarts;     /* Exon start positions */
    unsigned *exonEnds; /* Exon end positions */
    /* optional fields */
    unsigned optFields;             /* which optional fields are used */
    unsigned id;                    /* Numeric id of gene annotation */
    char *name2;                    /* Secondary name. (e.g. name of gene) */
    enum cdsStatus cdsStartStat;  /* Status of cdsStart annotation */
    enum cdsStatus cdsEndStat;    /* Status of cdsEnd annotation */
    int *exonFrames;                /* List of frame for each exon */
};
```

Figure 9.1 psl and genePred C structures. The structure fields are very similar to those used in psl and genePred file and table formats (see Appendix 2, Table A2.2, and Appendix 3, Table A3.1).

For tables with other formats, or if one wants more flexibility in data access, accessing the database table requires more than a single line but is still quite easy. For example, here is a segment of code to extract the start and end coordinates of CpG island annotations that overlap a query region:

```
struct sqlResult *sr = hExtendedRangeQuery(conn, "cpgIslandExt", chrom,
             start, end, NULL, FALSE, "chromStart,chromEnd", NULL);
while ((row = sqlNextRow(sr)) != NULL)
      {
      int cpgStart = sqlUnsigned(row[0]);
      int cpgEnd = sqlUnsigned(row[1]);
      /* Do something with the CpG data */
      }
sqlFreeResult(&sr);
```

Finally, one needs to perform any needed data post-processing (the "do something" in the code fragment here). In many cases, the necessary functions for performing the data manipulations will be defined in the same include file as that defining the structure holding the data. However, this is not always the case, and sometimes some searching is needed to find the library functions that one needs.

9.8 Finding what you need in the kent source code

The kent source code is very comprehensive, containing subroutines for performing most bioinformatics tasks one can imagine. However, this very comprehensiveness sometimes can make it challenging to determine precisely where within the code tree the needed subroutines are located.

The desired functionality can often be found simply by searching the inc and hg/inc subdirectories. In other cases, the necessary code can be found by scanning the list of programs in the utils directory. For example, if one is looking for code to sort or split FASTA sequence files, scanning the names of the files in the utils directory (or using a file search utility such as Unix "find") will quickly lead one to the files faSort.c and faSplit.c, respectively, from which the needed code can be extracted.

However, sometimes finding the location of a desired utility function or library function is not so easy. In Chapter 7, we described using a text-searching tool such as the Unix grep utility to find the appropriate Ensembl object for implementing a desired bioinformatics task. Similarly, using grep can be an effective way to locate functions in the UCSC code tree. Useful keywords or phrases for use in such searches may include genome browser track or table names (e.g., "CpG Islands," "multiz17way," or "phastCons") or simply words describing the data manipultion you need to perform (e.g., "reverse complement," "Smith Waterman," or "Hidden Markov"). In many cases, the returned source code lines will point you to the appropriate function and C structures in one or more of the lib or inc subdirectories.

Once the names of the C structures have been found, one can obtain more information about them and their associated functions from the include directory ".h" files in which they are defined. In addition, it is often useful to run grep again, this time with the name of the structure or function as the grep argument, to find examples illustrating correct structure and function usage.

Unfortunately, sometimes even using grep one cannot find any appropriate UCSC function. Of course, this could be because the desired functionality simply is not implemented in the kent source code. However, typically this is not the case. In particular, if the data manipulation is something you can do interactively on the UCSC Browser, then the functionality to perform that data manipulation must exist somewhere in the code. For example, say you need a subroutine that takes a genome location (chromosome and position) in the human genome and determines to what extent the nucleotide at that location is conserved in other vertebrate genomes. We may be certain that such capability exists somewhere in the code because we can answer the question for any single genomic position by setting the UCSC Browser to that position, selecting the "conservation" track, and zooming to "base level." However, if we cannot guess an appropriate keyword, grep will not be able to locate the relevant code.

In this case, we can try to find the relevant code by executing the desired function in the web browser and examining the resulting web address. In our example, when we display the conserved alignment in the browser, we see that the web address listed is http://genome.ucsc.edu/cgi-bin/hgTracks. This indicates that the browser has just executed the (CGI) program hgTracks. Therefore, hgTracks.c might be a good place to start our search. Locating hgTracks.c in the hg/hgTracks subdirectory in the code tree is straightforward using Unix find (or grep). However, hgTracks.c is around 14,000 lines long, so some additional searching is necessary. If we can identify some identifying keyword at this point, such as "conservation" or "maf" (if we remember that multiple alignments are stored in "maf" format in the database), scanning the file with a standard text editor may be sufficient for narrowing the search region.

However, sometimes the browser code is sufficiently long and complex that finding the code where the required data manipulation is being performed is still difficult. In this case, there may be another (simpler) web program that has the desired functionality. In our example, multiple alignments are also produced by the web browser using the Table Browser interface pointing to the multiz17way table with maf output selected. If we access the Table Browser in this manner, we notice that in this case the CGI program accessed is hgTables.c, which has 1,500 lines.

Although hgTables.c is shorter than hgTracks.c, it is still quite complex, and one may need to analyze the logic of the program further before one can isolate the code with the desired functionality. One way to carry out such an analysis is to run the program as a stand-alone command-line program. In fact, all of the UCSC Browser CGI programs can also be run in stand-alone mode. You just need to compile and link the program using its associated "make" file and pass the program the arguments it expects. Determining the arguments being used by a browser program can be accomplished with the cartDump utility, which is executed by loading the web address http://genome.cse.ucsc.edu/cgi-bin/cartDump. cartDump displays the most recent set of arguments used by the browser. Once the program has been compiled in standalone mode and appropriate input arguments have been determined from a cartDump, one can step through the program with an interactive debugger, such as gdb. In this way, one should be able to decipher the logic underlying even complex CGI browser programs.

The description in the preceding paragraphs may suggest that identifying the kent code that performs a specific task occasionally requires a considerable amount of detective work. However, this is actually rarely the case. Usually, "grepping" with one or two well-chosen keywords quickly identifies the appropriate code, and in most cases, the code is sufficiently clearly written that the underlying logic is apparent. Moreover, as should be apparent from the following examples, the advantages gained by incorporating the kent library functions and subroutines into one's own code are so great as to make the occasional additional detective work of locating the appropriate routines worthwhile.

9.9 UCSC API example programs

9.9.1 UCSC example 1: "Hello World"

Let us now illustrate how to use the UCSC API with some program examples. Our first program is a variant on the traditional beginner's "Hello World" computer program. Although we are assuming that the reader has C programming experience, we nevertheless begin with this simple program so we can illustrate some basic features of code development with the UCSC API.

The first issue we need to address is where to place the source file so that the compiler can locate any UCSC library routines that the program may need. The easiest approach uses the newProg utility and involves creating a new subdirectory within $KENTSRC/hg. We will call this subdirectory $KENTSRC/hg/ucscExamples. You can, of course, place your code in other places, but you will need to edit the UCSC makefiles or else make other modifications to ensure that the compiler and linker can find all the UCSC libraries.

We next need to build the "newProg" utility program, itself. newProg creates a new subdirectory for the program we want to write, along with skeleton source code and a makefile to compile the program and link it to the UCSC code libraries. We build "newProg" (we only need to do this once) by running the make command in the $KENTSRC/utils/newProg subdirectory.

Once newProg has been built, we can create a skeleton source file and a makefile for our "Hello World" program by executing the command:

```
$ newProg -jkhgap $KENTSRC/hg/ucscExamples/helloWorld Print Hello World
```

where -jkhgap tells the program to include the jkhgap.a library archive and, again, $KENTSRC specifies the location of the root directory of the kent/src tree. All the arguments to newProg after the name of the subdirectory to be created ("Print Hello World," in this case) are simply passed on to the program skeleton as a program description.

The code skeleton that newProg creates consists of a main program and two subroutines. The main program simply processes the program arguments and options and calls the principal subroutine, which has the same name as the program file (e.g., helloWorld, in the present case.) The other subroutine is called "usage" and contains a message to be displayed if the program is called with the wrong number of arguments.

We can now edit the skeleton so that the program prints "Hello World." The resulting code is shown in Figure 9.2 and, like all the code examples in the book, is available from the publisher's web site for the book. Note that we have retained some code generated by newProg that is not needed for printing "Hello World," such as the option-processing code.

Once we have completed editing and saving the source file, executing "make" in the directory $KENTSRC/hg/ucscExamples/helloWorld compiles and links the code. If all goes well and there are no error messages, we should be able to execute the program by typing "helloWorld" at the command line.

9.9.2 UCSC example 2: Accessing the UCSC public databases

For our second example, we will write a program called ucscDbConnTest that connects to one of the databases at the UCSC public mirror site (genome-mysql. cse.ucsc.edu), prints a list of the names of the tables in the database, and then prints the name of a gene from a specified gene table and in a specified genome region. Usage of the program will be, for example:

```
$ ucscDbConnTest sacCer1 sgdGene
```

where we are telling the program to access the sacCer1 *S. cerevisiae* database and the sgdGene gene table in that database. Running the completed program will produce output similar to that shown in Figure 9.3.

9.9.2.1 Program implementation

To write the program, we first create the skeleton program file, as in the previous example, by running the newProg utility, this time with

```
$ newProg -jkhgap $KENTSRC/hg/ucscExamples/ucscDbConnTest \
   Database Connection Test
```

We then fill in the skeleton with the code shown in Figure 9.4.

Let us now look at how the program is implemented. As in the helloWorld example, the main program, lines 63 through 73, processes arguments and options and calls the main subroutine, called ucscDbConnTest. Subroutine ucscDbConnTest (lines 30

```
1   /* helloWorld - Print Hello World. */
2   #include "common.h"
3   #include "linefile.h"
4   #include "hash.h"
5   #include "options.h"
6
7   void usage()
8   /* Explain usage and exit. */
9   {
10  errAbort(
11    "helloWorld - Print Hello World\n"
12    "usage:\n"
13    "   helloWorld XXX\n"
14    "options:\n"
15    "   -xxx=XXX\n"
16    );
17  }
18
19  static struct optionSpec options[] = {
20      {NULL, 0},
21  };
22
23  void helloWorld()
24  /* helloWorld - Print Hello World. */
25  {
26  printf("Hello World\n");
27  return;
28  }
29
30  int main(int argc, char *argv[])
31  /* Process command line. */
32  {
33  optionInit(&argc, argv, options);
34  if (argc != 1)
35      usage();
36  helloWorld();
37  return 0;
38  }
39
```

Figure 9.2 Source code for the "Hello World" program using the UCSC API.

```
Requesting connection
Made connection OK
Here are the names of the tables in the DB
all_est
all_mrna
...
transRegCodeMotif
transRegCodeProbe
yeastP2P
Gene name 1 = YAL034W-A
```

Figure 9.3 Part of the ucscDbConnTest program output.

```
1    /* ucscDbConnTest - Database Connection Test. */
2    #include "common.h"
3    #include "linefile.h"
4    #include "hash.h"
5    #include "options.h"
6    #include "jksql.h"
7    #include "genePred.h"
8    #include "genePredReader.h"
9    #include "hdb.h"
10
11   void usage()
12   /* Explain usage and exit. */
13   {
14   errAbort(
15     "ucscDbConnTest - Database Connection Test\n"
16     "usage:\n"
17     "   ucscDbConnTest db geneTable\n"
18     "   where db is the database (in the public mirror)\n"
19     "   to connect to and table is a table of genes in that database,eg\n"
20     "   ucscDbConnTest sacCer1 sgdGene\n"
21     "options:\n"
22     "   -xxx=XXX\n"
23     );
24   }
25
26   static struct optionSpec options[] = {
27      {NULL, 0},
28   };
29
30   void ucscDbConnTest(char *db, char *table)
31   /* ucscDbConnTest - Database Connection Test. */
32   {
33   struct slName *tableName, *allTables = NULL;
34   char *chr = "chr1"; int start = 80000; int end = 90000;
35   char* host = "genome-mysql.cse.ucsc.edu";
36   char *user = "genomep";
37   char *password = "password";
38   hSetDbConnect(host, db, user, password);
39
40   printf("Requesting connection\n");
41   struct sqlConnection *conn = sqlConnectRemote(host, user, password, db);
42   if (conn == NULL)
43           errAbort("Unable to establish connection to %s\n", db);
44   printf("Made connection OK\n");
45
46   allTables = sqlListTables(conn);
47   printf("Here are the names of the tables in the DB\n");
```

Figure 9.4 Source code for the ucscDbConnTest program for accessing the UCSC databases using the UCSC API.

```
48     for(tableName = allTables; tableName != NULL; tableName=tableName->next)
49             {
50             char *name = tableName->name;
51             printf("%s\n", name);
52             }
53
54     struct genePred *gp
55             = genePredReaderLoadRangeQuery(conn, table, chr, start, end, NULL);
56     if (gp == NULL)
57             errAbort("No gene found in %s overlapping %s:%d-%d\n",
58                             table, chr, start, end);
59     printf("Gene name 1 = %s\n", gp->name);
60     sqlDisconnect(&conn);
61     }
62
63     int main(int argc, char *argv[])
64     /* Process command line. */
65     {
66     optionInit(&argc, argv, options);
67     if (argc != 3)
68         usage();
69     char *db = argv[1];
70     char *table = argv[2];
71     ucscDbConnTest(db, table);
72     return 0;
73     }
74
```

Figure 9.4 (continued)

through 61), in turn, performs three tasks. First, it establishes the database connection (lines 34 through 38) using code we have already seen in Section 9.7. Next, in line 46, the program retrieves a C structure called allTables, which is an slName struct (defined in $KENTSRC/inc/common.h) consisting of a singly linked list of all the table names in the database:

```
allTables = sqlListTables(conn);
```

Then in the "for" loop (lines 48 through 52), the program cycles through the slName structure, printing out the name of each table. The program next (lines 54 and 55) acquires a singly linked list of genePred C structures for each of the genes in the specified gene table that overlaps the region of interest (hard-coded in this example to chromosome 1:80,000–90,000). Finally, in line 59, the name of one of the genes in the region (whichever happens to be first in the list) is printed. If we wanted to print the name of all the genes in the region, we would just loop through the list of genePred structures the same way we looped over the list of slName structures in lines 48 through 52.

```
Requesting connection
SQL_CONNECT 1078189 sacCer1 genome-mysql.cse.ucsc.edu genomep
SQL_TIME 1078189 sacCer1 0.804s
Made connection OK
SQL_QUERY 1078189 sacCer1 show tables
SQL_TIME 1078189 sacCer1 0.030s
Here are the names of the tables in the DB
```

Figure 9.5 Part of the JKSQL_TRACE output, verifying that a connection has been established to the sacCer1 database at the UCSC public mirror.

If all goes well, after compiling, linking, and executing the program with arguments sacCer1 and sgdGene, we will obtain a display similar to that in Figure 9.3. On the other hand, if we obtain an error message when we attempt to create the database connection, we will need to debug the program. For identifying such connection problems, in addition to the usual C debugging tools (e.g., gdb), it is useful to turn on the SQL trace function in the UCSC API by defining the environmental variable with (in Bash)

```
$ export JKSQL_TRACE='on'
```

The trace tool displays messages for all SQL transactions and can be quite useful for tracking down database connection problems. A part of the display from JKSQL_TRACE when running ucscDbConnTest is shown in Figure 9.5. When you are finished using the trace tool, you turn it off with

```
$ export JKSQL_TRACE='off'
```

If the cause of the database connection problem is still unclear, make sure that you can access genome-mysql.cse.ucsc.edu by directly using the MySQL program from the command line as illustrated in Section 4.5. If this fails as well, there is a problem in your overall MySQL configuration, or perhaps you have a local firewall preventing remote database access. However, if command-line access is successful, the problem is in the code and needs to be tracked down with gdb or some similar tool.

9.9.3 UCSC example 3: Intron length comparisons

For our third example, we consider a more realistic application – namely, our now familiar comparison of lengths among the introns of mammalian genes that contain embedded snoRNAs that we have previously investigated with both Galaxy and the Ensembl API. Recall that we wish to compare the median lengths of introns that have snoRNAs embedded in them and the remaining introns of those snoRNA host genes (see Section 5.4.6 for a detailed description of this application.) Here we determine the intron length distributions with a C program using the UCSC database and the UCSC API.

Let us write the program so that we can execute it with a command similar to the one we used in the analogous Ensembl program in Chapter 7. We will call this program ucscIntronLengths1 and invoke it with three arguments. The first argument is the genome database to use, the second argument is the name of the database gene table, and the third argument is the name of a file of (BED) locations, for example:

```
$ ucscIntronLengths1 hg18 ensGene snoBedFile
```

Here hg18 is the UCSC database, ensGene is the (Ensembl) gene table to be searched for genes that may have a snoRNA gene in one of their introns, and snoBedFile is a BED list of snoRNA gene coordinates in the (hg18) database.

Note that in contrast to our program based on the Ensembl API, with the UCSC API we need to explicitly specify which gene table to use. For example, with the UCSC hg18 human genome build, we could use the table for Ensembl genes, UCSC "Known Genes," NCBI RefSeq genes, GENSCAN, ACESCAN, or N-SCAN gene predictions, among others.

9.9.3.1 Program implementation

Code for ucscIntronLengths1 is shown in Figure 9.6. A flowchart of the program is shown in Figure 9.7. By comparing Figure 9.7 with the ensemblIntronLengths.pl flowchart in Figure 7.6, we see that the overall strategy of the program is almost identical to that of the Ensembl implementation. The only difference stems from the fact that, as described in Section 3.1.2, the UCSC system does not distinguish

```
1   /* ucscIntronLengths1 - Intron lengths from public database. */
2   /* ucscIntronLengths1 - illustrates accessing data from public UCSC
3    * database */
4   #include <time.h>
5   #include "common.h"
6   #include "options.h"
7   #include "jksql.h"
8   #include "bed.h"
9   #include "binRange.h"
10  #include "genePred.h"
11  #include "genePredReader.h"
12  #include "hdb.h"
13
14  struct hostGene
15  /* hostGene with associated bed records */
16      {
17      struct hostGene *next;  /* Next in singly linked list. */
18      struct genePred *hostGp; /* genePred list of overlapping genes */
19      struct bed *bedList; /* list of overlapping beds */
20      };
21
22
23  void usage()
24  /* Explain usage and exit. */
```

Figure 9.6 Source code for the ucscIntronLengths1 intron lengths distribution program.

```
25   {
26   errAbort(
27     "ucscIntronLengths1 - find median length of introns \n"
28     "                      overlapping ranges in input file\n"
29     "usage:\n"
30     "   ucscIntronLengths1 db dbTable myBedFile\n"
31     "       where db is the database name \n"
32     "       where dbTable is tableFileName in 'file' mode or else\n"
33     "           name of table to use in 'public' or 'localDb' modes \n"
34     "       where myBedFile is a bed file of genomic ranges \n"
35     " Options:\n"
36     "   -verbose print progress and warning messages\n"
37     "\n");
38   }
39
40   /**************************************/
41   void hostGeneHashFree(struct hash **hash)
42   /* Free hostGene Hash */
43   {
44   if (*hash != NULL)
45       {
46       struct hashEl *hashEl = NULL;
47       struct hashCookie cookie = hashFirst(*hash);
48       while ((hashEl = hashNext(&cookie)) != NULL)
49           {
50           struct hostGene *hg  = hashEl->val;
51           struct bed *hgBedlist = hg->bedList;
52           slFreeList(&hgBedlist);
53           freez(&hg);
54           }
55       hashFree(hash);
56       }
57   }
58
59   /****************************************/
60   int genePredMostExonsCmp(const void *va, const void *vb)
61   /* Compare to sort based on exon count and
62   * sizes of txEnd - txStart, most exons first. */
63   {
64   const struct genePred *a = *((struct genePred **)va);
65   const struct genePred *b = *((struct genePred **)vb);
66   int dif = b->exonCount - a->exonCount;
67   if (dif != 0)
68     return dif;
69   int lengthA = a->txEnd - a->txStart;
70   int lengthB = b->txEnd - b->txStart;
71   dif = lengthB - lengthA;
72   return dif;
73   }
74
```

Figure 9.6 (continued)

```
75   /*****************************************/
76   struct genePred *genePredWithMostExons(struct bed *bed,
77      struct sqlConnection *conn, char *geneTable, int *noHostGenePtr)
78   {
79   boolean verbose = optionExists("verbose");
80   int bStart = bed->chromStart; int bEnd = bed->chromEnd;
81   struct genePred *gp = NULL;
82   if (verbose)
83      printf("%s\t%d\t%d\t%s\t%d\t%c\n", bed->chrom, bStart, bEnd,
84          bed->name, bed->score, bed->strand[0]);
85   /* Create singly-linked list of gene prediction structures */
86      gp = genePredReaderLoadRangeQuery(conn, geneTable,
87                                  bed->chrom, bStart, bEnd, NULL);
88   if (gp == NULL)
89      {
90      if (verbose)
91          warn("#####No gene found in %s overlapping %s:%d-%d, skipping\n",
92                  geneTable, bed->chrom, bStart, bEnd);
93      *noHostGenePtr += 1;
94      return NULL;
95      }
96   /* Sort the gene structure list by number of exons and keep the
97    * structure for gene with most exons*/
98   slSort(&gp, genePredMostExonsCmp);
99   return gp;
100  }
101
102  /*****************************************/
103  boolean oneOverlappingIntron(struct genePred *gp, struct bed *bed,
104     int *noHostGenePtr, int *twoIntronPtr)
105  /* Check whether exactly one intron overlaps the bed region */
106  {
107  boolean verbose = optionExists("verbose");
108  int i, iStart, iEnd;
109  int foundOverlaps = 0; /* Number of introns overlapping bed region */
110  for (i=1; i< gp->exonCount; ++i)
111     {
112  /* For each intron in the gene, calculate its length. */
113     iStart = gp->exonEnds[i - 1];
114     iEnd = gp->exonStarts[i];
115     if (positiveRangeIntersection(bed->chromStart, bed->chromEnd,iStart,iEnd))
116      foundOverlaps++;
117     }
118  if (foundOverlaps == 0)
119     {
120     if (verbose)
121        warn("#####No overlapping intron found in %s  %s:%d-%d\n", gp->name,
122                bed->chrom, bed->chromStart, bed->chromEnd);
```

Figure 9.6 (continued)

```
123    *noHostGenePtr += 1;
124    return FALSE;
125    }
126   if (foundOverlaps > 1)
127     {
128     if (verbose)
129       warn("#####More than 1 overlapping introns found in %s   %s:%d-%d\n",
130              gp->name, bed->chrom, bed->chromStart, bed->chromEnd);
131     *twoIntronPtr += 1;
132     return FALSE;
133     }
134   return TRUE;
135   }
136
137
138   /**************************************/
139   struct hostGene *hostGeneNew(struct genePred *gp)
140   /* Create new hostGene struct */
141   {
142   struct hostGene *hostGene;
143   AllocVar(hostGene);
144   hostGene->hostGp = gp;
145   hostGene->bedList = NULL;
146   return hostGene;
147   }
148
149   /**************************************/
150   void hostGeneAddBed(struct hostGene *hg, struct bed *bed)
151   /* Add bed to list of overlapping regions */
152   {
153   struct bed *bedCopy = cloneBed(bed);
154   slAddTail(&hg->bedList, bedCopy);
155   }
156
157   /**************************************/
158   struct hash *makeHostGeneHash(struct bed *bedList,
159       struct sqlConnection *conn, char *geneTable,
160       int *noHostGenePtr, int *twoIntronPtr)
161   /* Create hash of structures of host gene genePred and contained
162   * bed regions */
163   {
164   struct bed *bed=NULL;
165   struct hostGene *hg = NULL;
166   struct hashEl *el;
167   struct hash *hostGeneHash = newHash(0);
168   struct genePred *gp = NULL;
169   for(bed = bedList; bed != NULL; bed = bed->next)
170     {
171     gp = genePredWithMostExons(bed, conn, geneTable, noHostGenePtr);
```

Figure 9.6 (continued)

```
172     if (gp == NULL)
173        continue;
174     if (!oneOverlappingIntron(gp,bed, noHostGenePtr, twoIntronPtr))
175        continue;
176     if (hashLookup(hostGeneHash, gp->name) == NULL)
177        {
178        hg = hostGeneNew(gp);
179        hashAdd(hostGeneHash, gp->name, hg);
180        }
181     hg = hashMustFindVal(hostGeneHash, gp->name);
182     hostGeneAddBed(hg, bed);
183     }
184  return hostGeneHash;
185  }
186
187  /*****************************************/
188  boolean overlapsAnyBed(struct bed *bedList, int start, int end)
189  /* */
190  {
191  struct bed *bed;
192  for(bed = bedList; bed != NULL; bed = bed->next)
193     {
194     if (positiveRangeIntersection(bed->chromStart, bed->chromEnd, start, end))
195            return TRUE;
196     }
197  return FALSE;
198  }
199
200  /*****************************************/
201  void intronLengthsForOneHostGene(struct hostGene *hostGene,
202     struct slDouble **overlapListPtr, struct slDouble **otherListPtr)
203  /* For each intron in the gene, calculate its length.
204  * Then append the length to a list of the lengths of
205  * either the introns that overlap the region specified by the
206  * 'bed' coordinates or the introns that don't overlap the 'bed' */
207  {
208  int i, intronStart, intronEnd;
209  struct genePred *gp = hostGene->hostGp;
210  struct bed *bedList = hostGene->bedList;
211  for (i=1; i< gp->exonCount; ++i)
212     {
213     intronStart = gp->exonEnds[i - 1];
214     intronEnd = gp->exonStarts[i];
215     double intronLength = (double) (intronEnd - intronStart);
216     struct slDouble *slIntronLength = slDoubleNew(intronLength);
217     if (overlapsAnyBed(bedList, intronStart, intronEnd))
218        {
219        slAddTail(overlapListPtr, slIntronLength);
220        }
```

Figure 9.6 (continued)

```
221     else
222         slAddTail(otherListPtr, slIntronLength);
223     }
224  }
225
226  /***************************************/
227  void processBedFile(char *bedFile, struct sqlConnection *conn,
228      char *geneTable)
229  // char *geneTable, struct hash *gpHash)
230  /* Main subroutine loop */
231  {
232  // List of lengths of introns containing a sno
233  struct slDouble *overlapList = NULL;
234  // List of lengths of host-gene introns NOT containing a sno
235  struct slDouble *otherList = NULL;
236  // Count of snos with no host-gene found
237  int noHostGeneWithIntron = 0;
238  // Count of snos overlapping two introns of host-gene
239  int twoIntronOverlapHost = 0;
240  struct bed *bedList = bedLoadAll(bedFile);
241  struct hostGene *hostGene = NULL;
242  struct hash *hostGeneHash = makeHostGeneHash(bedList, conn, geneTable,
243                       &noHostGeneWithIntron, &twoIntronOverlapHost);
244  struct hashEl *el;
245  struct hashEl *list = hashElListHash(hostGeneHash);
246  for(el=list;el;el=el->next)
247      {
248      hostGene = el->val;
249      intronLengthsForneHostGene(hostGene, &overlapList, &otherList);
250      }
251  printf("snos with no host gene containing intron = %d\n",noHostGeneWithIntron);
252  if (twoIntronOverlapHost !=0)
253      printf("Host genes with snos overlapping more than one intron = %d\n",
254          twoIntronOverlapHost);
255  printf("Host genes found = %d\n", slCount(list));
256  printf("Median value of lengths of %d overlapping introns = %f\n",
257      slCount(overlapList), slDoubleMedian(overlapList));
258  printf("Median value of lengths of %d other introns = %f\n",
259      slCount(otherList), slDoubleMedian(otherList));
260  hashElFreeList(&list);
261  bedFreeList(&bedList);
262  hostGeneHashFree(&hostGeneHash);
263  slFreeList(&overlapList);
264  slFreeList(&otherList);
265  }
266
267  /***************************************/
268  /* ucscIntronLengths1.c */
```

Figure 9.6 (continued)

```
269   int main(int argc, char *argv[])
270   /* Find median value of lengths of introns overlapping ranges in input file
271    * and compare with lengths of other introns in those genes
272    * Program reads 'bed fileŌ of genomic regions and
273    * extracts longest gene overlapping each region. For each
274    * gene, lengths of introns overlapping the region as well
275    * as those not overlapping the region are computed. Medians
276    * of each set of intron lengths is printed out.
277    * Once compiled and linked, the program is run as e.g.:
278    ucscIntronLengths1 sacCer1 sgdGene myYeastBedFile
279    * where the first argument is the db, the second argument is the name of
280    * the db or file gene table and the third program argument is the location
281    * file of (bed) locations to be screened for intron lengths,
282    */
283   {
284   time_t start_time, end_time;
285   double diff_time;
286   optionHash(&argc, argv);
287   char* host = "genome-mysql.cse.ucsc.edu";
288   char *user = "genomep";
289   char *password = "password";
290   char *db = argv[1];
291   char *geneTable = argv[2];
292   char *bedFile = argv[3];
293   if (argc != 4)
294       usage();
295   start_time = time(NULL);
296   struct sqlConnection *conn = NULL;
297   conn = sqlConnectRemote(host, user, password, db);
298   hSetDbConnect(host, db, user, password);
299   processBedFile(bedFile,conn, geneTable);
300   sqlDisconnect(&conn);
301   end_time = time(NULL);
302   diff_time = difftime(end_time, start_time);
303   printf("## Elapsed time (secs): %f \n", diff_time);
304   return 0;
305   }
306
307
```

Figure 9.6 (continued)

between "genes" and "transcripts." In the UCSC system, every transcript corresponds to a distinct gene. Consequently, whereas to retrieve transcript data with the Ensembl API we needed to perform two program loops – one to access every gene in a region and a second to retrieve every transcript of that gene – with the UCSC system, we will need only a single loop over every "gene."

Now let us look at the program implementation in more detail. ucscIntron-Lengths1 consists of a short main program and a number of subroutines that do

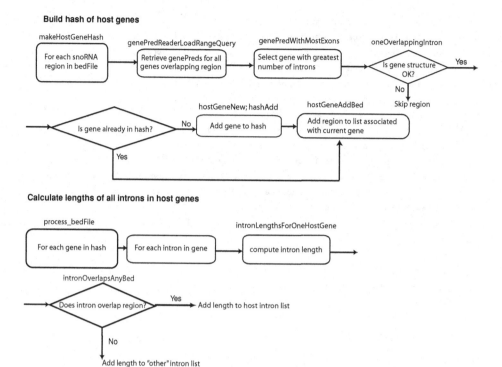

Build hash of host genes

Calculate lengths of all introns in host genes

Figure 9.7 Flowchart for the ucscIntronLengths1 program. The top part of the figure shows the construction of the hash associating host genes with their embedded snoRNAs. The bottom part of the figure outlines the steps involved in extracting all the host-gene introns and computing their lengths. The principal steps in the algorithm are indicated by rounded rectangles, with the program subroutine or UCSC API library function used to implement each step noted outside of the corresponding rectangle. Note the similarity between this flowchart and the outline of the Ensembl implementation in Figure 7.6.

most of the work. The main program first reads in the program arguments (lines 290 through 292). Next, in lines 297 through 298, the program creates an SQL connection to the database (just as in the previous example). Finally, the main program calls the "processBedFile" subroutine (line 299).

The processBedFile subroutine (lines 227 through 265) is the principal subroutine of the program. processBedFile first executes the single line (line 240):

```
bedList = bedLoadAll(bedFile);
```

which opens the file "bedFile," converts each line of the file into an element of a singly linked list of bed C structs (defined in bed.h and shown in Figure 9.8), and finally closes the file. Next, processBedFile calls makeHostGeneHash (lines 242 and 243) to create a hash associating each host gene with its embedded snoRNAs, just as we did in the Ensembl program. Once the host-gene hash has been constructed, the program cycles through each of the host genes in the hash (lines 246 through 250) and, for each one, calls the intronLengthsForOneHostGene subroutine to calculate the lengths of the

```
struct bed
/* Browser extensible data */
    {
    struct bed *next;  /* Next in singly linked list. */
    char *chrom;  /* Human chromosome or FPC contig */
    unsigned chromStart;  /* Start position in chromosome */
    unsigned chromEnd;  /* End position in chromosome */
    char *name;  /* Name of item */
    /* The following items are not loaded by the bedLoad routines. */
    int score; /* Score - 0-1000 */
    char strand[2];  /* + or -. */
    unsigned thickStart; /* Start of where display should be thick (start
                            codon for genes) */
    unsigned thickEnd;   /* End of where display should be thick (stop codon
                            for genes) */
    unsigned itemRgb;    /* RGB 8 bits each */
    unsigned blockCount; /* Number of blocks. */
    int *blockSizes;     /* Comma separated list of block sizes.  */
    int *chromStarts;    /* Start positions inside chromosome.  Relative to
                            chromStart*/
    };
```

Figure 9.8 BED C structure. The structure fields are very similar to those used in BED file and table formats (see Appendix 2, Table A2.1).

gene's introns. When the loop is completed, the median lengths are computed with the library function slDoubleMedian. Finally, processBedFile frees its allocated memory (lines 260 through 264). Explicitly freeing memory is not strictly necessary here because the program is terminating; however, freeing previously allocated memory is a good habit.

The subroutine makeHostGeneHash (lines 158 through 185) creates the hash that associates snoRNAs with host genes. Because C, unlike Perl, does not include hashes among its fundamental data types, we instead build the hash with library routines defined in the hash.h file in the kent/src/inc directory. Also, rather than create a second two-element hash to contain pointers to the associated gene structures and bed lists, as we did in the Ensembl example (compare with Figure 7.7), we instead define the hash value to be a new C struct called hostGene and defined in lines 14 through 20. The hostGene struct contains pointers to both the host gene's genePred struct as well as to a list of bed structs, one for each snoRNA embedded in the host gene's introns.

With these preliminaries, we build the hostGene hash as we did previously in the Ensembl example. For each bed location, makeHostGeneHash calls the subroutine genePredWithMostExons in line 171. Subroutine genePredWithMostExons (lines 76 through 100), in turn, first fetches the gene data from the database and stores the data in a list of genePred structures (lines 86 and 87), one for each of the genes that overlap the region. Next, genePredWithMostExons selects the overlapping gene with the most exons by sorting the gene list in terms of the number of exons and then returning the initial element of the sorted list (lines 98 and 99).

Next, makeHostGeneHash performs some error-checking by calling the oneOver-lappingIntron subroutine (lines 174 and 175), which determines whether a "believ-able" gene structure has been found, that is, by making sure that each snoRNA region is located within a single intron. If the gene passes the tests, the program checks whether a hostGene struct already exists for the gene in the hostGene hash. If the gene is not already in the hash, the subroutine creates a new hostGene struct by calling hostGeneNew and adds the new hostGene struct to the hostHash (lines 178 and 179). makeHostGeneHash then calls hostGeneAddBed (lines 181 and 182) to append the bed struct for the current region to the hostGene structure's bedList field. Note that in the subroutine hostGeneAddBed (lines 150 through 155), it is necessary to copy the entire bed struct using the cloneBed library function (line 153) before appending the bed to the bedList. This ensures that the "next" pointers in the singly linked list, bedList, all point appropriately to the next item in the list.

With the host-gene hash created, processBedFile can finally call intronLengths-ForOneHostGene for each host gene in hostGeneHash (lines 246 through 250) to implement the second phase of the program. Subroutine intronLengthsForOneHost-Gene (lines 201 through 224) cycles through the exons of the host gene, computes the length of each intron by subtraction of exon coordinates, and inserts the length into an slDouble struct so that the length can be added to a list of lengths (lines 213 through 216). The subroutine then determines whether the intron overlaps any of the input beds and appends the length to either the linked list overlapList or otherList using the library routine slAddTail (lines 217 through 223).

9.9.3.2 Running the program

Assuming we have created our program's makefile by using newProg, we can now compile and link the program simply by typing "make" at the command line. Once the program has been compiled and linked, we can try the program out with the command:

```
$ ucscIntronLengths1 hg18 ensGene hacaWgRna.hg18.bed
```

Here hacaWgRna.hg18.bed is our list of known H/ACA snoRNA locations, and we choose the ensGene table of Ensembl genes to facilitate comparison with our Ensembl API-based program. The results are

snos with no host gene containing intron = 10
Host genes with snos overlapping more than one intron = 1
Host genes found = 65
Median value of lengths of 86 overlapping introns = 1075.500000
Median value of lengths of of 857 other introns = 928.000000
Elapsed time (secs): 6.000000

The results are very similar to those we obtained in Chapter 7 using the Ensembl API, both in terms of the intron length distributions as well as the presence of the anomalous cases for which either no host intron is found or the snoRNA overlaps more than one intron. As we used the same human genome assembly and the same (Ensembl) set of gene structures, such agreement is not surprising.

However, there are also differences between the UCSC and Ensembl approaches. First, we notice that the results are not exactly the same. We have found 857 "other" introns, whereas in Section 7.7.3.3 we found 856. We can determine the cause of the difference by running each program with fewer snoRNA coordinates in the input BED file to isolate which input entry is causing the different results. If we do so, we find that for snoRNA AJ609441, UCSC and Ensembl pick different Ensembl transcripts (ENST00000255477 versus ENST00000379060, respectively). Checking the Ensembl web site, we find that "transcript ENST00000255477 is no longer in the Ensembl database, but it has been mapped to the following deprecated identifiers: ENST00000379055 ENST00000379060 ENST00000379056."

Another difference is that with the UCSC database, it is relatively easy to see how our results are sensitive to our choice of gene-structure annotation. For example, we can determine the effect of using NCBI RefSeq genes, or UCSC "Known Genes," or the results one of several gene structure prediction programs (e.g., GENSCAN or N-SCAN) rather than the Ensembl gene set. We would simply change the gene-table argument when executing our program. In contrast, with Ensembl's API, a "gene" can be only an Ensembl gene or, if we had queried the ensembl-vega database, a VEGA gene. We could also use the GENSCAN gene predictions in the Ensembl database; however, this would require modifying our program to retrieve "Gene Prediction" objects rather then "Gene" objects from the Ensembl database.

For example, to see the effect of using RefSeq gene annotations on our intron length distributions, we would run

```
$ ucscIntronLengths1 hg18 refGene hacaWgRna.hg18.bed
```

The result is

snos with no host gene containing intron = 26
Host genes found = 58
Median value of lengths of 72 overlapping introns = 958.000000
Median value of lengths of of 752 other introns = 920.000000
Elapsed time (secs): 19.000000

We see that the median values of the intron length distributions have not changed greatly. This is not surprising because in many cases, gene predictions and structures generated by Ensembl and NCBI RefSeq methods are identical. However, we do notice that the RefSeq gene set includes host genes for sixteen fewer snoRNAs. On the other hand, the RefSeq set does not include any anomalous host genes like ENST00000357861, with two introns overlapping a snoRNA. The point here is not that one gene data set is better than another; rather, it is that with genomic data, it is often helpful to check more than one annotation method.

The Ensembl API and UCSC API-based programs also differ in execution times. Using each program with its associated public mirror database, the UCSC-based C program required 6 seconds, whereas the Ensembl-based Perl program required 219 seconds. Admittedly, this is just a single program benchmark, and there are other factors that may contribute to this result. For example, the Ensembl public mirror may have been more heavily accessed than the UCSC mirror at the times the test programs were being run, and the tests were run from California (which is closer to the UCSC mirror than to the Ensembl mirror). That said, the almost forty-fold difference in performance is probably not entirely the result of such outside factors, and consequently, if one has a computation-intensive application, it may useful to consider execution time differences in deciding how one wants to implement a genome-database querying program.

Chapter summary

- The UCSC API includes stand-alone utilities and C language library functions addressing a wide variety of bioinformatics and general data-processing tasks.
- Essentially, any data manipulation that can be performed by the UCSC Genome Browser can be carried out programmatically with code from the UCSC API.
- Identifying which library functions and C structures one needs to use is often the crucial step in using the UCSC API, and can generally be accomplished by applying the grep utility to the source code tree.
- The UCSC core software is free for all uses; UCSC's database access and other specialized software are free for academic and research use but require licensing for commercial applications.

Exercises

Note that you will need to have the UCSC API and the MySQL client software installed to complete these exercises.

1. Find the code for the C structures and functions used to access the main "repeat" table in the UCSC databases. Write a program to read a set of BED regions and return a list of repeats that overlap each region that indicates the type of repeat and start and end coordinates for each one. The program should work for either the human, mouse, rat, or chicken genomes. Test the program against the public UCSC mirror database.

2. Modify the intron lengths program so that it selects the longest gene transcript as the representative gene rather than the transcript with the largest number of introns. Run the program with the hacaWgRna.hg18.bed dataset.

3. Write a program to read a list of BED regions and a UCSC database name (e.g., hg18) and output a list of the SwissProt protein IDs for each gene overlapping each region. This program should work for (at least) both the UCSC human and yeast genome databases.

4. Write a C program using the UCSC API to search for NMD candidates. Run the program against the human Ensembl gene set as well as against the human RefSeq gene set. Compare the percentage of genes that are NMD candidates in the two datasets. Are the results similar in mouse?

More Advanced Applications Using the UCSC API

In the present chapter, we continue our description of the UCSC API and its application in programmed querying of the UCSC databases. In particular, we consider applications for which database access via the public UCSC mirror is insufficient or inconvenient, and describe alternative methods for accessing the UCSC databases. For each method, we detail the steps required and the relative advantages and disadvantages of the approach. We also present three complete programming examples illustrating each of these methods. Finally, we describe the procedure for mirroring all or part of a UCSC database.

10.1 Alternate methods for accessing the UCSC databases

In Chapter 8, we noted that there are reasons – for example, performance, data security, or the need to customize a database – why one might want to create a private, local mirror of one or more of the Ensembl databases. We noted that this was true, despite the fact that any program written using the Ensembl API works equally well with the Ensembl public database as with a private mirror (aside from issues relating to program execution speed).

Performance, security, and customization considerations may also lead one to want to install a local mirror of one or more of the UCSC genome databases. However, in the case of the UCSC databases, there is an additional reason for not relying exclusively on the UCSC public mirror. The reason is that although most code developed with the UCSC API will run properly on the UCSC public mirror, there is one important exception. The issue is that remote public access to genome-mysql.cse.ucsc.edu is via the MySQL client-server interface. Consequently, remote users can only access data that is explicitly stored within the MySQL databases. However, in the UCSC system, some commonly used data, such as sequence-alignment data and genomic and GenBank sequence data, are stored in auxiliary flat files and not in the MySQL databases. (A schematic view of the UCSC database architecture is shown in Figure 10.1.) Because sequence and alignment data can only be accessed locally, these data cannot be retrieved via the UCSC public mirror.

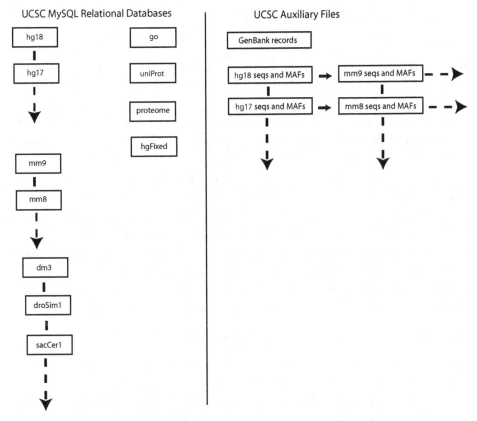

Figure 10.1 Highly simplified schematic overview of the UCSC database architecture. The UCSC architecture includes both multiple MySQL databases, shown on the left side of the figure, and multiple auxiliary flat-file databases, outlined on the right. The MySQL databases include a separate database corresponding to each species' genome assembly covered by the system. These databases are labeled by the corresponding UCSC database-build ID (hg18, hg17, hg16 . . . for human; mm9, mm8, mm7 for mouse; etc.). In addition there are MySQL databases that are shared among multiple species, such as the GO, UniProt, and proteome databases. Genomic sequences and multiple sequence alignments corresponding to each build are stored in a flat-file subdirectory of the /gbdb directory (e.g., /gbdb/hg18/nib for human genomic sequences and /gbdb/hg18/multiz17way for human MAF alignments). Other flat-file data, such as GenBank and Visigene data, are stored in separate /gbdb subdirectories.

To address these issues, there are essentially three distinct (though related) means for accessing the data in the UCSC databases without relying on the public mirror. The first method involves downloading individual database tables as flat files and subsequently reading the flat files directly into one's application programs. The second approach involves downloading one, or a few, database tables or auxiliary flat files from the UCSC databases and then installing them into a skeleton UCSC mirror. Finally, one may decide to mirror an entire UCSC database (e.g., hg18), or even multiple UCSC databases. We now describe each of these approaches in more detail.[1]

---

[1] For completeness, there is yet one other method, namely, via a web robot or "bot." A *bot* is a computer program that sends requests to an interactive web resource and processes the

10.2 Using flat files containing individual tables

If one only needs data from one or a few database tables, one option is to simply download the needed tables. For the UCSC databases, the individual tables can be downloaded using either the Table Browser, the UCSC FTP site (e.g., http://hgdownload.cse.ucsc.edu/goldenPath/hg18/database), or the UCSC DAS interface.

Once the UCSC table data have been downloaded, the data files can be read directly into computer memory by one's application program. If the tables are stored in one of the common UCSC table formats, such as BED, PSL, genePred, or MAF, reading the flat files in this manner can be accomplished easily using the library functions and subroutines available in the UCSC API for this purpose. In fact, in the last chapter, we already used one such function, bedLoadAll, which read a flat file in BED format and converted it into a singly linked list of bed C structures (line 240 in Figure 9.6). Similarly, in the next example, the library function genePredLoad (defined in hg/inc/genePred.h) is used to read a line in a data file in genePred format and convert it into a genePred C structure. In this way, the application program can access the data in one or more UCSC database tables without having to directly connect to any database at all.

10.2.1 UCSC example 4: Intron lengths using three data sources

We now illustrate the flat file, table-access approach (as well as the local mirror database access method) with the program ucscIntronLengths2. This program has the same purpose as ucscIntronLengths1, which we examined in the previous chapter, and is implemented with largely the same code. However, with ucscIntronLengths2, we can access the required gene table data in any one of three ways – via the public database (as in Chapter 9), via a downloaded flat file version of the table data, or via a local mirror.

The program is called in a manner similar to ucscIntronLengths1, except that now there is one more program argument, the "method," which can be "public," "file," or "localDb." That is, one calls the program using one of these three commands:

1. Using the public UCSC mirror

```
$ ucscIntronLengths2 hg18 ensGene hacaWgRna.hg18.bed public
```

2. Using a private mirror

```
$ ucscIntronLengths2 hg18 ensGene hacaWgRna.hg18.bed localDb
```

responses automatically as well. The requests are transmitted as though they were coming from a person using the web interface (in our case, a genome browser). In principle, a bot could be written to automate any task that is performed interactively by human biologists. However, the genome browsers have not been designed to handle the volume of queries generated by bots and, for this reason, are generally configured to deny service to Internet requests that they determine are coming from bots. Consequently, in practice, bots are not a feasible means for establishing programmatic access to a genome database.

3. Using a downloaded gene table

```
$ ucscIntronLengths2 hg18 ensGene.hg18.txt \
    hacaWgRna.hg18.bed file
```

As in example 3 in Chapter 9, hacaWgRna.hg18.bed is the list of snoRNA locations. Also, when using either the public or local database access methods, the second program argument (e.g., ensGene) is the name of the database table to use. However, if the "file" method is selected, the second program argument is the name of a file that contains the downloaded data from the required table (e.g., ensGene.hg18.txt). Of course, using the "file" method assumes that the required file with the gene table data has previously been downloaded, for example, with the Table Browser. Similarly, using the localDb method only works if one has installed a private mirror of at least part of the specified UCSC database (in this case, at least the ensGene table of the hg18 database).

10.2.1.1 Program implementation

Since most of the implementation of ucscIntronLengths2 is identical to ucscIntronLengths1, we will not repeat the analysis of the entire program here. Rather, we will just describe the new parts of the program necessary to implement the additional data-input features. The flowchart of ucscIntronLengths2 is essentially the same as that for ucscIntronLengths1, shown in Figure 9.7. The only difference is in the implementation of the "Retrieve genePreds for all genes overlapping region" box in the flowchart. In ucscIntronLengths1, this step is implemented exclusively by accessing the UCSC public database. In contrast, in ucscIntronLengths2 the genePred retrieval is implemented in one of three ways, depending on whether the "method" option has been set to "public," "file," or "localDb."

Let us now look at the differences between ucscIntronLengths1 and ucscIntronLengths2 in more detail. (The modified and new subroutines are shown in Figure 10.2 and the entire program is available from the publisher's web site for the book.) The first difference is in the ucscIntronLengths2 main program, where we see (Figure 10.2a, lines 18 through 29)

```
if (sameWord(method, "file"))
        gpHash = readGpToBinKeeper(geneTable);
  else
        {
        if (sameWord(method, "public"))
                {
                conn = sqlConnectRemote(host, user, password, db);
                hSetDbConnect(host, db, user, password);
                }
        else
                conn = sqlConnect(db);
        }
```

```
1    int main(int argc, char *argv[])
2    {
3    time_t start_time, end_time;
4    double diff_time;
5    optionHash(&argc, argv);
6    char* host = "genome-mysql.cse.ucsc.edu";
7    char *user = "genomep";
8    char *password = "password";
9    char *db = argv[1];
10   char *geneTable = argv[2];
11   char *bedFile = argv[3];
12   char *method = argv[4];
13   if (argc != 5)
14       usage();
15   start_time = time(NULL);
16   struct sqlConnection *conn = NULL;
17   struct hash *gpHash = NULL;
18   if (sameWord(method, "file"))
19           gpHash = readGpToBinKeeper(geneTable);
20   else
21           {
22           if (sameWord(method, "public"))
23                   {
24                   conn = sqlConnectRemote(host, user, password, db);
25                   hSetDbConnect(host, db, user, password);
26                   }
27           else
28                   conn = sqlConnect(db);
29           }
30   processBedFile(bedFile,conn, geneTable, gpHash);
31   sqlDisconnect(&conn);
32   binKeeperGpHashFree(&gpHash);
33   end_time = time(NULL);
34   diff_time = difftime(end_time, start_time);
35   printf("## Elapsed time (secs): %f \n", diff_time);
36   return 0;
37   }
```

Figure 10.2 Main program of ucscIntronLengths2 and ucscIntronLengths2 subroutines, which differ from corresponding subroutines in the ucscIntronLengths1 program. (a) Main program. (b) Subroutine genePredWithMostExons. (c) Subroutine readGpToBinKeeper, which reads the file containing gene data into a hash of binKeeper structures. (d) Subroutine bkToGenePreds, which extracts gene data from the binKeeper-structure hash to a list of genePred structures. For a program flowchart and listing of other subroutines, see Figures 9.6 and 9.7. Only the implementation of the "Retrieve genePreds for all genes overlapping region" step is changed from the ucscIntronLengths1 flowchart in Figure 9.7. See text for details.

```
1    struct genePred *genePredWithMostExons(struct bed *bed,
2           struct sqlConnection *conn, char *geneTable,
3           struct hash *gpHash, int *noHostGeneWithIntronPtr)
4    {
5    boolean verbose = optionExists("verbose");
6    int bStart = bed->chromStart; int bEnd = bed->chromEnd;
7    struct genePred *gp = NULL;
8    if (verbose)
9          printf("%s\t%d\t%d\t%s\t%d\t%c\n", bed->chrom, bStart, bEnd,
10               bed->name, bed->score, bed->strand[0]);
11   /* Create singly-linked list of gene prediction structures
12    * from either mySQL database or in-memory hash loaded from
13    * flat-file */
14   if (gpHash == NULL)
15         gp = genePredReaderLoadRangeQuery(conn, geneTable,
16                                  bed->chrom, bStart, bEnd, NULL);
17   else
18         gp = bkToGenePreds(gpHash, bed->chrom, bStart, bEnd);
19   if (gp == NULL)
20         {
21         if (verbose)
22               warn("#####No gene found in %s overlapping %s:%d-%d, skipping\n",
23                     geneTable, bed->chrom, bStart, bEnd);
24         *noHostGeneWithIntronPtr += 1;
25         return NULL;
26         }
27   /* Sort the gene structure list by number of exons and keep the
28    * structure for gene with most exons*/
29   slSort(&gp, genePredMostExonsCmp);
30   return gp;
31   }
32
```

Figure 10.2 (b)

We see that if the "method" is "public,"[2] a database connection is established to the public database, exactly as in ucscIntronLengths1. If the method is "file," the gene data is read from the file "geneTable" into memory using the subroutine readGpTo-BinKeeper, which we will describe shortly. Finally, if the method is neither "public" nor "file," we connect to the local database using the command:

```
conn = sqlConnect(db);
```

Note that sqlConnect automatically determines the correct MySQL arguments from the user's configuration file (called .hg.conf). Later in this chapter, we describe how this configuration file is created as part of the local mirror database installation.

---

[2] For the string comparison, we apply the useful "sameWord" library function, defined in kent/source/inc/common.h.

```
1     struct hash *readGpToBinKeeper(char *gpFileName)
2     /* adapted from readPslToBinKeeper in psl.c */
3     {
4     #define MAX_CHROM_SIZE 400000000
5     struct binKeeper *bk;
6     struct genePred *gp;
7     struct lineFile *pf = lineFileOpen(gpFileName , TRUE);
8     struct hash *hash = newHash(0);
9     char *row[21] ;
10    int genePredLineCtMin = 10;
11    while (lineFileNextRow(pf, row, genePredLineCtMin))
12        {
13        gp = genePredLoad(row);
14        if (hashLookup(hash, gp->chrom) == NULL)
15           {
16            bk = binKeeperNew(0, MAX_CHROM_SIZE);
17            hashAdd(hash, gp->chrom, bk);
18           }
19        bk = hashMustFindVal(hash, gp->chrom);
20        binKeeperAdd(bk, gp->txStart, gp->txEnd, gp);
21        }
22    lineFileClose(&pf);
23    return hash;
24    }
```

Figure 10.2 (c)

```
1     /*****************************************/
2     struct genePred *bkToGenePreds(struct hash *gpHash,
3                           char *chrom, int start, int end)
4     /* Retrieve all genePreds from binkeeper overlapping region */
5     {
6     struct genePred *gpList = NULL;
7     struct genePred *gp;
8     struct binKeeper *bk = hashFindVal(gpHash, chrom);
9     struct binElement *el, *elist = binKeeperFind(bk, start, end) ;
10    for (el = elist; el != NULL ; el = el->next)
11        {
12        gp = el->val;
13        if (gp != NULL)
14               {
15               slSafeAddHead(&gpList, gp);
16               }
17        }
18    slFreeList(&elist);
19    return gpList;
20    }
```

Figure 10.2 (d)

The main program then calls the principal subroutine, processBedFile (Figure 10.2a, line 30), passing it both "conn" (which contains a pointer to the database-connection structure, if we are using local or public database access) and "gpHash" (which points to a hash structure containing the gene table data, if we are in "file" mode).

Subroutine processBedFile handles the BED file of snoRNA locations in exactly the same manner as in ucscIntronLengths1, with one exception – the subroutine makeHostGeneHash is passed one additional parameter, gpHash, the pointer to the gene data hash (compare with Figure 9.6, lines 242 and 243):

```
hostGeneHash = makeHostGeneHash(bedList, conn, geneTable,
            gpHash, &noHostGeneWithIntron, &twoIntronOverlapHost);
```

In turn, makeHostGeneHash is also almost identical to the subroutine of the same name in ucscIntronLengths1. The sole difference is that the subroutine genePredWith-MostExons, which returns a genePred structure for the overlapping transcript with the most exons, is also called with one additional argument, gpHash (see Figure 9.6, line 171, for comparison):

```
gp = genePredWithMostExons(bed, conn, geneTable, gpHash,
    noHostGeneWithIntronPtr);
```

Finally, in the genePredWithMostExons subroutine, we see the following code (Figure 10.2b, lines 14 through 18):

```
if (gpHash == NULL)
        gp = genePredReaderLoadRangeQuery(conn, geneTable,
                        bed->chrom, bStart, bEnd, NULL);
else
        gp = bkToGenePreds(gpHash, bed->chrom, bStart, bEnd);
```

If gpHash was set to "NULL" in the main program, we connect to a database and we retrieve the gene data directly from the database via the call to genePredReaderLoad-RangeQuery, just as in ucscIntronLengths1. However, depending on the data stored in the "conn" data structure, here we may be accessing either the public database or a local mirror.

In contrast, if gpHash is not "NULL," we are in "file" mode. In this case, the program needs to retrieve the gene data from the gpHash structure using the new subroutine bkToGenePreds. Once the gene data has been retrieved as a list of genePred structures pointed to by the variable "gp" by one of these methods, the remainder of the ucscIntronLengths2 is identical to ucscIntronLengths1.

The only new subroutines[3] are readGpToBinKeeper (Figure 10.2c) and bkTo-GenePreds (Figure 10.2d) for reading the gene file into a data structure (gpHash) and then for extracting data from that structure, respectively. The point is that with

---

[3] There is also a third new routine, binKeeperGpHashFree, which simply frees the gpHash data structure at the end of program execution, as shown in Figure 10.2a, line 32.

```
struct binElement
/* An element in a bin. */
    {
    struct binElement *next;
    int start, end;        /* 0 based, half open range */
    void *val;             /* Actual bin item. */
    };
struct binKeeper
/* This keeps things in bins in memory */
    {
    struct binKeeper *next;
    int minPos;       /* Minimum position to bin. */
    int maxPos;       /* Maximum position to bin. */
    int binCount;     /* Count of bins. */
    struct binElement **binLists; /* A list for each bin. */
    };
```

Figure 10.3 UCSC data structures, defined in binRange.h, used to build a binKeeperHash struct. The binKeeperHash itself is a hash whose keys are chromosome IDs and whose values are pointers to binKeeper structs that store the gene data for the corresponding chromosome.

the "file" method, we need have a way to load an entire gene data table (e.g., the Ensembl gene table or the RefSeq table) into memory and then to efficiently retrieve records that overlap a specified genomic region.

Developing an efficient data retrieval method, keyed on chromosomal location, is actually a nontrivial programming exercise. However, the kent code has a ready-made solution in the form of the binKeeperHash structure. The binKeeperHash structure, of which gpHash is an example, is a two-level data structure in which the top level is a simple hash in which the keys are chromosome names. The values of the hash are pointers to the structures at the second level, called binKeeper structures. This data structure segments each chromosome into numerous "bins," and data associated with any specific genomic coordinate range are stored in the smallest bin that completely contains the specified region. This implementation enables very fast data retrieval. The kent code structs that are used to build a binKeeperHash are shown in Figure 10.3 and are defined in binRange.h in $KENTSRC/inc.

However, the important point is that we do not need to understand the details of the binKeeper implementation to take advantage of its capabilities. Rather, because reading files into, and accessing records from, hashes and binKeeper structs are functions that the UCSC system does already, writing the necessary subroutines is just a matter of finding these routines in the kent code and copying or modifying them to perform the tasks we need. In particular, the UCSC subroutine pslToBinKeeper (which is found in $KENTSRC/lib/psl.c) reads PSL-formatted alignments from a file into a binKeeperHash of psl C structs, and contains code very similar to what we need. Modifying pslToBinKeeper into our subroutine readGpToBinKeeper (Figure 10.2c), so that we can read in genePred data rather than PSL data, just involves replacing the psl data-loading functions and psl structures with the corresponding genePred functions and structures.

The data-retrieval subroutine bkToGenePreds performs the reverse operation – retrieving a list of genePred structs for all the genes in a specified genomic region from the binKeeperHash struct. bkToGenePreds performs this data retrieval by first finding the binKeeper structure corresponding to the specified chromosome in the binKeeperHash struct (Figure 10.2d, line 8). The subroutine then retrieves from the binKeeper struct a linked list of the binElement structures that overlap the specified coordinates, that is, binElement structs for each of the genes in the region (line 9). Finally, the list of binElement structs are converted into a list of genePred structs (lines 10 through 17) and returned to the calling program (line 19). By now, the linked list looping syntax should be familiar and, in any case, by using grep with the keyword "binKeeperFind," we can find several examples in the kent code with correct syntax to copy.

10.2.1.2 Running the program

If we have a file of downloaded gene table data (such a file for the hg18 ensGene table is included among the files for the book on the publisher's web site), we can now execute the program ucscIntronLengths2 in "file" mode. If we have installed a skeleton hg18 mirror database – consisting of the single table, ensGene (as described in Section 10.3) – we can also test the program in "localDb" mode. We carry out these tests by using the commands shown at the beginning of Section 10.2.1. Performing these tests, we will find that with all three methods, we obtain the same result that we obtained with ucscIntronLengths1 in Section 9.9.3.2 (except, of course, for the execution times.)

10.3 Using a skeleton database

As we have already noted, there are some important applications of the UCSC API for which neither the method of public database access nor that of downloading MySQL tables as files is sufficient. These applications include ones in which one needs to access sequence data, such as genome sequences, mRNA/EST sequences, or sequence alignments. For these applications, we need to locally mirror a "skeleton" of one of the UCSC databases, consisting of, at a minimum, one of the UCSC auxiliary data files. In our final two examples using the UCSC API, we illustrate this approach. In the examples, we will assume that the installation of the required skeleton mirror has been completed, deferring the description of the installation procedure to the end of the chapter.

10.3.1 UCSC example5: Displaying local multiple-sequence alignments

For our next example program, mafWriteRegions, we revisit the task of displaying multiple alignments that overlap a set of genomic regions, which we previously addressed using Galaxy and Ensembl. We will assume that we have a local mirror of the hg18 database, including at least the auxiliary MAF files located at /gbdb/hg18, as well as the hg18 MySQL database with the MAF index table, multiz17way. Note that you will need an understanding of UCSC's MAF file and table formats, described in Appendices 2 and 3, for this example.

The program, mafWriteRegions, takes a set of genomic regions and, for each region, retrieves all MSAs that overlap at least a part of the region. For those alignments that meet minimum alignment-score and alignment-length cutoffs, the program outputs the part of the alignment that overlaps the input coordinates either as an alignment (in MAF format) or as a set of individual sequences (in FASTA format).

mafWriteRegions' functionality is similar to that of the ensemblComparaExample.pl from Section 8.1.1. However, since we are using the UCSC databases rather than Ensembl's, the set of species included in the MSAs will be different and, in addition, different alignment algorithms will be used (MULTIZ with UCSC versus PECAN with Ensembl). We note that as with the Ensembl program, it would be straightforward to modify mafWriteRegions to perform additional analyses on the retrieved alignments, such as determining the alignment consensus sequence or counting the number of the sequences having the consensus nucleotide at a specified position.

We will execute our program with the command:

```
$ mafWriteRegions [options] hg18 multiz17way regionLocations.bed outFile
```

Here hg18 is the species database we are using, multiz17way is the name of the MAF index table to use, regionLocations.bed is a BED file containing a list of the genomic regions for which we want alignments, and outFile is the desired destination of the program output. The program includes several options, including ones to change the minimum alignment length and score necessary to display the alignment and options selecting whether the output format should be FASTA or MAF.

10.3.1.1 Program implementation

The code for the mafWriteRegions program is shown in Figure 10.4 and a flowchart is shown in Figure 10.5. The basic approach for accessing and displaying alignments in mafWriteRegions is similar to ensemblComparaExample.pl, as can be seen by comparing Figures 8.3b and 10.5. However, one important difference in the implementations stems from the fact that UCSC stores all data for MSAs in the database of the "query" species. For example, even if we want to access alignments for sixteen vertebrate genomes to the human genome, we only need to access a single UCSC genome database (in this case, the human genome database). mafWriteRegions retrieves the alignment data in the form of C data structures. Specifically, the UCSC

```
1    /* mafWriteRegions - write all maf alignments in specified region. */
2    /* mafWriteRegions takes a list of genomic regions in BED format and
3     * for each one retrieves all overlapping alignments */
4
5    #include "common.h"
6    #include "linefile.h"
7    #include "hash.h"
8    #include "options.h"
9    #include "hdb.h"
10   #include "maf.h"
11   #include "fa.h"
12
13   void usage()
14   /* Explain usage and exit. */
15   {
16   errAbort(
17     "mafWriteRegions - write all maf alignments in multiple specified regions\n"
18     "usage:\n"
19     "   mafWriteRegions db table bedFile mafOutputFilename\n"
20     "options:\n"
21     "   -public use public ucsc mirror for relational data\n"
22     "        (maf data **MUST** still be local)\n"
23     "   -hitAny output all of maf if any of it overlaps specified region\n"
24     "        (default is to output intersection region only) \n"
25     "   -fasta output fasta files (default is maf file)\n"
26     "   -minLen=integer minimum length of intersected aligned seq \n"
27     "        (ie including dashes) to output\n"
28     "   -minScore=integer minimum score of intersected alignment to output\n"
29     "   -outDb=outDbID used with -fasta to specify which species seq to output\n"
30     "        (default is to output all species) \n"
31     " maf alignments are loaded using coordinates from the positive (W) strand\n"
32     "eg mafWriteRegions hg18 multiz17way myBedFile stdout\n"
33     "or mafWriteRegions -fasta -outDb=mm8 hg18 multiz17way myBedFile stdout\n"
34     );
35   }
36   /*********Globals***********/
37   time_t start_time, end_time;
38   double diff_time;
39   static struct optionSpec options[] = {
40   {NULL, 0},
41   };
42   /***************************************/
43   void removeDashes(char *out, char *in, int size)
44   /* removeDashes - copy size non-dash characters from in to out,
45    * skipping dashes and zero terminating */
46   {
47   int count = size;
```

Figure 10.4 Source code of the mafWriteRegions program for displaying a multiple-sequence alignment using the UCSC API.

```
48  while (count > 0)
49      {
50      if (*in != '-')
51          {
52          *out++ = *in;
53          count--;
54          }
55      in++;
56  }
57  *out = '\0';
58  }
59  /****************************************/
60  void mafWriteComponentFa(FILE *f, struct mafComp *mc)
61  /* mafWriteComponentFa - write component sequence of maf alignment
62   * in fasta, trimming dashes */
63  {
64  char buf[512];
65  char *s = needMem(mc->size + 1);
66  removeDashes(s, mc->text, mc->size);
67  sprintf(buf, "%s:%d:%d:%c", mc->src, mc->start, mc->size, mc->strand);
68  faWriteNext(f, buf, s, mc->size);
69  freez(&s);
70  }
71  /****************************************/
72  void mafWriteAllFa(FILE *f, struct mafAli *ali)
73  /* mafWriteAllFa - write all components of maf alignment in fasta,
74   * trimming dashes */
75  {
76  struct mafComp *mc;
77  for (mc = ali->components; mc != NULL; mc = mc->next)
78      mafWriteComponentFa(f, mc) ;
79  mafCompFree(&mc);
80  }
81  /****************************************/
82  struct mafComp *mafMayFindDbComponent(struct mafAli *ali, char *dbName)
83  /* Find component of given source with specified dbName or NULL if not found.*/
84  {
85  struct mafComp *mc;
86  for (mc = ali->components; mc != NULL; mc = mc->next)
87      {
88      if ( startsWith(dbName, mc->src) )
89          return mc;
90      }
91  return NULL;
92  }
93  /****************************************/
94  void mafWriteGeneric(FILE *f, struct mafAli *ali)
95  /* mafWriteGeneric - output alignment data either as maf or fasta */
```

Figure 10.4 (continued)

```
96    {
97    char *outDb;
98    struct mafComp *mc;
99
100   outDb = optionVal("outDb", "all");
101   if (!optionExists("fasta"))
102       mafWrite(f, ali);
103   else if (sameWord(outDb, "all") )
104       mafWriteAllFa(f, ali);
105   else
106       {
107       mc = mafMayFindDbComponent(ali, outDb );
108       if (mc)
109           mafWriteComponentFa(f, mc);
110       }
111   }
112   /*****************************************/
113   void mafWriteSubset(FILE *f, struct mafAli *ali,
114       char *dbChrom, int start, int end)
115   /* mafWriteSubset - write intersecting subset of single maf alignment   */
116   {
117   int minLen = optionInt("minLen", 0);
118   int minScore = optionInt("minScore", 0);
119
120   struct mafAli *subset = mafSubset(ali, dbChrom, start, end);
121   if (subset == NULL)
122       return;
123   if (  (minLen != 0)  && (subset->textSize < minLen)  )
124       {
125       mafAliFree(&subset);
126       return;
127       }
128   subset->score = mafScoreMultiz(subset);
129   if (  (minScore != 0)  && (subset->score < minScore)  )
130       {
131       mafAliFree(&subset);
132       return;
133       }
134   mafWriteGeneric(f, subset);
135   mafAliFree(&subset);
136   }
137
138   /*****************************************/
139   void mafWriteOneRegion(char *db, struct sqlConnection *conn, char *table,
140       char *chrom, int start, int end, FILE *f)
141   /* mafWriteRegions - write all maf alignments in specified region. */
142   {
143   char dbChrom[64];
```

Figure 10.4 (continued)

```
144    struct mafAli *ali, *mafList;
145    if (start >= end)
146        errAbort("Start %d greater than end  %d in bedFile\n", start, end);
147    mafList = mafLoadInRegion(conn, table, chrom, start, end);
148    if (!optionExists("fasta"))
149        mafWriteStart(f, "multiz");
150    for (ali = mafList; ali != NULL; ali = ali->next)
151        {
152        if (optionExists("fasta"))
153            {
154            mafWriteGeneric(f, ali); /* Write full alignment to file. */
155            }
156        else
157            {
158            safef(dbChrom, sizeof(dbChrom), "%s.%s", db, chrom);
159            /* Write maf of intersection to file. */
160            mafWriteSubset(f, ali, dbChrom, start, end);
161            }
162        }
163    mafAliFreeList(&mafList);
164    }
165
166    /****************************************/
167    void mafWriteRegions(char *db, char *table, char *bedFile, char *outName)
168    /* Read file and process */
169    {
170    FILE *f = mustOpen(outName, "w");
171    struct sqlConnection *conn = NULL;
172    if (optionExists("public"))
173        {
174        char *host = "genome-mysql.cse.ucsc.edu";
175        char *user = "genomep";
176        char *password = "password";
177        hSetDbConnect(host, db, user, password);
178        conn = sqlConnectRemote(host, user, password, db);
179        }
180    else
181        conn = sqlConnect(db);
182    struct bed *bedList=NULL, *bed=NULL;
183    bedList = bedLoadAll(bedFile);
184    for(bed = bedList; bed != NULL; bed = bed->next)
185        {
186        mafWriteOneRegion(db, conn, table, bed->chrom, bed->chromStart,
187            bed->chromEnd, f);
188        }
189    bedFreeList(&bedList);
190    if (!optionExists("fasta"))
191        mafWriteEnd(f);
```

Figure 10.4 (continued)

```
192   carefulClose(&f);
193   sqlDisconnect(&conn);
194   end_time = time(NULL);
195   diff_time = difftime(end_time, start_time);
196   printf("## Elapsed time (secs): %f \n", diff_time);
197   }
198
199   /***************************************/
200   int main(int argc, char *argv[])
201   /* Process command line. */
202   {
203   optionHash(&argc, argv);
204   start_time = time(NULL);
205   if (argc != 5)
206     usage();
207   mafWriteRegions(argv[1], argv[2], argv[3], argv[4]);
208   return 0;
209   }
210
```

Figure 10.4 (continued)

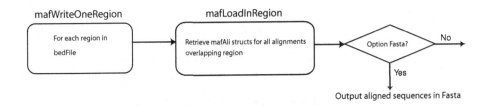

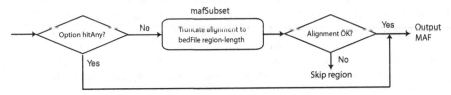

Figure 10.5 Flowchart for the mafWriteRegions program. Principal steps in the algorithm are indicated by rounded rectangles, with the program subroutine or UCSC API library function used to implement each step noted outside of the corresponding program block. The MAF data retrieval step assumes that the necessary tables and auxiliary files have been mirrored locally.

API uses mafAli C structures for storing an entire MAF alignment and mafComp C structures for storing a single component (i.e., a single sequence) of the alignment. These C structures are defined in inc/maf.h and are illustrated in Figure 10.6. We note that, as usual, these structures are defined with "next" pointer fields, enabling mafAli and mafComp structs to be concatenated together as singly linked lists.

Using these structures, the program is implemented as follows. After processing the command-line options, the main program calls subroutine mafWriteRegions.

```
struct mafComp
/* A component of a multiple alignment. */
    {
    struct mafComp *next;
    char *src;  /* Name of sequence source.  */
    int srcSize; /* Size of sequence source.  */
    char strand; /* Strand of sequence.  Either + or -*/
    int start;  /* Start within sequence. Zero based. If strand is - is
        relative to src end. */
    int size;  /* Size in sequence (does not include dashes).  */
    char *text;  /* The sequence including dashes. */
    char leftStatus; /* the syntenic status of the alignment before us */
    int leftLen;     /* length related information for the previous
        alignment */
    char rightStatus; /* the syntenic status of the alignment after us  */
    int rightLen;     /* length related information for the following
        alignment */
    };
struct mafAli
/* A multiple alignment. */
    {
    struct mafAli *next;
    double score;    /* Score.  Meaning depends on mafFile.scoring.
        0.0 if no scoring. */
    struct mafComp *components;  /* List of components of alignment */
    int textSize;             /* Size of text in each component. */
    };
```

Figure 10.6 C structs for multiple-alignment files (MAFs). The mafAli struct corresponds to a single alignment and simply stores the score and number of characters in each sequence of the alignment, along with pointers to a list of the sequence components in the alignment and to the (optional) next alignment in a list of alignments. The mafComp struct stores the data for each sequence in the alignment. Note that in contrast to bed, psl, and genePred structs – which use formats that are very similar to BED, PSL, and genePred file and table formats – mafAli amd mafComp structs have quite different formats from the MAF file and table formats described in Sections A2.4 and A3.3, respectively. Note that the leftStatus, leftLen, rightStatus, and rightLen fields are not needed by most applications.

Subroutine mafWriteRegions begins by setting up a connection to the database at line 181.[4] mafWriteRegions then reads in the entire BED file (line 183) and for each region in the BED file, the program calls the subroutine mafWriteOneRegion, at lines 186 and 187. The mafWriteOneRegion subroutine (lines 139 through 164) first loads all the alignments that overlap each region of interest into a linked list of mafAli structures (at line 147) using the mafLoadInRegion library function (defined in hg/inc/hdb.h):

```
mafList = mafLoadInRegion(conn, table, chrom, start, end);
```

[4] At this point, we are assuming that the program has been called without the -public option.

mafWriteOneRegion then checks whether FASTA or MAF output is requested (lines 148 and 152). For FASTA output, mafWriteOneRegion calls mafWriteGeneric at line 154, which writes the entire alignment as a set of FASTA sequences using the faWriteNext library function (defined in inc/fa.h), at line 68 in subroutine mafWriteComponentFa.

On the other hand, for MAF output, mafWriteOneRegion truncates the alignment to the size of the region[5] and then outputs the truncated alignment as a MAF file. mafWriteOneRegion performs these tasks by first concatenating the "db" and "chrom" strings using the UCSC safef library function[6] defined in inc/common.h (line 158). This is necessary because the mafSubset library function, which truncates the alignment (at line 120) requires the database and chromosome information as a concatenated string. Finally, mafWriteOneRegion calls subroutine mafWriteSubset (at line 160) to print out all the alignments that overlap the region.

Subroutine mafWriteSubset (lines 113 through 136) first extracts the subset of each alignment that overlaps the specified region by calling the mafSubset library function, defined in inc/maf.h, at line 120:

```
subset = mafSubset(ali, dbChrom, start, end);
```

Next, mafWriteSubset skips alignments that do not meet the alignment-score and alignment overlap-length minima (lines 123 through 133). Lastly, mafWriteSubset calls mafWriteGeneric (at line 134) to output the alignments that pass the length and score tests. mafWriteGeneric outputs the MAF alignments using the mafWrite library function (at line 102).

10.3.1.2 Executing the program

As in the previous examples, we create a makefile to compile and link the program using newProg. Once the program is compiled and linked, we can run it with the command:

```
$ mafWriteRegions hg18 multiz17way ensemblCompara.test.bed stdout \
  | egrep -vw 'e|i'
```

(The grep command serves to remove unwanted lines from the MAF format output.) To compare our results with those obtained from Ensembl, we use the same input file that we used with the ensemblCompara program in Section 8.1.1. We recall that this particular test file has just a single input line:

```
chrX   100162141   100162165   cxorf34   0   -
```

The program output is shown in Figure 10.7. Comparing Figure 10.7 and Figure 8.1, we see that in this particular case, Ensembl and UCSC produce the same results for

---

[5] This condition applies unless the option -hitAny is selected.

[6] The safef function allows one to concatenate strings in a buffer, like the standard C sprintf function, but also provides for buffer-overflow checking.

```
##maf version=1 scoring=multiz
a score=133374.000000
s hg18.chrX                100162141 24 + 154913754 TGCAGTCCATC---TTGCATCCTCCAC
s panTro1.chrX             103325023 24 + 160174553 TGCAGTCCATC---TTGCATCCTCCAC
s rheMac2.chrX              99725831 24 + 153947521 TGCAGTCCATC---TTGCATCCTCCAC
s rn4.chrX                 121709624 24 + 160699376 TGCACTCCACC---TGGCATCCTCTAC
s mm8.chrX                 129586438 24 + 165556469 TGCAGTCCACC---TGGCATCCTCTAC
s oryCun1.scaffold_113276       4423 24 +      8378 AGCCGTCCACC---TGGCATCCTCTAC
s bosTau2.scaffold2252         17861 24 +    126872 TGCTGTCCACC---GGGCATCCTCTGC
s canFam2.chrX              78022477 24 + 126883977 TGCAGTCCACC---TGGCATCCTCTAC
s dasNov1.scaffold_45247       11824 23 -     12890 TG-AATGCACC---AGGCATCCTCTAC
s loxAfr1.scaffold_20375        8297 24 -     34054 TGCAGTCCACC---TGGCATCCTCTAC
s echTel1.scaffold_313318      36025 27 -    111094 TGCAGTCCACCACATAGCATCCTCTAG
s monDom4.chr3             437629003 24 + 526135210 AGCAGTCCATT---TAGCATCCTCTAG
s galGal2.chr4               1939414 24 +  90634903 TGCATTCCACC---TGGCATCCTCTAT
s danRer3.chr13            31321733 24 +  47719189 TGCATTAAACT---TGGCATCCTCCAC
```

Figure 10.7 Part of the output generated by the mafWriteRegions program using the test data file shown in the text.

the species that occur in both alignments. In general, UCSC and Ensembl alignments will not be identical. First, sometimes UCSC and Ensembl may use different genomic assemblies.[7] Second, even if the same genome assemblies are used, UCSC and Ensembl might yield different results because UCSC uses MULTIZ alignments whereas Ensembl uses the PECAN program, as described in Appendix 4. Apparently, in this case, neither using the different chicken assemblies nor different algorithms changes the resulting alignment.

A couple of additional comments regarding the MafWriteRegions program. First, although mafWriteRegions does not currently have an option to restrict which species in the alignment to display (as the ensemblCompara program does), it would be straightforward to use the UCSC maf subroutine library routines to implement such an option. Second, we note that mafWriteRegions has an option -public. Selecting this option causes the program to use the public MySQL database for accessing the database *tables*, for example, multiz17way. However, the MAF files themselves must still be accessed locally. In fact, if we run

```
$ mafWriteRegions -public hg18 multiz17way ensemblCompara.test.bed stdout
```

we obtain the same answer as before. However, in this case we have accessed the public database for the table information. To convince ourselves that we have actually

---

[7] In fact, close inspection of Figures 8.1 and 10.7 show that, for example, the chicken assemblies used in the two alignments are not the same. From Figure 8.1, we see that Ensembl used the "WASHUC2" chicken sequence, which, as indicated on the Ensembl web site, is from the May 2006 chicken assembly. In contrast, from Figure 10.7 we see that the UCSC alignment used UCSC's "galGal2" database, which corresponds, according to the UCSC documentation, to the February 2004 chicken assembly. Note that the May 2006 chicken assembly is also available now from the UCSC databases as UCSC build galGal3; however, at the time I downloaded the MAF alignment data, UCSC's hg18 MAF alignments were based on the earlier galGal2 sequences.

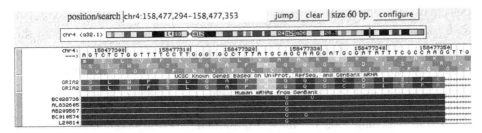

Figure 10.8 Evidence of ADAR sites in human glutamate receptor gene, GRIA2, as shown by comparison of genomic and mRNA sequences. Note the presence of guanine (G) residues in mRNAs at locations where adenines (A) are found in the genomic sequence.

accessed the public database, we can turn on the JKSQL_TRACE function (described in Section 9.9.2.1) and run the program again, both with the -public option and without it. With the -public option, the TRACE output of the SQL connect command displays

```
SQL_CONNECT 1078192 hg18 genome-mysql.cse.ucsc.edu genomep
```

whereas with the default (local) database selected, the TRACE displays

```
SQL_CONNECT 7 hg18 localhost ucscDbUser
```

showing that the SQL queries are indeed going to different hosts in the two cases.

These observations may lead you to suspect that we did not need to include the multiz17way table in our local skeleton database mirror. And you would be correct. The reason we do not need the multiiz17way table locally (and hence, that the -public option works) is that the MAF files and multiiz17way MAF indexing table in human database hg18 are quite stable. Consequently, the (current) public MAF indexing table is compatible with the (older) downloaded MAF files. However, in general, this approach of combining local files and public index tables will not work because the public index will not correspond to the downloaded indexed files.

10.3.2 UCSC example 6: Comparing genomic and mRNA/EST sequences

For our last UCSC API example, we illustrate comparing genomic and mRNA/EST-transcript sequences. Comparing transcript and genome sequences arises in many important applications, including searching for novel SNPs and RNA editing sites and detecting genomic sequencing, assembly, and alignment errors. For example, in RNA editing, adenosine deaminase enzymes (ADARs) convert specific adenosines in RNA to inosine. The precise sequence motifs that result in adenosine targeting by an ADAR are not yet known. One method to screen for potential ADAR target sites is to search for genomic locations that code for an "A" while a "G" has been observed at the corresponding location in an mRNA or EST (inosine is generally interpreted as guanosine by both reverse transcriptase and the ribosome). For example, Figure 10.8 shows how ADAR sites in the GRIA2 gene appear in the UCSC Genome Browser.

The program pslDisplaySeqs demonstrates this application. The program first reads in a BED file of genomic regions and the name of the UCSC transcript-alignment table to use. By default, the program queries the hg18 human genome database. For each region, the program displays the alignment between the transcript and the genome, highlighting places where the transcript and genomic sequence differ by displaying them with capital letters. The program is invoked with the command:

```
$ pslDisplaySeqs [options] transcriptTable regions.hg18.bed outFile
```

where transcriptTable indicates the transcript table to use (e.g., xenoEst if one wants alignments of ESTs from other species) and regions.hg18.bed contains a BED file of the genomic regions to compare. Various options are available, including ones that filter alignments by the transcript's name or that enable one to extend the regions beyond the ranges specified in the input BED file.

10.3.2.1 Program implementation

The program code is shown in Figure 10.9 and is outlined in flowchart form in Figure 10.10. As with the mafWriteRegions program, pslDisplaySeqs requires that at least a skeleton local UCSC mirror has been installed because the program accesses genome sequences and the GenBank mRNA and EST sequences. For example, for use with the UCSC Human Genome Database hg18, the local system needs to include at least the UCSC data directories /gbdb/hg18 and /gbdb/genbank. The local mirror also needs to include a skeleton of the hg18 MySQL database, including the GenBank and genome-sequence index tables – specifically, chromInfo, for indexing the genomic sequence data files, and gbSeq and gbExtFile, for indexing the GenBank transcript files. Note that in the present example, accessing the indexing tables via the public database (illustrated by the "-public" option in the mafWriteRegions program) will almost definitely not work. This is because the UCSC GenBank files are updated frequently and, consequently, the index tables and the downloaded files will almost certainly no longer be in sync.

The overall structure of the pslDisplaySeqs program and many of the library functions and structures used in its implementation are similar to what we have seen in previous examples. However, the current program also needs to access transcript alignments. Transcript alignments are stored in PSL format, which can be somewhat confusing (especially in the case of negative-strand alignments). The main subtleties of negative-strand PSLs are described in Section A1.4 of Appendix 1 and Section A2.3 of Appendix 2. The reader who has not read this material previously is advised to do so now.

Let us now look at the program in more detail. As usual, the main program processes arguments and options (lines 184 through 189) and then calls a principal subroutine that has the same name as the overall program, that is, pslDisplaySeqs (line 190). pslDisplaySeqs (lines 162 through 177) first sets up database access (lines 165 and 166). The subroutine next opens the output file (line 167) and reads in the input BED file (line 169). pslDisplaySeqs then cycles through the list of genomic regions from the BED file and, for each region, calls the subroutine doOneBed (line 172).

```
1    /* pslDisplaySeqs - align transcripts with genome for region. */
2    #include "common.h"
3    #include "bed.h"
4    #include "options.h"
5    #include "jksql.h"
6    #include "hdb.h"
7    #include "dnautil.h"
8    #include "genbank.h"
9    #include "linefile.h"
10   #include "pslReader.h"
11
12   void usage()
13   /* Explain usage and exit. */
14   {
15   errAbort(
16     "pslDisplaySeqs - display genome and EST/mrna seqs of psls \n"
17     "                   overlapping input beds \n"
18     "usage:\n"
19     "   pslDisplaySeqs [options]  table inFile outFile\n"
20     "       where infile is a bedFile of genome locations \n"
21     "       and table is db table to use (eg all_mrna or \n"
22     "       all_est or xenoMrna or xenoEst)\n"
23     "Options:\n"
24     "   -db - database [hg18]  \n"
25     "   -filter - only use psls whose qName matches filter \n"
26     "   -bedExtend=int - increase size of bed by int nt at both ends \n"
27     "\n");
28   }
29
30   /* command line option specifications */
31   static struct optionSpec optionSpecs[] = {
32      {"db", OPTION_STRING},
33      {"filter", OPTION_STRING},
34      {"bedExtend", OPTION_INT},
35      {NULL, 0}
36   };
37
38   /* globals*/
00   FILE *f; /* output file handle */
40
41   /****************************************/
42   boolean isTranslatedAlignment(struct psl *psl)
43   /*  Determine whether psl is for translated alignment
44    *  from whether strand field has 1 or 2 characters */
45   {
46   if (strlen(psl->strand) == 1)
47      return FALSE;
48   return TRUE;
49   }
50
```

Figure 10.9 Source code of the pslDisplaySeqs program for displaying transcript to genome alignments.

```
51   /******************************************/
52   void doOneBlock(int bStart, int bEnd, struct psl *psl,
53      int blockIx, struct dnaSeq *qSeq, struct dnaSeq *tSeq)
54   /* extract both subsequences for each alignment
55    * block. Any mismatching transcript characters to
56    * capitals and then print out alignment block.*/
57   {
58   int blockSize = psl->blockSizes[blockIx];
59   int ix = 0; /* index within part of block within bed */
60   /* Truncate block to size of BED if necessary */
61   int overlapSize = positiveRangeIntersection(bStart, bEnd,
62              psl->tStarts[blockIx], psl->tStarts[blockIx] + blockSize);
63   if (overlapSize <= 0)
64      return;
65   int qOffset = psl->qStarts[blockIx];
66   int tOffset = psl->tStarts[blockIx] - psl->tStarts[0];
67   if (bStart > psl->tStarts[blockIx])
68      {
69      tOffset += (bStart - psl->tStarts[blockIx]);
70      qOffset += (bStart - psl->tStarts[blockIx]);
71      }
72   DNA *qSeqStart = qSeq->dna + qOffset;
73   DNA *tSeqStart = tSeq->dna + tOffset;
74   /* Convert mismatches to caps */
75   for (ix = 0; ix < overlapSize; ix++)
76      {
77      char tchar = tSeqStart[ix];
78      char qchar = qSeqStart[ix];
79      if (tchar == qchar)
80         continue;
81      qSeqStart[ix] = toupper(qSeqStart[ix]);
82      }
83   fprintf(f, "mRNA/EST block %d: %.*s\n", blockIx, overlapSize, qSeqStart);
84   fprintf(f, "genome   block %d: %.*s\n\n", blockIx, overlapSize, tSeqStart);
85   }
86
87   /******************************************/
88   void doOnePsl(int bStart, int bEnd, struct psl *psl,
89      struct sqlConnection *conn)
90   /* Retrieve transcript and genome DNA, reverse comp
91    * if on negative strand, then cycle through each
92    * gapless block of alignment */
93   {
94   int blockIx;
95   int blockCount = psl->blockCount;
96   fprintf(f, "mrna/est: %s\n", psl->qName);
97   if (psl->strand[1] == '-')
98      reverseIntRange(&bStart, &bEnd, psl->tSize);
```

Figure 10.9 (continued)

```
99   struct dnaSeq *tSeq = hChromSeq(psl->tName, psl->tStart, psl->tEnd);
100  if (psl->strand[1] == '-')
101      reverseComplement(tSeq->dna, strlen(tSeq->dna));
102  struct dnaSeq *qSeq = hExtSeq(psl->qName);
103  if (psl->strand[0] == '-')
104      reverseComplement(qSeq->dna, strlen(qSeq->dna));
105  for (blockIx = 0; blockIx < blockCount; blockIx++)
106      {
107      doOneBlock(bStart, bEnd, psl, blockIx, qSeq, tSeq);
108      }
109  dnaSeqFree(&tSeq);
110  dnaSeqFree(&qSeq);
111  }
112
113  /*****************************************/
114  boolean bedAndTargetSameStrand(struct psl *psl, char bStrand)
115  /*  Check whether transcript aligns to strand that bed is on
116   * Assumes negative strand, nucleotide alignment has been
117   * reverse complemented to make query in forward orientation */
118  {
119  char pslTStrand = '+'; /* true for all nucleotide alignments */
120  if (isTranslatedAlignment(psl))
121      pslTStrand = psl->strand[1];
122  return (pslTStrand == bStrand);
123  }
124
125  /*****************************************/
126  void doOneBed(struct bed *bed,
127      struct sqlConnection *conn, char *table)
128  /* Run BED region for requested transcript table(s) */
129  {
130  bedTabOutN(bed, 6, f);
131  int bedExtend = optionInt("bedExtend", 0);
132  int bStart = bed->chromStart - bedExtend;
133  int bEnd = bed->chromEnd + bedExtend;
134  if (bStart >= bEnd)
135      return;
136  char *chrom = bed->chrom;
137  char bStrand = bed->strand[0];
138  struct psl *pslList = NULL, *psl;
139  char *filter = optionVal("filter", NULL);
140  pslList =
141      pslReaderLoadRangeQuery(conn, table, chrom, bStart, bEnd, NULL);
142  if (pslList == NULL)
143      return;
144  for(psl = pslList; psl != NULL; psl = psl->next)
```

Figure 10.9 (continued)

```
145        {
146        if (filter && !wildMatch(filter, psl->qName))
147            continue;
148        if (!isTranslatedAlignment(psl) && (psl->strand[0] == '-'))
149            pslRc(psl);
150        if (bedAndTargetSameStrand(psl, bStrand) == FALSE)
151            {
152            fprintf(f, "mRNA/EST on opposite strand for psl %s, skipping\n",
153              psl->qName);
154            continue;
155            }
156        doOnePsl(bStart, bEnd, psl, conn);
157        }
158    pslFreeList(&pslList);
159    }
160
161    /****************************************/
162    void pslDisplaySeqs(char *table, char *inFile, char *outFile)
163    /* Load regions and call doOneBed for each region */
164    {
165    char *db = optionVal("db", "hg18");
166    struct sqlConnection *conn = sqlConnect(db);
167    f = mustOpen(outFile, "w");
168    struct bed *bedList=NULL, *bed=NULL;
169    bedList = bedLoadAll(inFile);
170    for(bed = bedList; bed != NULL; bed = bed->next)
171        {
172        doOneBed(bed, conn, table);
173        }
174    bedFreeList(&bedList);
175    carefulClose(&f);
176    sqlDisconnect(&conn);
177    }
178
179    /****************************************/
180    int main(int argc, char *argv[])
181    /* Process command line. */
182    {
183    char *outFile, *inFile, *table;
184    optionInit(&argc, argv, optionSpecs);
185    if (argc != 4)
186        usage();
187    table = argv[1];
188    inFile = argv[2];
189    outFile = argv[3];
190    pslDisplaySeqs(table, inFile, outFile);
191    return 0;
192    }
193
```

Figure 10.9 (continued)

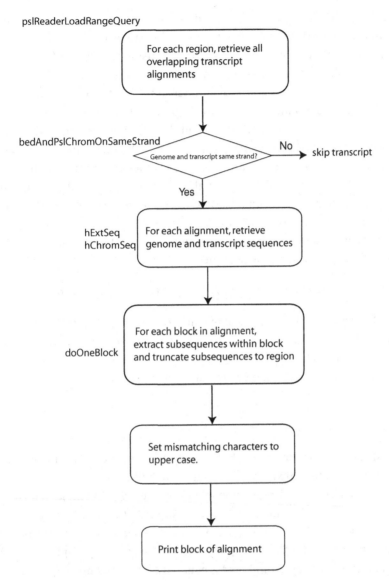

pslReaderLoadRangeQuery

For each region, retrieve all overlapping transcript alignments

bedAndPslChromOnSameStrand

Genome and transcript same strand?

No → skip transcript

Yes

hExtSeq
hChromSeq

For each alignment, retrieve genome and transcript sequences

doOneBlock

For each block in alignment, extract subsequences within block and truncate subsequences to region

Set mismatching characters to upper case.

Print block of alignment

Figure 10.10 Flowchart for the pslDisplaySeqs program. The program subroutine or UCSC API library function used to implement each step is noted outside of the corresponding block. The data retrieval step assumes that the necessary tables and auxiliary files have been mirrored locally.

The doOneBed subroutine (lines 126 through 159) first prints the BED region coordinates (line 130) and then optionally extends the range of the BED region (lines 131 through 133). Next, doOneBed calls the library function pslReaderLoadRangeQuery (defined in hg/inc/pslReader.h) to retrieve all overlapping transcript alignments as a linked list of PSL data structures (recall the PSL C structure from Figure 9.1a) at lines 140 and 141.

Next, doOneBed cycles through each transcript alignment. For each alignment, doOneBed first performs optional filtering on the transcript name using the library function wildMatch from inc/common.h (lines 146 and 147). Next, because we want our alignments to be displayed with the transcript in positive orientation, the subroutine reverse-complements the PSL alignment (using the library function pslRc, defined in inc/psl.h) if the alignment is a nucleotide alignment and the PSL strand field is equal to "-" (lines 148 and 149). doOneBed then determines if the transcript aligns to the same genomic strand as the region of interest in the BED record by calling subroutine bedAndTargetSameStrand (line 150). If the transcript is on the specified strand, doOneBed then calls subroutine doOnePsl at line 156. Otherwise, the alignment is skipped with a warning message (lines 152 through 154).

Subroutine doOnePsl (lines 88 through 111) first converts the start and end positions of the BED region from absolute to strand coordinates (lines 97 and 98) using the library function reverseIntRange from inc/dnautil.h.[8] The subroutine then retrieves the chromosomal and transcript sequences. The chromosomal subsequence is retrieved from the auxiliary hg18 sequence files at line 99:

```
tSeq = hChromSeq(psl->tName, psl->tStart, psl->tEnd);
```

where "psl->tName" is the chromosome name and the library function hChromSeq is defined in hg/inc/hdb.h. Note that the sequence is retrieved as a dnaSeq C struct, defined in inc/dnaSeq.h.

Similarly, the sequence of the transcript is retrieved from the auxiliary GenBank files at line 102:

```
cSeq = hExtSeq(psl->qName);
```

where "psl->qName" is the GenBank accession ID for the sequence to be retrieved, and the library function hExtSeq is also defined in hg/inc/hdb.h.

Next, the transcript and/or genomic sequences are reverse-complemented, if necessary, as specified by the values of the PSL strand field (lines 100 and 101, and 103 and 104). As described in Section A2.3, if the first character of the strand field is equal to "-," the transcript (i.e., the "query") needs to be reverse-complemented. If the second character of the psl strand field is equal to "-," the genomic sequence (i.e., the "target") needs to be reverse-complemented. Sequence reverse-complementation is implemented with the reverseComplement library function defined in inc/dnautil.h. Finally, doOnePsl loops through the alignment (lines 105 through 108), calling the subroutine doOneBlock for each (gapless) block of the alignment.

Subroutine doOneBlock (lines 52 through 85) first computes the number of bases of overlap between the current block and the BED region of interest, and skips the block unless the amount of overlap is greater than zero (lines 61 through 64). The library function positiveRangeIntersection, defined in inc/common.h, is used to perform the range intersection.

---

[8] This is necessary because we will later be comparing the BED region to the alignment region on the chromosome, which is in strand coordinates in PSL format.

Next, doOneBlock needs to compute the offsets into the transcript and chromo-somal sequences to retrieve the subsequences corresponding to the current block. If the start of the block in the genomic sequence is located within the BED region, the offsets are given by lines 65 and 66:

```
qOffset = psl->qStarts[blockIx];
tOffset = psl->tStarts[blockIx] - psl->tStarts[0];
```

Note that hExtSeq retrieves the *entire* transcript sequence in line 102, so that the tran-script offset is simply "psl–>qStarts[blockIx]." In contrast, hChromSeq only retrieves the part of the chromosomal sequence between "psl–>tStart" and "psl–>tEnd" in line 99, so "psl–>tStarts[0]" needs to be subtracted from "psl–>tStarts[blockIx]" in the computation of tOffset. Consequently, if the start of the BED is within the genomic region specified by the current block, we need to offset into the current block by the distance between the block start and the start of the BED. The offsets for this case are computed in lines 69 and 70.

With the proper offsets calculated, doOneBlock can now retrieve the subsequence for the current block, one nucleotide pair at a time (lines 72 through 78). The sub-routine then checks if the nucleotides are the same at each position and converts the transcript character to uppercase, if the genome and transcriptnucleotides differ (lines 79 through 81). Finally, the subroutine prints out the aligned sequence block (lines 83 and 84).

10.3.2.2 Executing the program

As before, we create the source code file and makefile using the newProg utility. We can then compile and link the program using make. pslDisplaySeqs can now be executed with the command:

```
$ pslDisplaySeqs all_mrna pslDisplay.hg18.bed stdout
```

where pslDisplay.hg18.bed is a BED file of genome locations and all_mrna is the database table used for locating the same-species mRNA PSL transcript-alignment records. A portion of the program output is

```
chr4 158477314 158477333 GRIA2       0     +
mrna/est: BC028736
mRNA/EST block 10: gcctttatgcGgcaGggat
genome    block 10: gcctttatgcagcaaggat
chr2 47650962 47650974 KCNK12        0     -
mrna/est: AF287302
mRNA/EST block 0: atgtcctcccgc
genome    block 0: atgtcctcccgc
```

The first alignment is for part of the coding region of the GRIA2 gene, and illustrates the identification of two positions where a genomic "A" has been replaced by a

"G" in the mRNA transcript as a result of RNA editing. We see that the program highlights places where the transcript and genome sequences differ by capitalizing the nucleotide in the transcript sequence.

The second alignment is for the initial part of the coding region of the KCNK12 gene, which is found on the negative strand (as indicated by the "-" in the BED record for KCNK12). Note that even though KCNK12 aligns to the negative strand, the alignment is correctly oriented, as shown by its starting with a start codon (ATG).

As we have emphasized, implementing pslDisplaySeqs requires that you have installed a skeleton mirror of at least one UCSC database. That said, we note that the computational comparison of genomic and transcript sequences is important for many applications and cannot be accomplished using any of the tools that we have described previously, including Galaxy or Ensembl.

10.4 Installing a private UCSC genome database mirror

The last two examples indicated why mirroring at least a small part of the UCSC databases is necessary if one wants to fully exploit the capabilities of the UCSC databases and API. In addition, reasons of performance, data security, and database customization may also lead one to want to create a local private UCSC mirror.

That said, mirroring large parts of the UCSC genome database, such as the entire human hg18 database, is not a trivial task, and also requires a large amount of disk space. For example, mirroring the MySQL tables for hg18 requires about thirteen gigabytes, and mirroring the entire UCSC genome databases currently requires over one terabyte. Moreover, the size of the UCSC databases is continually increasing. Current information on disk space requirements for UCSC database mirroring can be found at http://genome.ucsc.edu/admin/mirror.html.

The task of mirror installation will generally be easier if one's target system configuration is similar to UCSC's own configuration (e.g., linux and MySQL). In my experience, downloading and installing a large UCSC database, such as hg18, can be successfully completed on a Macintosh G5 running OS X 10.3 without too many complications. However, with any large software installation, it is difficult to predict exactly what sorts of complications will arise because these problems often depend on the specific hardware and software configuration being used. In case of difficulties, you may need to get help from someone with Unix database administration experience (if you do not have it) or else seek assistance from the UCSC support team via the UCSC mailing list (genome-mirror@soe.ucsc.edu). In addition, it is important to remember that even when downloading and installation are executing completely smoothly, they are both slow processes. For example, downloading and installing the database and auxiliary files for hg18 each required approximately twenty hours on a 1.8-Ghz Mac G5 with a wide-band cable modem network connection. Downloading the GenBank auxiliary files – which are needed for the mRNA and EST sequence comparisons – required about another twenty hours of data transfer time.

Consequently, before performing a large mirror installation, it is highly recommended to first install a small test database to identify potential installation problems. One possible choice would be to use a small-genome database such as the yeast sacCer1 database. Installation of the yeast genome database follows essentially the same procedure as that for the human genome described here (in fact, simply replacing each reference to hg18 with sacCer1 should work) but will be approximately 100 times faster.

An alternate choice of test database for initial installation testing would be to install a small subset of the database tables and files from the target database. For example, to carry out the local database implementation of ucscIntronLengths2, we would actually only need to download and install a single table, the ensGene table, into our mirror database. To compare the length-distribution results for the Ensembl and RefSeq gene sets, as also described in example 4, we would only need to have the ensGene and refGene tables installed. Similarly, to display the multiple alignments of example 5, we would need to install the multiz17way table and the MAF auxiliary files in our local mirror. In fact, one can even create a working skeleton of the UCSC Genome Browser by only downloading and installing five tables (grp, chromInfo, trackDb, hgFindSpec, and any other table that represents a single browser track) into one's local mirror database. This minimal browser installation is described in http://genomewiki.ucsc.edu/index.php/Browser_installation.

Whether one installs a skeleton database with a single table or a complete mirror, the necessary sequence of commands described in the following three subsections are almost identical, with only a single command (as noted in Section 10.4.1) changing. However, because the skeleton database installation is several orders of magnitude faster, beginning with a test database installation has significant advantages. In particular, if there are going to be installation problems, you will be confronted with them much more quickly, and testing potential solutions can be carried out more rapidly as well. Moreover, with this approach, any subsequent problems during a large-database installation will most likely be specifically a result of manipulating large data files.

In the following sections, we outline the steps required to install the UCSC hg18 database (or a skeleton subset of hg18) with some of its auxiliary databases and files. The steps are also shown schematically in Figure 10.11. For additional details on UCSC database installation, the reader is referred to the mirror installation documentation at http://genome.ucsc.edu/admin/mirror.html.

10.4.1 Data download

Before performing the database installation, you need to have downloaded and installed the UCSC software code, as described in Chapter 9. You will also need to have the MySQL Server software installed and have privileged access to MySQL and to the host computer system. You will also need to ensure that any local firewalls that might prevent you from downloading large data files have been disabled.

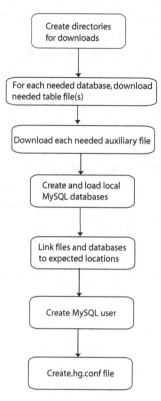

Figure 10.11 Flowchart outlining the principal steps required in the installation of mirror UCSC databases.

Assuming these preliminary steps have been completed, you can now download the data files from the UCSC FTP site. Precisely which files you need to download will depend on how much of the UCSC system you want to mirror. As an example, here we will describe installing the entire human database (hg18) along with its associated sequence and alignment files, the UniProt and GO (gene ontology) databases and the auxiliary files containing the GenBank EST and mRNA records. However, as we have emphasized, you are strongly advised to initially just install one or two tables and auxiliary files.

We first need to create and specify a root directory on a local disk drive with sufficient free disk space and make this our working directory. The commands are as follows (where "diskWithFreeSpace" is replaced by the actual disk directory location to be used):

```
$ mkdir −p /diskWithFreeSpace/goldenPath/hg18/
$ cd /diskWithFreeSpace/goldenPath/hg18/
```

Next, we execute the actual data download using a data-transfer utility program such as rsync or ncftp. This is the only step that varies between skeleton-database and full-database installation. If we build a skeleton database with only a single table (e.g., ensGene) or a few tables, as suggested previously, the download command using rsync would be

```
$ rsync -avzP --delete --max-delete=20 \
rsync://hgdownload.cse.ucsc.edu/genome/goldenPath/hg18/database/ensGene* \
/diskWithFreeSpace/goldenPath/hg18/database/
```

To obtain the data for more than one table, we would need to repeat this command for each table or else write a shell script for this purpose.

Alternatively, if we want to mirror the entire hg18 database, we would execute a command to download the data for the entire database, such as the following (this time illustrated with ncftp):

```
$ ncftp ftp://hgdownload.cse.ucsc.edu/goldenPath/hg18/
> get -R *
```

In contrast to the single-table download, which should complete within a few seconds, the download of the entire hg18 database files will require approximately fifteen to twenty hours. To minimize the likelihood of losing the connection during data downloading, it is helpful to disable any automatic sleep timer on one's local machine during this time. If, nevertheless, the network link is lost during the download, or the download does not complete successfully for some other reason, one can complete the download by simply re-executing the rsync or ncftp-get command.

To create mirrors of the UniProt and GO databases, we need to use ncftp for the corresponding files for these databases as well. Commands for these downloads are identical to those for the hg18 download, with the terms "uniProt" (or "go") substituted for "hg18" in the prior example.

Once all the MySQL files are downloaded, we need to create a root directory for the auxiliary sequence and alignment data, and we then need to download the corresponding data files. In particular, to perform the analyses illustrated in our programming examples, we need to download the hg18 and GenBank auxiliary files. The commands for this are

```
$ mkdir -p /diskWithFreeSpace/gbdb/hg18/
$ rsync -avzP rsync://hgdownload.cse.ucsc.edu/gbdb/hg18/ \
  /diskWithFreeSpace/gbdb/hg18/
$ mkdir -p /diskWithFreeSpace/gbdb/genbank/
$ rsync -avzP rsync://hgdownload.cse.ucsc.edu/gbdb/genbank/ \
  /diskWithFreeSpace/gbdb/genbank/
```

Again, if the connection is lost before the download completes successfully, we can complete the transfer later by simply repeating the rsync command.

Finally, because the UCSC software expects that the sequence and alignment files are located in a directory named /gbdb, we need to create a link from the actual file directory to /gbdb. As these directories may well be on different disks, this needs to be a symbolic link:

```
$ ln -s /diskWithFreeSpace/gbdb//gbdb
```

10.4.2 Installation

Once all of the required files have been successfully downloaded, the next steps are to create the local databases, uncompress the database files, and load the data into the local databases. To create the databases, we need to log on to the MySQL server with MySQL administrator privileges. We can then initialize each database we need as follows:

```
mysql> create database hg18;
mysql> create database uniProt;
mysql> create database go;
 etc.
```

In addition, if you are installing a large database like hg18, you will probably need to increase the value of the "max_allowed_packet" parameter (otherwise, the database loading script may abort). This can be done at the MySQL command line with

```
mysql> set global max_allowed_packet=25165824;
```

(If you plan to be frequently installing large databases, you may prefer to include this command in your MySQL configuration file, for example, /etc/mysql/my.cnf or /etc/my.cnf.)

Next, we need to address the fact that MySQL generally assumes that its databases are on the same hard drive as the MySQL server. However, typically this will not be the case with the large genome databases. Consequently, we need to configure appropriate links so that MySQL knows where to locate its databases. There are several different ways for accomplishing this – see, for example, (Dubois, 2005) Chapter 10, for more details. One simple approach is to create a separate symbolic link for each new database. Note that you will typically need to be system superuser (using the "su" command under Unix/linux or "sudo su" under Mac OS X) to create these links. For example, on the Macintosh, assuming that the expected location of the MySQL data directory is /usr/local/mysql/data:

```
peter$ mkdir -p /diskWithFreeSpace/var/mysql/data/hg18
peter$ mkdir -p /diskWithFreeSpace/var/mysql/data/uniProt
peter$ mkdir -p /diskWithFreeSpace/var/mysql/data/go
(etc.)
peter$ sudo su
root# ln -s /diskWithFreeSpace/var/mysql/data/hg18 /usr/local/mysql/data/hg18
```

```
#!/bin/sh
DB="hg18"
WEBROOT="/Volumes/LaCieDisk/ucscDbRoot"
cd $WEBROOT/goldenPath/${DB}/database
for SQL in *.sql
do
  T_NAME=${SQL%%.sql}
  echo "loading table ${T_NAME}"
  mysql -uAdmin -pAdminPw -e "DROP TABLE ${T_NAME};" \
      ${DB} > /dev/null 2>/dev/null
  mysql -uAdmin -pAdminPw ${DB} < ${SQL}
  zcat "${T_NAME}.txt.gz" | mysql -uAdmin -pAdminPw --local-infile=1 \
      -e "LOAD DATA LOCAL INFILE \"/dev/stdin\" INTO TABLE ${T_NAME};" ${DB}
done
```

Figure 10.12 Template installation script to load data downloaded from the UCSC FTP site into MySQL tables of a local UCSC database mirror. See text for template modifications required before the script can be used for table loading.

```
root# ln -s /diskWithFreeSpace/var/mysql/data/uniProt \
      /usr/local/mysql/data/uniProt
root# ln -s /diskWithFreeSpace/var/mysql/data/go /usr/local/mysql/data/go
(etc.)
```

Now we are almost ready to load the database tables. The UCSC installation documentation provides a script that automates this process, shown here in Figure 10.12. However, we first need to edit the script in Figure 10.12 so that

1. /Volumes/LaCieDisk/ucscDbRoot is replaced by the location where you previously loaded the compressed UCSC tables.
2. "-uAdmin -pAdminPw" is substituted with the appropriate administrator user name and password.
3. hg18 is replaced with the database that you are loading.

We now need to run this script separately for each database that must be loaded, or else modify the script to load multiple databases sequentially. Note that loading all the tables of a large database, such as hg18, using the script in Figure 10.12 may require fifteen hours or more, depending on your hardware. (If we are only creating a skeleton database with one or a few tables, the database table-loading script will finish within a few seconds or less.) Once the downloaded data have been loaded into the local MySQL database, the download files can be deleted, if desired, to free up disk space. Finally, for security, if you use this type of script, you will want to remove the administrator MySQL password from the script once it has completed.

10.4.3 Configuration

Once all the databases have been loaded, the final steps are to create a MySQL user account through which one's software can access the UCSC databases, and to set up either a configuration file or environmental variables so that the software can find the database account information.

A simple method for granting read access to the databases is with the following single command, executed by the MySQL administrator:

```
mysql> GRANT SELECT ON *.* TO ucscDbUser@localhost IDENTIFIED BY "ucscDbPw";
```

Alternatively, more complex (but more secure) grant options can be configured – see DuBois (2005), chapters 11 and 12, for examples.

Finally, we need to set up a configuration file for each user who needs programmatic access to the database via the kent code. The file needs to have the name ".hg.conf" (note the initial dot), to be located in the user's home directory, and to contain the following three lines of code (with "db.host," "db.user," and "db.password" values substituted as appropriate):

```
db.host=localhost
db.user=ucscDbUser
db.password=ucscDbPw
```

Alternately, a user can define the following three environmental variables, typically in one of their login files (syntax given is for the Bash shell):

```
export HGDB_HOST="localhost"
export HGDB_USER="ucscDbUser"
export HGDB_PASSWORD="ucscDbPw"
```

Once the database is set up and configuration is complete, one should run tests to confirm that everything has been properly configured. As an initial step, one should confirm the ability to log in to the different databases and the ability to query them interactively, for example:

```
$ mysql -u ucscDbUser -p -A hg18
password:
mysql > show tables;
```

If direct access works, the next tests might be to run the demonstration programs described in this chapter.

10.4.4 Maintaining UCSC code and databases

In contrast to Ensembl, UCSC's code revisions are not explicitly linked to its database releases. Usually, the most current UCSC code will function properly both with current as well as with older database builds. In many cases, older versions of the UCSC API – in particular, the library functions that one principally uses – will continue to function properly with newer UCSC database builds. That said, it is recommended – and easy – to periodically update one's copy of the UCSC API either by using the CVS updating system or by re-downloading the entire API from http://hgdownload. cse.ucsc.edu/admin/jksrc.zip, which is updated by UCSC every two weeks.

Keeping the data in UCSC mirror databases and auxiliary files current is a more substantial task. Essentially, one needs to update (via the rsync program) every one of the database tables and auxiliary files that one is mirroring, and then to replace each changed database table with one containing the current data. Because a mirror database may well contain numerous UCSC databases (e.g., from many genomes) and because some of the tables and files are modified on a daily or weekly basis by UCSC, this is a nontrivial task.

To assist one in automating these updating tasks, two Unix command scripts are available from the UCSC genome wiki site. The first script, called doDownloads.sh, handles the updating of the auxiliary files and compressed database table files. The second script, doUpdateDb.sh, automates the process of loading the decompressed updated tables into the appropriate database. Each of the scripts can read a list of database names that can be customized so that just the databases present in one's local mirror are updated.

A subtlety with which one is confronted in the updating process is the removal of obsolete database tables. If one simply mirrors the UCSC databases without adding any custom data tables, one can simply include the --delete option to the rsync command and the rsync program will automatically delete the corresponding tables in the mirror database. However, this simple approach will fail if one has added custom tables to one's local mirror. In this case, one needs to customize the arguments to rsync so that the rsync program does not inadvertently remove one's custom tables. For more information on updating UCSC mirror sites, the reader is referred to the discussion of partial mirrors at the UCSC genome wiki site (http://genomewiki.ucsc. edu/index.php/Browser_Mirrors).

Chapter summary

- UCSC API access of data from the UCSC databases – without using the UCSC public mirror databases – is possible by either downloading individual database tables as flat files or by installing a local mirror of all or part of one or more UCSC databases.
- Programs using sequence and alignment data can only be executed if the relevant components of the non-MySQL data from the UCSC database have been mirrored locally.
- For many applications, only one or a small number of tables or auxiliary files need to be mirrored locally.

- Even if one wants to eventually mirror an entire UCSC database, it is usually advisable to first install a skeleton mirror to ensure that the installation procedure has been configured properly for one's local system.
- Although mirroring parts of the UCSC databases is not trivial, creating such a private mirror enables one to carry out genomic analyses – such as base-level comparisons of genomic and transcript data – which are quite difficult to accomplish by other means.

Exercises

Note that you will need to have the UCSC API and the MySQL client and server software installed to complete these exercises.

1. Download a copy of the hg18 refSeq gene table from UCSC (e.g., using the Table Browser). Use this data file with the ucscIntronLengths2 program using the "file" method option. Compare your results to those you obtain if you use the hg18 refSeq gene table from the UCSC public databases while running ucscIntronLengths2 with the "public" method option.
2. Follow the procedure outlined in the text to install a local copy of a skeleton of the hg18 database consisting just of the ensGene table. Test your installation by directly querying the database with SQL and by executing the ucscIntronLengths2 program against your local database.
3. Install a local copy of the UCSC yeast database using the procedures described in the text. Test your installation by directly querying the database with SQL and by executing the program from Chapter 9, Exercise 3, against your local yeast database.

11

Customized Genome Databases

Biological research often involves combining publicly available data from genome (and other) databases with custom data that has been identified in one's own research. So far, we have primarily focused on techniques for accessing public data, especially in the integrated formats used by the major genome database projects. We have also seen how to incorporate one's own data using the method of custom tracks offered by the UCSC and Ensembl Browsers and how to use the tools for including custom data in batch queries via the UCSC Table Browser and Galaxy interfaces.

We now turn our attention to additional techniques for integrating custom data with data in the public databases. These range from quite simple methods for navigating among features on a custom track to quite complex approaches for creating entirely new genome databases and browsers for newly sequenced genomes. As we will see, some of these methods are applicable to both the Ensembl or UCSC architectures. Others are significantly easier or only possible in one of the two systems, or are best performed outside of either the UCSC or Ensembl framework using the CMOD tools. In particular, we will restrict ourselves to approaches that do not require writing new software for either the UCSC or Ensembl APIs.

The tasks described in the present chapter are more specialized and are needed less frequently by the typical researcher than the database querying and data analysis tasks on which we have focused in previous chapters. Moreover, some of these methods require significantly more computer system skills (e.g., Unix systems-administrator experience) than have been needed in other parts of this book. Also, the requirements of custom database configuration vary widely with the specific application and, consequently, describing all the issues that may arise in a database customization is difficult. For all of these reasons, our descriptions of genome database customization are less detailed than our descriptions of genome database querying have been. In particular, in some cases, we will restrict ourselves to briefly describing the various tools available, noting the tradeoffs among these tools, and pointing the reader toward the detailed documentation, which describe the methods more fully.

11.1 Overview of genome database customization

Despite the capabilities of the custom-track tools provided by both the Ensembl and UCSC systems, there are several situations in which one might want custom-data integration capabilities beyond those provided by conventional custom tracks. First, custom-track data are kept on the UCSC and Ensembl sites for only a limited time period. Consequently, custom tracks need to be continually re-uploaded if they are needed on a regular basis. Second, if one's custom data set is large, it may exceed the space allocation for custom data provided by UCSC or Ensembl. In addition, if the custom tracks only include a small number of features that are sparsely distributed across the genome, navigating among the custom features can be awkward. Also, if the custom data consist of numerous, interrelated tracks, batch querying with the UCSC Table Browser may be difficult or inconvenient, whereas in Ensembl BioMart, batch querying using custom data is not possible at all. Another limitation of custom tracks is that it may not be possible to express all of the relationships among one's data using one of the available custom-track formats (e.g., BED or PSL). Also, there may be security concerns with depositing one's private data as custom tracks on a public genome database site (even though both the UCSC and Ensembl systems are designed so that custom data can only be accessed by the machines that initially uploaded the data). Finally, if one's data comes from a species whose genome is not in the UCSC or Ensembl databases at all, then clearly one cannot incorporate one's data into the standard databases and one needs to create one's own genome database.

Various methods have been developed to address the limitations of conventional custom tracks. Each has its advantages and its limitations. For example, custom frames are simple and easy to implement, but only address the issue of displaying sparse datasets. Using DAS server software enables one to display large and permanent custom datasets on the public Ensembl Browser while representing one's data with any desired data model. With a local DAS server, one can also make batch queries of one's data using the Bio::DAS API. However, one cannot easily make integrated batch queries (either interactively or via API) that involve both local data and annotations in the Ensembl database.

In contrast to the DAS server approach, UCSC provides tools that enable one to add custom data in a permanent manner to a local mirror of the UCSC database. These tools facilitate the creation of a UCSC database mirror in which one can browse and batch query both custom data and public genome annotations in a truly integrated way. The disadvantage of this approach is that setting up such a system can be technically challenging, particularly if your machine architecture differs from the UCSC's.[1]

---

[1] In principle, one can apply the approach of adding custom tables to a local database mirror to an Ensembl mirror as well. However, adding tables to an Ensembl database requires adding software to the Ensembl API and, because of this added complexity, will not be further described here.

Databases for genomes not present in UCSC or Ensembl's databases can be designed by "cloning" or copying the architecture of an already existing UCSC or Ensembl database or by using the genome-database architecture of the GMOD (Generic Model Organism Database) project as a template. Cloning an Ensembl or UCSC database has the advantage that when one is finished, one will have access to the entire array of database-access tools developed by UCSC or Ensembl. However, the disadvantage of this approach is that cloning an Ensembl or UCSC database schema is quite challenging technically and is often most appropriate for large-scale projects similar to those for which the UCSC and Ensembl systems were designed. In contrast, using the GMOD genome database construction strategy enables one to build a genome database and browser interface for a small- or medium-size genome with a relatively modest investment of development time. The disadvantage is that the resulting system will not be as full-featured for either genome browsing or batch querying as the Ensembl or UCSC systems.

11.2 Custom frames

A custom frame is a simple tool, available in the UCSC system, for displaying sparse custom data. In the next section, we will illustrate custom frames via their use within the UCSC Genome Graphs tool. In the following section, we will describe custom frames more generally, showing how they can be used to display other types of sparse custom data.

11.2.1 Genome Graphs

The Genome Graphs tool was developed to aid in the display of custom genetic association data, such as those generated by genotyping arrays. Genetic association studies are designed to identify genomic regions for which there is a statistically significant correlation between a specific genetic variant (typically a SNP) and some phenotype, such as the susceptibility to a disease. Identifying such regions is important because they are likely to include genes whose variations contribute to the observed range of phenotypes.

To display one's genetic association data using Genome Graphs, one needs to format one's data slightly differently than for a standard UCSC custom track. First, as each value in a genetic association study is associated with a SNP, a location in the custom data file is specified by a single coordinate rather than by a start and end coordinate. Also, this location may be specified indirectly via a dbSNP rsID or an Affymetrix or Illumina chip location rather than explicitly specifying its genomic coordinate. The other addition to the input format is the inclusion of a user-adjustable threshold for data significance so that one can specify precisely which records in the input file are included in the data display.

As shown in Figure 11.1, a Genome Graphs display is similar to a standard UCSC custom data track, however, with some new features. First, we see the appearance

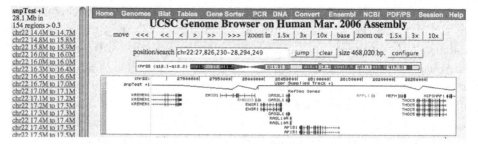

Figure 11.1 Example of a custom frame display obtained after selecting the "Browse regions" option from the UCSC Genome Graphs page. The left side of the display contains a list of genomic coordinates. Clicking on any one of them causes the browser display on the right side to show the region of the genome specified by the selected coordinates.

of line segments connecting the genetic association probabilities of adjacent SNPs, where "adjacency" is determined by a user-specified maximum distance. The other new feature of the Genome Graphs display is the use of a custom frame. The custom frame includes a list of all of the regions that contain custom data. The list appears as a separate frame on the left side of the display. By clicking on any one of the regions in the list on the left side of the display, the main browser window is refreshed to display the specified region. In this way, it is easy to navigate among the various genomic regions containing the custom data.

11.2.2 General custom frames

The custom frame display shown in Figure 11.1 is useful for applications besides genetic association studies. In particular, custom frames can be helpful for displaying any type of sparse custom data (i.e., data from a relatively small number of regions sparsely distributed throughout the genome).

Assuming the UCSC API has been downloaded and installed as described in Section 9.4, the bedToFrames program (located in the hg/orthoMap subdirectory) can be used to easily create a custom frame of one's own data. After the bedToFrames program has been compiled and installed, a list of genomic regions in BED-file format can be converted into custom-frames format with a command like

```
$ bedToFrames hg18 bedFileIn.bed htmlTableFile.html htmlFrameFile.html ""
```

In this command, bedFileIn.bed is the file of input regions and htmlTableFile.html and htmlFrameFile.html are the output HTML pages implementing the custom frame. The final command-line argument to bedToFrames specifies an optional URL; because we do not require this option, we replace this argument with an empty pair of double quotes.

After executing the program, pointing one's web browser to the local file html-FrameFile.html will create a custom frame display analogous to that shown in Figure 11.1, with the left part of the display showing the list of regions in the original

BED file with their browser links and the right side of the display showing a standard view of the UCSC Genome Browser for the selected region.

We note that the custom frame display is complementary to the custom track displays that we have described previously. In fact, it is often useful to also upload a custom track of one's custom-frame data to the UCSC Browser. However, creating a custom track at UCSC is not necessary for creating a custom frame and, in particular, the custom frame (which uses data located entirely on one's own computer) will continue to function even after one's custom track has been removed from the UCSC database.

It is also worth noting that there is an even easier, albeit slightly less convenient, means for navigating among a sparse set of selected genomic regions. One simply uploads the set of regions as a custom track and then selects the custom track in the Table Browser. If one then chooses "hyperlinks" as the output data format, the Table Browser will return a list of links of one's regions of interest to the genome browser, though not in the form of a separate frame.

11.3 User-generated DAS tracks in Ensembl

In Chapter 3, we saw how to load custom-track data via Ensembl's internal DAS server. Ensembl also has a custom-track tool that uses the client computer's web-serving capability to create custom tracks (see http://www.ensembl.org/info/using/external_data). These approaches work well for relatively small custom datasets that only need to be temporarily integrated into the Ensembl Browser. For situations where one wants to integrate a large amount of custom data or wants the data to be available relatively permanently, Ensembl provides a third mechanism for custom tracks – using a local DAS server.

The DAS protocol enables computers to add annotation tracks to a remote genome browser. DAS includes specifications for annotation formats – described in Appendix 2 – as well as for communications between the DAS annotation server and the main sequence-server of the genome browser. Ensembl has been designed to be fully compliant with the DAS protocol and, consequently, a DAS server can create annotations that will be properly incorporated into the Ensembl Browser as a standard Ensembl (DAS) track.

Setting up a local DAS annotation server is more complex than using other means of creating Ensembl custom tracks, such as using Ensembl's internal DAS server. However, the installation procedure is relatively straightforward and well documented (see http://www.ensembl.org/info/using/external_data/das/das_server.html and links therein), and will not be detailed further here. Once a DAS server has been installed and configured on one's local computer, one can easily and stably view one's custom data as tracks on the Ensembl Browser. However, as the DAS data is not incorporated into the Ensembl BioMart database, the data is not available for access via MartView, nor is the data accessible via the Ensembl API. (The data can, however, be accessed via the Bio::DAS API described in Chapter 8.)

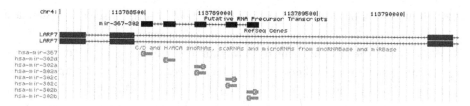

Figure 11.2 Custom track of putative miRNA precursor on the UCSC Genome Browser.

11.4 Adding tables to the UCSC database

The UCSC system does not currently support the integration of custom tracks onto its public browser via a remote DAS server. Instead, the UCSC software provides tools for integrating one's data into a local mirror of the UCSC database. This approach has both advantages and disadvantages compared to integrating DAS server tracks. It is more complex than using DAS, and one needs to install a UCSC database mirror and – if one wants to display the custom data visually on the browser – also a mirror of the UCSC Genome Browser software. On the other hand, with this approach one's data is truly integrated with the public genome data and is accessible for batch querying via the Table Browser and the UCSC API, as well as visually via the Genome Browser.

Just how difficult integrating one's custom data into the UCSC database will be, as well as the extent that one will be able to access that data via the UCSC toolset, will depend on the nature of the data being added. If the custom data consists solely of additional tables having exactly the same table formats as ones that already exist in the database (e.g., tables containing gene predictions from some new gene-prediction program), then adding the data into the database is easy. If the new data requires a new type of database table, then integrating the data with the database and the UCSC API requires more effort.

We illustrate both of these scenarios with a simple example involving annotations for microRNA-precursor sequences. *microRNAs* (also known as miRNAs) are small ncRNAs that have recently been found to play important roles in eukaryotic gene regulation. microRNAs are coded in the genome by short (20 to 22 nt) regions of DNA that are located in larger (60 to 80 nt) *miRNA-primary sequences*. These primary miRNAs may themselves be located within larger transcripts known as *miRNA-precursor sequences*, which may include two or more primary miRNA sequences. Figure 11.2 shows annotations for several human miRNA primary sequences on the UCSC Genome Browser, as well as a putative precursor sequence that contains them.

11.4.1 Adding standard-format database tables to a UCSC mirror

Let us now imagine that we have experimentally or computationally identified the locations of a set of miRNA-precursor sequences. An example of one such putative precursor miRNA is shown as a custom track in Figure 11.2. Now there is no standard UCSC table format that completely captures all of the relationships among

Table 11.1 Creating a new database table with standard format (e.g., BED-12)

Step	Task	Utility or library function	Comments
1	Create SQL code for table generation	Text editor	Could also use autoSql
2	Create MySQL table	mysql or hgsql	
3	Convert data format for database loading	Custom parser using gff.c autoXml	Data in GFF Data in XML
4	Load table into database	hgLoadBed, hgLoadPsl, hgLoadMaf, or other dataloading program	

miRNA-precursor data. However, BED-12 table format (see Appendix 2, Table A2.1) can describe most of the data if we set chromStart and chromEnd equal to the precursor-transcript start and end coordinates, and set the blockStarts and block-Sizes fields to the start coordinates and lengths of the primary miRNA sequences. In fact, the "putative precursor" track shown in Figure 11.2 was input to the UCSC Genome Browser as a custom track using BED-12 format.

If we store our miRNA-precursor data in BED-12 formatted tables, we can integrate our data into a mirror database in just a few steps. These steps are outlined in Table 11.1. First, we need to create a new BED-12 table in our mirror database (assumed here to be hg18). We can do this by editing the table name in the fullBed.sql file in the hg/lib directory to rnaPrecursor and renaming the file rnaPrecursor.sql. Alternatively, we could create the rnaPrecursor.sql file directly with the autoSql utility, as we describe in the following section. In either case, once we have created the rnaPrecursor.sql file, we can create the new database table using the command:

```
$ hqsql hg18 < rnaPrecursor.sql
```

where "hgsql" is a simple UCSC API wrapper program (located in the hg subdirectory of the UCSC API code) for the MySQL program, which eliminates the need to explicitly enter the database user name and password. Note that for this command or the following hgLoadBed command to work, you need to ensure that the database "user," defined by either $HGDB_USER or .hg.conf, has MySQL privileges to create and drop tables. (How to grant such privileges to a database user was described in the database configuration section of Chapter 10.)

Next, we need to load the precursor data from a file in computer memory into the rnaPrecursor table. If we have generated the data ourselves, it is generally easy to store our data in BED-12 file format. In this case, we can load the entire table with the hgLoadBed program (located in the hg/makeDb subdirectory) using the command:

```
$ hgLoadBed hg18 rnaPrecursor rnaPrecursor.bed
```

where "rnaPrecursor.bed" is the BED file with the (hg18 human-genome) precursor locations and "rnaPrecursor" is the hg18 database table name.

On the other hand, if our data is available only in another format, such as XML or tab-delimited GFF, we will need a program to parse the input file and convert it into a format compatible with one of the UCSC database-loader functions. However, here as well there are kent code programs and library functions available to facilitate the task. For some database table formats, a parser that can convert a tab-delimited GFF-formatted file into the required database format is already available (see, for example, gtfToGenePred in the utils subdirectory). For other GFF-formatted files, one can use the parsing routines available in the gff.c file in the lib subdirectory to write a GFF parser.

If our data is in XML, we can use the UCSC xmlToSql, autoDtd, and autoXml programs to directly generate the necessary SQL code as well as to generate an XML parser program to read the XML input file into memory. See the articles "autoSql and autoXml: Code generators from the Genome Project" and "XML, SQL and C" (which are available online; see the references in Appendix 7), as well as the embedded documentation in the autoXml, autoDtd, xmlToSql, and sqlToXml subdirectories of the hg directory for more information on autoXml.

Finally, once our data has been loaded into the UCSC database, it can be accessed either directly via SQL or via the UCSC API using the library functions described in the file bed.h of the hg/inc subdirectory.

11.4.2 Creating new types of database tables

So far, we have seen how by storing our miRNA-precursor annotations in BED-12 format in one of the UCSC databases, we can access them via the UCSC API. However, by using BED-12 format we have lost some of the information contained in our annotation. Specifically, although each table record contains the precursor and primary transcript coordinates, it does not contain the coordinates of the miRNAs themselves.

One way of including this additional information would be to add a field to each record of our rnaPrecursor table consisting of a comma-separated list of names of the embedded miRNAs. This field could then be used to derive keys to look up information on the constituent miRNAs stored in some other table, for example, the wgRna table. To implement this approach, we need to first design the structure of our modified table and then execute an SQL command to create the table in the database. Next, we need to load the table with our data. Finally, we need to write C code to access the data from the database and to store the data in memory in appropriate C structures. The tasks required to construct such a new database table are shown in outline form in Table 11.2.

As in the previous example, these tasks are made much easier by using UCSC utility programs, in particular, the autoSql program. autoSql is located in the UCSC code tree in subdirectory hg/autoSql. autoSql is described in detail in the "autoSql

Table 11.2 Creating a database table with new structure (e.g., rnaPrecursor)

Step	Task	Utility or library function	Comments
1	Create SQL + C code for table generation and access	autoSql	Creates rnaPrecursor.sql, rnaPrecursor.c, and rnaPrecursor.h files
2	Create MySQL table	mysql or hgsql	
3	Convert data format for database loading	Custom parser using rnaPrecursorCommaIn	Comma-separated data
		Custom parser using gff.c	GFF Data
		autoXml	XML Data
4	Load table into database	Custom loader using rnaPrecursorSaveToDb	

and autoXml" article noted in Section 11.4.1, as well as in the autoSql.doc file in the hg/autoSql subdirectory. With autoSql, we need only specify the structure of the annotation data. The autoSql program then generates the C and SQL code necessary to create and load the new MySQL tables and to access these tables via the UCSC API.

Figure 11.3a shows a sample autoSql specification file for miRNA-precursor annotations. In the specification file, we see two new fields. miRnaCount stores the number of miRNAs contained within the precursor sequence, whereas miRnaNames is a comma-separated string with the names of the miRNAs (which are also the primary keys of the miRNA records in the wgRna table). We have also changed some of the BED-12 field names (e.g., chromStart has become precursorStart, and thickStart has become primaryStart) to more accurately correspond to their usage in our example.

Once the autoSql program has been compiled and linked, we can run it against the template file of Figure 11.3a with the command:

```
$ autoSql rnaPrecursor.as rnaPrecursor -dbLink
```

where rnaPrecursor.as is the name of the autoSql template file. This will create three output files, rnaPrecursor.sql, rnaPrecursor.c, and rnaPrecursor.h. Creating the database table can now be accomplished using the same hgsql command indicated in the previous section. The rnaPrecursor.c and rnaPrecursor.h files provide the C code for loading and accessing the rnaPrecursor table from a C program (see Figure 11.3c). In particular, the rnaPrecursorLoad function loads a row from the rnaPrecursor table into a C structure, whereas rnaPrecursorSaveToDb performs the reverse task – saving the data from an rnaPrecursor C structure into the database.

The final step before we can access our data is to actually load the data into the database. Because our data table has a new format, we can no longer use a standard UCSC loader program, such as hgLoadBed or hgLoadPsl. However, if we format our data records as strings of comma-separated fields, we can use the autoSql-generated

a)
```
table rnaPrecursor
"A cluster of cotranscribed ncRNAs (eg miRNAs or snoRNAs)."
    (
    string name;   "Name of RNA cluster"
    string chrom;  "Reference sequence chromosome or scaffold"
    char[1] strand;      "+ or - for strand"
    uint precursorStart;      "Precursor transcript start position"
    uint precursorEnd;         "Precursor transcript end position"
    uint primaryStart;   "Primary transcript region start"
    uint primaryEnd;        "Primary transcript region end"
    uint primaryCount;      "Number of primary transcripts"
    uint miRnaCount;      "Number of processed RNAs"
    uint[primaryCount] primaryStarts; "Primary transcript start positions"
    uint[primaryCount] primaryEnds;  "Primary transcript end positions"
    string[miRnaCount] miRnaNames;   "Processed RNA names"
    )
```

b)
```
CREATE TABLE rnaPrecursor (
    name varchar(255) not null,  # Name of RNA cluster
    chrom varchar(255) not null, # Reference sequence chromosome or scaffold
    strand char(1) not null,   # + or - for strand
    precursorStart int unsigned not null, # Precursor transcript start position
    precursorEnd int unsigned not null,   # Precursor transcript end position
    primaryStart int unsigned not null,   # Primary transcript region start
    primaryEnd int unsigned not null,     # Primary transcript region end
    primaryCount int unsigned not null,   # Number of primary transcripts
    miRnaCount int unsigned not null,     # Number of processed RNAs
    primaryStarts longblob not null,      # Primary transcript start positions
    primaryEnds longblob not null,        # Primary transcript end positions
    miRnaNames longblob not null,         # Processed RNA names
              #Indices
    PRIMARY KEY(name)
);
```

Figure 11.3 Use of autoSQL program to add database table and C code for RNA precursor annotations to the UCSC database. (a) Input file for autoSql program specifying the precursor-transcript structure. (b) autoSql-generated SQL code for creating the database table for RNA-precursor data. (c) Part of autoSql-generated C struct and subroutine definitions for manipulating RNA-precursor data within the UCSC API.

functions rnaPrecursorCommaIn and rnaPrecursorSaveToDb to read the data into memory and then write it out to the database table. In contrast, if the data is only available in XML or tab-separated GFF format, we will need a parser to read the data into memory, as in the previous example. For XML data, we can again use the autoXml program to automatically generate an XML data parser. Similarly, for tab-delimited GFF files, we can again use the GFF parsing routines from the gff.c library to write a GFF parser. In addition, for XML, we can automatically generate a database table

```
c)
struct rnaPrecursor
/* A cluster of cotranscribed ncRNAs (eg miRNAs or snoRNAs). */
    {
    struct rnaPrecursor *next;  /* Next in singly linked list. */
    char *name;    /* Name of RNA cluster */
    char *chrom;   /* Reference sequence chromosome or scaffold */
    char strand[2];      /* + or - for strand */
    unsigned precursorStart;  /* Precursor transcript start position */
    unsigned precursorEnd;    /* Precursor transcript end position */
    unsigned primaryStart;    /* Primary transcript region start */
    unsigned primaryEnd;      /* Primary transcript region end */
    unsigned primaryCount;    /* Number of primary transcripts */
    unsigned miRnaCount;      /* Number of processed RNAs */
    unsigned *primaryStarts;  /* Primary transcript start positions */
    unsigned *primaryEnds;    /* Primary transcript end positions */
    char **miRnaNames;  /* Processed RNA names */
    };

struct rnaPrecursor *rnaPrecursorLoadByQuery(struct sqlConnection *conn, char
    *query);
/* Load all rnaPrecursor from table that satisfy the query given. */

void rnaPrecursorSaveToDb(struct sqlConnection *conn, struct rnaPrecursor *el,
    char *tableName, int updateSize);
/* Save rnaPrecursor as a row to the table specified by tableName. */

struct rnaPrecursor *rnaPrecursorCommaIn(char **pS, struct rnaPrecursor *ret);
/* Create a rnaPrecursor out of a comma separated string. */

void rnaPrecursorOutput(struct rnaPrecursor *el, FILE *f, char sep, char
    lastSep);
/* Print out rnaPrecursor.  Separate fields with sep. Follow last field with
    lastSep. */
```

Figure 11.3 (continued)

schema using autoXml (rather than autoSql) that will be compatible with the XML
data representation.

Once our input data file has been parsed and loaded into the database, the data
can be programmatically accessed with the UCSC API using the autoSql-generated
rnaPrecursorLoadByQuery function.

11.5 Adding tracks to the UCSC Browser

In the previous sections, we have seen how we can add custom data to a local UCSC
mirror database and access that data via the UCSC API in an integrated manner.
However, we are not yet able to view this data as a track on the Genome Browser. To

visualize the new data with the Genome Browser, we must first install a mirror of the UCSC Browser software itself and then configure the browser so that it displays the data in our new database tables as browser tracks. In the following sections, we outline the procedure for performing these tasks assuming that our custom data can be represented using one of the standard UCSC table formats (e.g., BED or PSL). In contrast, displaying data from a new type of database table requires adding new code to the UCSC Browser software itself, which is beyond the scope of the present chapter.

11.5.1 Installing a mirror of the UCSC Browser

In Chapter 10, when we described installing a mirror of the UCSC databases, we did not discuss creating a mirror of the Genome Browser itself. The reason is that unless one is creating new database tables, it is just as effective to visualize the results of any batch query using the public Genome Browser at UCSC. However, if one wants to visualize annotations corresponding to our custom data, it is necessary to install a mirror of the UCSC Genome Browser itself (or else to continually upload custom tracks to the UCSC site).

Assuming that we have already installed a mirror of one or more UCSC databases, it is straightforward – though not trivial – to install a mirror of the UCSC Genome Browser. Detailed browser-installation procedure is located in the UCSC mirror installation guide at http://genome.ucsc.edu/admin/mirror.html. Additional useful documentation for installing a browser mirror can be found at http://genomewiki.ucsc.edu/index.php/Browser_installation and in the ex.installExample and README. install files in the kent/src/product subdirectory of the kent code distribution. If your target system uses RedHat or a similar linux distribution, the documentation at http://genomewiki.ucsc.edu/index.php/Browser_ Installation (note capitalization of "Installation") is also helpful.

The main steps in the browser installation procedure are

- Install and configure the Apache web server.
- Create a new MySQL user and grant that user read and write privileges on the UCSC annotation databases.
- Create an additional database, called hgcentral, which contains overall browser-configuration tables.
- Create html, cgi-bin, and trash subdirectories in the main Apache document directory.
- Copy static HTML pages from the UCSC download site to the html subdirectory.
- Either copy precompiled CGI binaries to your system's main Apache CGI binary directory (for linux systems) or compile the CGI binaries from source by executing "make compile" and "make install" commands in the kent/src/hg subdirectory. In the latter case, when you run "make install," you will need to define the DESTDIR and CGI_BIN variables so that the output binary is written to the main Apache CGI binary directory.

```
track rnaPrecursor
shortLabel  RNA Precursors
longLabel snoRNA and miRNA Precursors
group genes
priority 50
type bed 12
visibility hide
```

Figure 11.4 A section of a trackDb.ra file, which is configured to display a BED-12 formatted track called rnaPrecursor.

- Load the hgcentral database using the (edited) command script ex.hgcentral.sql located at kent/src/product and described in the README.install file.
- Modify the ex.hg.conf configuration file located at kent/src/product and copy it to the main cgi-bin directory, again as described in the README.install file.

In practice, the effort required to install a browser mirror depends on how similar the target hardware and software configurations are to those used at UCSC. If the target system uses linux, the Apache web server, and MySQL, installation will be less difficult. If the target system uses another version of Unix, such as Mac OS X, or some other web server or relational database, installation and system configuration will require more time and effort because installation on such systems has been much less tested.

In any case, it is advisable to make sure that one's web server properly executes basic CGI programs before one undertakes Genome Browser installation because otherwise, it is difficult to distinguish browser installation problems from other web server configuration issues. It is also important to install all of the browser html and cgi-bin subdirectories in exactly the location expected by the browser, that is, in the primary web server directory, because parts of the browser software are hard-coded to expect them in those locations.

11.5.2 Displaying new tracks on the UCSC Browser

Once one has a genome browser mirror installed and running, any custom tables that you have installed in your local database will be viewable via the Table Browser. However, if one also wants to view the data as tracks on the Genome Browser, one has to modify and reload one of the track-configuration trackDb.ra files in hg/makeDb/trackDB or one of its subdirectories. These files specify which tracks to display on the browser, which table to associate with each track, and how to display each track.

In a trackDb.ra file, configuration data for each track is contained in a multi-line record separated by a blank line. A small portion of a sample trackDb.ra file, illustrating the track-configuration file format for an rnaPrecursor track, is shown in Figure 11.4. As can be seen, the trackDb.ra format includes multiple fields and options for customizing the track display. These options are described in detail in the README file in the hg/makeDb/trackDB. However, if one only needs to display tracks from tables with standard UCSC table formats, one does not need to learn all of these options. Instead, it is sufficient to simply copy a record from another preexisting, track that uses the same database format (e.g., BED-12) and edit the copied record with the new track and table names, as shown in Figure 11.4.

The various track-configuration files are located in multiple subdirectories of the hg/makeDb/trackDb directory hierarchy, and all have the same file name (trackDb.ra). All trackDb.ra files have the same function and format. Which trackDb.ra file you need to modify depends on in which genome assemblies you want your new track to appear. For example, tracks listed in the trackDb.ra file of the human/hg18 subdirectory of hg/makeDb/trackDb will be displayed only if the user selects the hg18 build of the human genome. In contrast, tracks in the trackDb.ra file of the human subdirectory will appear in all assemblies of the human genome. Similarly, those in the main trackDb directory will be displayed in the browser for all species.

After one has added the record for the new track to one of the trackDb.ra files, one has only to load the modified trackDb.ra file for the changes to appear in the browser. Loading the trackDb.ra files is accomplished by executing the following command in the hg/makeDb/trackDb directory:

```
$ make alpha DBS=hg18
```

where the argument of DBS (in this case, hg18) indicates which browser builds should be updated. After this command has been completed, any new tracks in the trackDb.ra files in the trackDb, trackDb/human, or trackDb/human/hg18 subdirectories will be visible the next time the hg18 human genome build is selected in the Genome Browser.

In the case that our custom data requires the use of a nonstandard table format, the previous procedure for displaying the data as a browser track will not work. This is because there is no standard track type associated with our new type of table and, consequently, we cannot configure the track in a trackDb.ra file. Instead, we would need to create a new type of track. This would involve writing software to display the new track in the browser code (e.g., in the display program hgTracks.c) and will not be described further here. For more information on adding tracks to a UCSC mirror, see http://genomewiki.ucsc.edu/index.php/Adding_New_Tracks_to_a_browser_installation.

11.6 Creating new genome databases

So far, we have described situations where we already have a genome browser or database and we want to add annotations to it. However, there are cases where one needs to create an entirely new genome database. The typical scenario would be for annotating a genome that has been newly sequenced. To be sure, this is not a task with which most researchers are often confronted. However, as the time and cost required for sequencing and assembling genomes continues to drop, the number of sequenced genomes that have not yet become available on the Ensembl, MapViewer, or UCSC systems is likely to increase.

One approach could be to design a new database and associated browser interface from scratch. First, one would need to select and install the underlying database management system (DBMS) and define the database-table specifications (i.e., the database schema). Next, one would need to format one's data so that it can be loaded into the database and execute the SQL code to create the database and load it with the data. Finally, one needs to develop an API and user interface to facilitate accessing the data via a browser or with querying programs.

Building a genome database from scratch is a very major undertaking and is almost never actually done. Instead, one generally copies (or "clones") the database schema and tools from a genome-database architecture that already exists. Obvious choices for architectures to clone include either the UCSC or Ensembl systems (or some subset of one of them). The resulting cloned systems could then be customized to meet the specific requirements of the new genome database. However, the designs of the Ensembl and UCSC systems are in certain ways specific to the requirements of the Ensembl and UCSC projects, and are implemented using hardware that is much more powerful than may be available to projects for smaller genomes. Consequently, an attractive alternative approach for building a new genome database is to use the generic database-building tools provided by the GMOD project. These tools are relatively easy to use for small- to medium-size databases, while maintaining multiple customization options so that they will be appropriate for a wide range of genome database projects.

11.6.1 Cloning a UCSC or Ensembl database

Database cloning is the approach used by the Ensembl and UCSC Browser teams when they are building a database for a newly sequenced genome, or for a new assembly of a previously sequenced genome. However, cloning a database for a new genome is much more complex than simply mirroring an already existing Ensembl or UCSC database, as described in Chapters 8 and 10.

To clone a database for a newly sequenced genome, UCSC and Ensembl use different approaches, both in terms of the extent of integration of the annotation generation with the database-building process and in the level of automation used. UCSC uses a partially automated procedure in which annotation generation and database construction are largely separated. The steps involved in these procedures are documented in files in the hg/makeDb/doc subdirectory, with names like "DatabaseName.txt," where "DatabaseName" is the name of the database being built (e.g., the hg18 build documentation is in the file hg18.txt).

At first glance, the makeDb/doc files look like Unix shell scripts that could be executed automatically. However, in fact, the contents of these files are a set of individual Unix commands, many of which are currently not set up for completely automated execution. Although the makeDb files are generally well documented, they are long and complex – hg18.txt, for example, is over 12,000 lines. Consequently modifying such a command list for use with a new genome or on a computer configuration that is not likely to be identical to UCSC's will inevitably uncover installation challenges.

To be sure, one does not need to load all the tables used in the hg18 database to configure a UCSC database clone for a newly sequenced genome. In fact, to clone a UCSC database and browser for a new genome, one could (and probably should) start by simply creating a skeleton database of five tables. This approach was mentioned in Chapter 10 and is described in http://genomewiki.ucsc.edu/index.php/Browser_installation. After one has successfully installed and configured the skeleton database and browser for the new genome, one can add additional tables and tracks as needed. Although gradually building a UCSC clone database is much easier than attempting to create a new UCSC database with large numbers of tables and tracks all at once, it is still challenging. In fact, to date only one public genome database, the Archaea and Bacterial Genome Browser (http://archaea.ucsc.edu), has been developed by cloning the UCSC architecture.

In contrast to UCSC's approach, Ensembl's database-construction procedure is more integrated and automated. Ensembl uses a database-construction program suite called the Ensembl pipeline (Potter et al., 2004), which not only performs the creation and loading of the new database but also executes the programs (e.g., BLAST, GEN-SCAN, repeatmasker) that create many of the database annotations. As a result, the procedures in the Ensembl pipeline are tightly connected with the specific annotations that are to be included in the Ensembl database. Consequently, the strategy of cloning the Ensembl database architecture has been primarily used in the annotation of large, newly sequenced genomes, similar to those in Ensembl itself. Examples of the use of Ensembl's software for implementing genome databases include the genome browsers for agricultural and other grains used by the Gramene Project (http://www.gramene.org/genome_browser) and the farm-animal genome browsers used by SIGNAE Project (http://public-contigbrowser.sigenae.org:9090).

Some customization of the Ensembl pipeline is possible using a configuration program known as the RulesManager. In addition, a more generic form of the Ensembl pipeline, called Biopipe (Hoon et al., 2003) exists. However, even using Biopipe, creating an Ensembl-like genome database for a newly sequenced genome is a complex task requiring significant programming and Unix system-administration experience, as well as access to considerable computer hardware (e.g., computer clusters). In general, genome database designers considering the use of the Ensembl pipeline and database architecture for new genome database construction are well advised to first contact the Ensembl developer's mailing list for support and guidance. The developer's mailing list can be accessed via Ensembl's general mailing list page at http://www.ensembl.org/info/about/contact.html.

11.6.2 Database construction using the GMOD tools

For the reasons described in the previous section, cloning and adapting the UCSC or Ensembl database structure for annotating a new genome is a challenging task.

At a minimum, a genome database designer should have a year of Unix system-administration experience, or the equivalent, before attempting such database cloning. An alternative, and often easier, approach to creating databases and browsers for new genomes is to use the genome-database architecture and tools provided by the GMOD project. Using GMOD software components, one can build a genome database system that provides genome browsing, interactive and programmed batch querying, and the display of genetic maps and biochemical pathways with relatively modest effort. Moreover, the resulting genome databases will have uniform user interfaces, so that someone familiar with one GMOD database can easily navigate another. FlyBase, WormBase, the Mouse Genome Database, FleaBase, and BeetleBase are some of the widely used genome databases that have been built at least partly using the GMOD tools.

Descriptions of all of the tools in the GMOD toolset can be found at the GMOD web site (http://www.gmod.org). At first glance, the tool descriptions at the GMOD web site may seem a little confusing, in part because the tools have been developed by independent groups and, in some cases, have redundant or overlapping capabilities. For several steps in the genome database construction process, multiple tools are available – generally including both simple tools, which are more thoroughly tested and documented, and more powerful ones that are in earlier stages of development. In general, the GMOD project is still somewhat of a work in progress, with some tools that are well developed and documented and others that are much less so. In particular, many of the tools are well developed for stand-alone use, whereas the mechanisms for combining these tools into an integrated genome database system are less well developed and documented.

11.6.2.1 Genome browsing with GBrowse

The central component of a typical GMOD database is the web-based genome browser, GBrowse. GBrowse is relatively easy to set up and yet offers considerable flexibility. GBrowse can handle genomic data in varying formats and multiple different database schemas and database-management systems.

Setting up a GBrowse system is essentially a three-step process. First, one chooses the DBMS and database schema. Second, one formats one's data so that it can be loaded into the database (in the case of relational database implementations) and so that it can accessed for genome browsing and batch querying. The simplest GBrowse database configuration is a flat-file (i.e., nonrelational) database, using GMOD's Bio::DB::SeqFeature::Store database schema and with data in the GFF3 format.[2] The last step in GBrowse database installation is that of installing and configuring the GBrowse program itself. In general, the procedure is similar to that for installing and configuring the UCSC Browser. One first needs to install and configure a web server program (i.e., Apache). One then downloads and installs the GBrowse software and

---

[2] See Appendix 2 for a description of the GFF3 data formats.

copies the GBrowse HTML and binary files to the web server directory. Finally, one edits the GBrowse configuration files so that the browser communicates properly with the web server and with the underlying database. We will not cover the installation procedure further because it is described in detail in the GBrowse tutorial (available at http://www.gmod.org/nondrupal/tutorial/tutorial.html) and the GBrowse installation guide (see http://www.gmod.org/wiki/index.php/GBrowse_Install_HOWTO). Once one's data has been loaded into the database and GBrowse has been configured to point to this database, one can visualize the data using one's web browser. The resulting genome browser display will be similar to that shown in Figure 3.14 in Chapter 3.

So far, we have described setting up GBrowse with a flat-file database using GFF3-formatted input data and the Bio::DB::SeqFeature::Store database schema. However, the GMOD architecture supports several other database-management systems, schema, and file formats. For example, if one wanted to implement the database as a relational database, one could modify the GMOD/GBrowse installation procedure to specify RDBMS data storage (currently, GMOD supports MySQL, PostgreSQL, and Oracle, among others). For use with, say, MySQL, one would need to create a MySQL database and create a user with read and write privileges on that database. One then would load the MySQL database with one's data in GFF3 format. Loading the database is easily performed with the bp_seqfeature_load.pl load script (available in the BioPerl 1.5.2 distribution). If, at a later point, one wants to incorporate additional data into the database, one can simply run the bp_seqfeature_load.pl script again with the new data, as long as the new data is also GFF3-compatible. In contrast, if our data is only available in GFF format rather than in GFF3, we could configure our MySQL database to use the Bio:DB::GFF schema instead of the Bio::DB::SeqFeature::Store schema by using the BioPerl GFF loading script bp_load_gff.pl instead of bp_seqfeature_load.pl.

11.6.2.2 Batch querying with GMOD's Perl API

In addition to visualizing one's data with GBrowse, batch queries can be executed against a GMOD database using either standard SQL or one of several available APIs. For example, if you are programming in Perl, you can use the library routines from the BioPerl Toolkit (you will need BioPerl, version 1.5.2 or later). BioPerl database-access adaptors are available for several GMOD-supported database schemas including Bio::DB::SeqFeature::Store, Bio:DB::GFF, and Chado. The necessary syntax for calling the database adaptors for the various APIs is very similar to the Ensembl Perl API syntax. For example, retrieving sequence-feature objects on chromosome 3 from a flat-file database (using, say, a GFF3-formatted file at /var/databases/test.gff3) with Bio::DB::SeqFeature::Store could be performed with Perl code like this:

```
$db = Bio::DB::SeqFeature::Store->new(-adaptor => 'memory',
                              -dsn => '/var/databases/test.gff3');
@features = $db->get_features_by_location(-seq_id=>'chr3');
```

Similarly, if you have stored the GFF3 data in a local MySQL database using the Bio::DB::SeqFeature::Store schema, you could retrieve sequence-feature objects associated with gene "ZK909" or sequence-feature objects of type "mRNA," "match," and "repeat region" from Chromosome1:5000–6000 with Perl code like this:

```
use Bio::DB::SeqFeature::Store;
$db = Bio::DB::SeqFeature::Store->new( -adaptor => 'DBI::mysql',
                             -dsn  => 'dbi:mysql:test');
@features = $db->get_features_by_name('ZK909');
# or
@features = $db->features(-seq_id=>'Chr1', -start=>5000,
             -end=>6000, -types=>['mRNA','match','repeat_region']);
```

Syntax for accessing a Bio:DB::GFF-based database would be similar, except that the lines in the code here referring to "Bio::DB::SeqFeature::Store" would be replaced by "Bio:DB::GFF."

11.6.2.3 Cmap, Pathway Tools, and BioMart

Although running GBrowse as a stand-alone genome browser is quite useful by itself, GMOD also includes several other tools for genome data analysis. Three of the most useful tools are Cmap (http://www.gmod.org/wiki/index.php/Cmap), Pathway Tools (http://www.gmod.org/wiki/index.php/Pathway_Tools), and BioMart (http://www.biomart.org).

With Cmap, one can display genetic maps (such as maps of quantitative trait loci), physical maps (e.g., chromosomal markers), and sequence assemblies. In addition, pairs of maps can be compared to identify syntenic regions (i.e., sets of related markers, such as homologous sequences or regions associated with similar phenotypes, which occur in the same order in both maps). Such comparisons can be made between maps of any two species within the Cmap database. By means of such comparisons, Cmap can help identify genes that influence different phenotypes even in cases where the underlying genomic sequence has not yet been sequenced or assembled.

GMOD's Pathway Tools provide a mechanism for storing and graphically displaying biochemical pathway data (e.g., metabolic pathways or signaling pathways). In this case as well, the database need not include the sequence of the underlying genome. In addition, metabolic or signaling pathways can be compared between related species. In this way, Pathway Tools can help, for example, identify enzymes that are missing (or perhaps have simply not yet been identified) in specific species.

Although GBrowse, Cmap, and Pathway Tools provide powerful features for the browsing of genomic and other biological data, they have limited batch-querying functionality. Instead, batch-querying support is provided by the BioMart component of the GMOD project. As we have seen in Chapter 4, BioMart is a data-management system that simplifies the task of creating web-based batch-query interfaces to complex

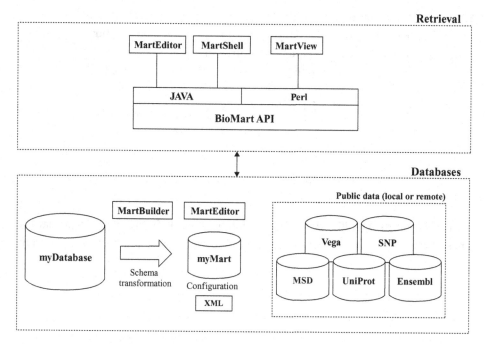

Figure 11.5 General BioMart architecture.

biological data. With BioMart installed, a GMOD database will have interactive batch-querying capability very similar to Ensembl's BioMart interface. In addition, as noted in Chapter 8, BioMart provides its own API, which can be used as an alternative to GMOD's APIs for programmatic access to one's GMOD data. BioMart uses a relational database backend and can currently support the Oracle, PostgreSQL, or MySQL relational database platforms. BioMart can be constructed from an existing relational schema (e.g., Bio::DB::SeqFeature::Store, Bio:DB::GFF, or Chado) or can be built from nonrelational data sources. Figure 11.5 shows a schematic of the data-mart architecture as implemented by BioMart.

Setting up a BioMart for a GMOD database is described in detail in the BioMart documentation at http://www.biomart.org/user-docs.pdf. Essentially, one first downloads and installs the BioMart software itself. One then needs to use BioMart's MartBuilder and MartEditor tools to generate the SQL for creating one or more new databases (the data marts). One then runs the SQL code to create the data marts and fill them with data from the GMOD database. Finally, one needs to install the BioMart Perl API and configure the MartView BioMart application to run as a web-based program under one's Apache installation.

11.6.2.4 Integrated GMOD databases with Chado

So far, we have considered the various GMOD components – GBrowse, Cmaps, Pathway Tools, and BioMart – as separate, stand-alone tools. However, the GMOD architecture is designed so that multiple GMOD tools can be incorporated into a single integrated genome database. But implementing such tool integration requires using a more complex database schema called Chado (Mungall and Emmert, 2007).

Chado differs from the database schemas that we have seen before – such as the Bio::DB::SeqFeature::Store and the Bio::DB::GFF schemas, as well as the Ensembl and UCSC database schemas – in that the relationships among the data are not completely described by the table definitions in the Chado database. Rather, Chado uses an additional layer of data description that constrains its data specifications to follow a "controlled vocabulary" of biological data types. In particular, a Chado implementation can only use biological-sequence terms defined by the Sequence Ontology Project (Eilbeck et al., 2005). Forcing data within a Chado database to follow the rules of the controlled vocabulary is generally implemented via the use of trigger functions defined within the underlying relational database management system itself. Consequently, all data that is to be loaded into a Chado database must be in a format, such as GFF3, that is already compliant with the data-type constraints of the Sequence Ontology Project.

Although controlled vocabularies impose constraints on the data formats that can be handled, the result is a genome-database architecture that is very flexible and customizable. At the same time, the controlled vocabulary ensures that tools developed by one Chado-based project will be interoperable with tools developed by any other Chado-compliant project. For example, with a fully installed GMOD database implementing the Chado database schema, one could view genome annotations with GBrowse, edit genome annotations with the Apollo genome-annotation editor (http://www.fruitfly.org/annot/apollo; Lewis et al., 2002), compare maps with Cmap, make data mining batch queries using the BioMart API, and browse the entire database using GMODWeb (http://www.gmod.org/wiki/index.php/GMODWeb).

Chado is a relatively new genome-database architecture. Several of its components and features – for example, those for incorporating BioMart into the Chado framework, or for supporting Chado under RDBMS other than PostgreSQL – are currently under active development. Moreover, documentation and tutorial material for some Chado tools are still somewhat limited. In particular, the details for implementing Chado may well change in the near future and, as a result, we will not describe them further here. For these same reasons, at present, installing and maintaining a database using the Chado schema requires considerably more effort and Unix system administrator and Perl programming experience (probably at least a year's experience) than using a database with the Bio::DB::SeqFeature::Store schema. However, this situation is likely to change in the near future, so the reader who is interested in deploying the Chado schema for integrated genome-database development is encouraged to consult the Chado web pages (http://www.gmod.org/chado) to determine the current status of the Chado toolset and its documentation.

Chapter summary

- Although custom tracks work well for the temporary integration of private data with genome annotations, for reasons of security, convenience, or enhanced data manipulation, it is sometimes desirable to integrate one's data with a public genome database in a more permanent manner.
- Adding permanent tracks to the Ensembl databases can be performed by setting up a DAS server system.
- If one has installed a local mirror of one or more UCSC databases, it is straightforward to add additional tables to the database using code generated with the autoSql program.
- To visualize the new tables on the UCSC Browser requires that a UCSC Browser mirror has been installed in addition to a database mirror. If this has been done, the data in the new tables can be displayed by editing the database's trackDb.ra configuration files.
- Creating an entirely new genome database, for example, to support the sequencing of a new genome, can be accomplished using the tools of the GMOD project.

Exercises

1. In the text, we described custom frames in terms of linking a set of custom regions of interest to the UCSC Browser. However, custom frames can also be used to move quickly between a set of related regions that are not necessarily based on custom data. Make a BED list of the coordinates of all genes whose descriptions include the term "Fanconi anemia" (e.g., by using the "Position and search term" input on the UCSC Browser). Convert the BED list into a custom frame as described in the text. Now by selecting different regions from the region list, one can easily compare genomic regions containing regions that have been implicated with this single disease. (To create the custom frame using the UCSC bedToFrames program, you will need to install the UCSC API and compile the bedToFrames program.)

2. Add a tRNA table to the UCSC *S. cerevisiae* database. (Actually, this information is already in the database as part of the sgdOther table, but this exercise asks you to make a separate tRNA-only table.)

   a. Create an autoSql template for a tRNA table.

   b. Execute autoSql to create SQL code to create the table and the associated C code.

   c. Obtain the locations of yeast tRNAs as a BED file (this can be done, for example, with the Table Browser by filtering the sgdOther table).

   d. Assuming you have installed a local yeast UCSC mirror (Chapter 10, Exercise 10.3), load the new table with hgLoadBed.

   e. Finally, confirm that you can access the newly loaded data either directly using SQL or else using the UCSC API with the C functions that you have generated with autoSql.

3. Create a list of all protein genes in the *S. cerevisiae* genome in GFF format (this can be done, for example, by using the GTF output option for the sgdGene table in the Table Browser). Create a new GMOD yeast database and load the database using the BioPerl bp_load_gff.pl program. Confirm that you can access the newly loaded data by writing a BioPerl script, as described in the text. Install GBrowse and display the gene track.

12

___

Genomes, Browsers, Databases –
The Future

With the acquisition of complete sequence data from multiple organisms, and the development of genome browsers and genome databases, researchers are now able to investigate fundamental biological questions in ways that were impossible to address in the past. Moreover, the rate at which useful sequence and annotation data is accumulating is not showing any signs of slowing down – even with the completion of the Human Genome Project. Instead, as technical innovation and economies of scale continue to drive down the cost per base pair of sequencing DNA, the quantity of available data increases ever more rapidly.

One may well wonder what new opportunities and applications may emerge from this new data, and what sorts of new tools will be required to analyze them. Of course, it is impossible to answer these questions with any certainty. However, in this final chapter, we will speculate regarding future trends in genomic data from the perspective of genome browsers and databases. In addition, we introduce web sites that provide an early glimpse of some of the new genomic-data resources being developed.

12.1 Glimpses to the near future

At least in the near term, development of new tools for analyzing and displaying genomic data are likely to come from the very research teams whose work we have already described in this book – UCSC, Ensembl, NCBI MapViewer, Galaxy, and GMOD. Most of these groups carry out a portion of their development work on public web sites that can be viewed by all. Consequently, if one is interested in glimpsing the future of genome databases and browsers, periodically checking these sites can be illuminating.

The UCSC "genome-test" web site (http://genome-test.cse.ucsc.edu) displays work in progress on the UCSC Genome Browser. The layout of genome-test is essentially identical to the main UCSC site and, in fact, all features on the main UCSC site initially appear on the test site. Only after they have remained on the test site for a period of weeks or even months and been determined to be useful and bug-free are they transferred to the main UCSC Genome Browser. In particular, by looking at the test site, one can determine which species and genome assemblies may soon be added to the main Genome Browser. For example, at the time of this writing genome assemblies for fourteen eukaryotic species were available on the test site but not on the main UCSC Browser. In most cases, annotations for these new species are limited. Nevertheless, if one is studying one of these species and integrated genomic information is not yet available elsewhere (e.g., at Ensembl or MapViewer), then being able to access the genome data on the UCSC test site may be advantageous.

In addition to browsers for newly sequenced genomes, the UCSC test site contains annotations and browser features that are not yet implemented on the main UCSC site. For example, at the time of this writing, there are some 80 tracks on the hg18 build of the human genome on the main UCSC site, whereas there are almost 200 hg18 tracks on the UCSC test site, including ones for VEGA genes, structural variations, and Eponine-predicted transcription start sites.

To be sure, many of the UCSC test-site annotations are experimental, or are relevant primarily for internal UCSC research projects, and are of limited general interest. Moreover, these tracks often have had only limited testing and may fail, or even display invalid data. As the test-site documentation indicates, using data from the test site should be done with caution, and generally only if there is no equivalent data available on the main UCSC site. On the other hand, if one simply wants a glimpse at what features and annotations will be coming to the main UCSC site in the future, examining the UCSC test site can be quite useful.

One can also glimpse the future of the UCSC Browser via the "ENCODE" tracks, which have been developed out of a collaboration between UCSC and the ENCODE Project (Birney et al., 2007). The ENCODE annotations are the result of multiple experimental collaborations that have acquired transcriptional data, chromosomal immuno-precipitation information, CpG methylation data, and many other annotations for characterizing genomic regions. Some twenty-five separate annotation tracks available only for the ENCODE regions can be found on the hg18 build of the UCSC Genome Browser. Even more annotations of the ENCODE regions are available on the older hg17 build of the UCSC Browser. At present, the ENCODE annotations are available for only about one percent of the genome. However, in the future, similar data should become available for the rest of the human genome and, quite possibly, for the genomes of some of the model organisms as well. The UCSC ENCODE tracks provide a useful preview of the how this data will appear in the genome browsers of the future.

Another source of previews of future developments by the UCSC group is the "Browser Development" section of the UCSC genome wiki web site (http://genomewiki. ucsc.edu/index.php/Main_Page#Browser_Development). On these pages, one can find proposals for new features and tools to be incorporated into the UCSC system. Descriptions on the Browser Development pages are sometimes sketchy and may well never actually be incorporated into the UCSC site. However, they do provide a glimpse of future features being considered within the UCSC developer community.

Ensembl and Galaxy also provide opportunities for learning more about their development plans. The Ensembl Pre! site (http://pre.ensembl.org) includes browsers for newly sequenced genomes that are not yet available via the main Ensembl site. For example, at the time of writing, six genome assemblies were available on the Ensembl Pre! site. The level of annotation on the Ensembl Pre! site is generally quite limited. In contrast to the UCSC test site, Ensembl Pre! does not include test browsers for upcoming builds for species that are already represented in the main Ensembl site. Ensembl also does not have a public web site for posting proposals for new Ensembl features, analogous to the Browser Development pages of the UCSC genome wiki site. Instead, if one wants a glimpse of future Ensembl plans, one can subscribe to the Ensembl developer's mailing list or view the mailing list archives, both of which are accessible at http://www.ensembl.org/info/about/contact.html.

Galaxy provides a view of its upcoming features via its test server at http://test.g2. bx.psu.edu, from which, every few weeks, software is transferred to the main Galaxy server. In addition, users interested in future Galaxy enhancements can subscribe to the Galaxy developer's mailing list at http://mail.bx.psu.edu/cgi-bin/mailman/ listinfo/galaxy-user. At the present, NCBI does not provide a preview web site or a public mailing list for discussion of upcoming MapViewer features.

12.2 Future genome database features

We now look at the kinds of data and features we may find in genome browsers and databases in the coming years. My guess – and, to be sure, it is just a guess – is that the most significant new features will be in the areas of regulatory region and epigenetic annotations, genome-wide RNA expression data, population-variation data, integrated multigenome querying, ancestral genome data, and databases describing microbial communities.

Acquiring many of these new types of data requires the sequencing of large amounts of genomic data. At the same time, new technologies are emerging that greatly lower the costs and increase the speed for the acquisition of such sequence data. Consequently, to better appreciate the range of new data types that are being added to the genome databases, we first briefly describe these new technologies, which are currently revolutionizing genomic sequencing.

12.2.1 New sequencing technologies and genome databases

Until recently, most sequence data in the genome databases have been generated by the Sanger sequencing technique. Sanger sequencing works well; its costs have decreased significantly and its throughput has greatly increased over the thirty years that it has been in use. Nevertheless, the costs and throughput of Sanger sequencing still limit its application for many important biological questions. To address these limitations, two other classes of sequencing technologies – referred to as *sequencing by hybridization* and *sequencing by synthesis* – are becoming increasingly important. These technologies promise dramatically lower sequencing cost and increased sequencing speed. For a more detailed introduction to these sequencing technologies, the reader is referred to the literature on sequencing technologies, such as chapter 7 of Primrose and Twyman (2006), as well as to the specific references indicated here.

Sequencing by hybridization (SBH) refers to a class of technologies in which the sequence of a ("target") DNA is determined by its pattern of hybridization to a large number of short DNA elements ("probes") of known sequence. In most SBH implementations, the probes are deposited in a grid pattern on a micro-array chip, similar to the micro-arrays used for RNA-expression measurements. Although SBH was proposed in the late 1980s and early 1990s (Drmanac et al., 1989; Southern et al., 1992), its use has been limited until recently. This is because SBH is limited primarily to *resequencing*, that is, sequencing of genomic regions that have been sequenced previously – for example, in other individuals of the species or of a closely related species. However, with the completion of the sequencing of numerous genomes (and especially the human genome) and the increasing interest in determining sequence variants within populations, resequencing by SBH has become increasingly important. In particular, as improved technology has increased the number of oligonucleotides that can be placed on an array and has lowered the costs of producing such arrays, SBH has enabled the detection of large numbers of novel SNPs. In addition, variants of the SBH approach, such as array-CGH (Sebat et al., 2004), have enabled the cost-effective detection of genomic structural variations, including some classes of genomic insertions, deletions, and copy-number variations. Moreover, when combined with biochemical assays, SBH can be used for genome-wide identification of DNA binding sites or DNA- and histone-methylation patterns.

However, except for very short (e.g., mitochondrial) genomes, *de novo* DNA sequencing generally can still not be performed by SBH. In addition, determining certain important genomic variations, such as finding rare SNPs, identifying chromosomal inversions, and determining high-resolution maps of CNV breakpoints, is still difficult or impossible to accomplish with SBH. To address these applications, sequencing by synthesis (SBS) technologies are being developed. Several different SBS technologies have recently been commercialized or are likely to become available in the near future – for example, see Leamon and Rothberg (2007) and Kartalov and Quake (2004). Although these technologies differ among one another in important details, they all involve modifying the DNA replication process so that a detectible signal (e.g., fluorescent or enzymatic light emission) is generated as each nucleotide is incorporated into nascent replicated DNA. Consequently, if the target DNA is used as a replication template for the DNA polymerase, the sequence of light signals emitted during the replication process can be used to determine the sequence of the target DNA.

As of 2007, SBS technologies are already capable of determining DNA sequence at rates that are about 100 times faster than Sanger sequencing and have about 100 times lower per-base cost – for example, see Rogers and Venter (2005), and Olson (2007). Moreover, as these technologies are still in relatively early stages of development, it is likely that improvements in cost and speed will continue. In fact, the National Institutes of Health (NIH) currently has established programs for funding research with the goal of reducing sequencing costs so that an entire human genome can be sequenced for $1,000 or less.

The main limitations of SBS technologies are that they produce shorter sequencing "reads" than Sanger sequencing and that they have higher error rates. However, for many important applications, these limitations are not critical, and methods for addressing them exist. For example, by using a fraction of the increased throughput available from SBS to replicate each sequencing run, one can significantly decrease the error rate with only a minor decrease in throughput. Also, as the main sources of SBS sequencing errors are usually known (e.g., several of these technologies have increased error rates when sequencing regions consisting of a single repeated nucleotide, or "homopolymer regions"), one can selectively resequence the error-prone regions using Sanger sequencing to decrease the overall error rate with only a small increase in sequencing cost.

Because of its short read lengths, SBS has to date been focused on resequencing applications. Such applications include searching for rare, but medically important, genomic variants, identifying genetic and epigenetic sequence variations in different cell types, and sequencing entire genomes of single individuals. For such applications, short read-length sequencing (which range from about 500 base pair reads for the most mature approaches to less than 50 base pairs for techniques with the highest speed and lowest cost) is just as effective as Sanger sequencing with its 1,000 base pair read length. For *de novo* sequencing of small (e.g., microbial) genomes, the limitations of short read lengths have also been shown to be quite surmountable. In contrast, *de novo* sequencing of large genomes is not yet feasible with short read-length technology. However, even here, the situation is improving. For example, new computational approaches show promise for assembling mammalian-size genomes from reads as short as 200 base pairs (Sundquist et al., 2007).

In short, there appears to be little doubt that SBH and SBS technologies will enable the acquisition of far more biologically important sequence data than could be previously acquired with Sanger techniques. This new data will greatly enhance the range of data that will be available for integrated querying from the genome databases. We now look at some areas where the availability of this new data is likely to have the greatest impact on the genome browsers and databases, and on biological research in general.

12.2.2 Regulatory regions and epigenetics

It has been evident for a number of years that genomic regulatory features, which control gene expression, are likely to have as significant an impact on phenotypic variation as protein-coding gene sequences. Such regulatory features include both properties of the sequence itself – for example, the occurrence of CpG islands or pro-moter sequences – as well as epigenetic genomic markers – for example, nucleotide methylation patterns or variations in the distribution of chromatin that binds to the genomic DNA.

In most cases, identifying regulatory regions in the genomic sequence is still much more difficult than locating protein-coding genes. Determining regions of epigenetic variation is typically even more difficult. In many cases, it is not even known to what extent epigenetic markers are determined by the underlying sequence or by other factors, such as the cellular environment, developmental stage, or the parental origin of the chromosome. However, recently the identification of regulatory regions in the genome has become more tractable via the use of techniques such as affinity assays to identify transcription-factor binding sites and chromatin immuno-precipitation assays to locate sites of chromatin-DNA association.

As new regulatory regions are identified, new approaches may be required to dis-play this data in the genome browsers. One the one hand, regulatory regions that are simply specific genome sequences can be displayed on a standard genome-browser annotation track. In contrast, regulatory regions including epigenetic markers that vary with cell type or developmental stage will require a more complex display – such as a separate annotation track showing the epigenetic modifications (e.g., methyla-tion state) for each different cellular environment.

Annotations for epigenetic modifications and genome regulatory regions can already be found in the current genome browsers. One example is in the ENCODE regions of the human genome. For these regions, the ENCODE Project has generated a variety of gene-regulation data including annotations for DNaseI-hypersensitive sites, locations of unmethylated CpGs, and regions of DNA-chromatin association as deter-mined by ChIP/chip experiments. Tracks describing many of these annotations can be visualized in the ENCODE tracks of the UCSC Human Genome Browser.[1] As such reg-ulatory data becomes available for entire genomes, annotation tracks similar to the ENCODE tracks are likely to become incorporated into all of the genome databases, providing important clues as to the detailed regulation of gene transcription.

12.2.3 Genome-wide micro-array transcription data

Until recently, most micro-array transcription data – that is, data measuring the transcription level of thousands of RNAs as a function of cell type and cellular envi-ronment – has been limited to the mRNA transcription of protein-coding genes. As a result, these data, though of considerable biological significance, have only been incorporated to a limited extent into the genome databases (e.g., primarily via the Gene Sorter or Known Gene Details pages of the UCSC Browser or via third party DAS tracks in Ensembl). Instead, these data are principally stored in specialized

---

[1] Some ENCODE annotations are currently only available in the hg17 build of the UCSC Genome Browser. Others are also available in hg18.

micro-array databases (e.g., GEO or ArrayExpress), where the data are accessed via gene name or keyword rather than by genomic location.

However, in the last few years, the development of "tiling" micro-arrays, which measure transcription from all locations in an entire chromosome or even an entire genome, has made the incorporation of micro-array data into the genome databases more attractive. Tiling-array experiments show that, at least for mammals, a large portion of the non-protein-coding portion of the genome is transcribed and that some, or perhaps even most, of this transcription is biologically functional. However, unraveling the extent of functional transcriptional activity is not easy. To address this important question, integrating transcriptional data with other genomic annotations should provide valuable insights. To date, viewing tiling-array data via a genome browser is available primarily for the ENCODE regions, for example, using the ENCODE Affymetrix-transfrag and Yale transcriptionally active region tracks on the UCSC Human Genome Browser. In the near future, genome browsers are likely to become important tools for asking questions such as which transcriptionally active regions are highly conserved, or which ones overlap gene candidates from gene-prediction programs or are found in loci associated with known diseases.

Because of the large amounts of data acquired from tiling-array experiments, incorporating such expression data into the genome browsers will introduce considerable challenges in data storage and display. A separate annotation track may well be needed for each cell-type, developmental-stage, or cellular environment for which tiling-array expression data is available. Besides the demands imposed by the sheer volume of data, it will also be necessary to provide flexible user interfaces so that the user can access the expression data that is relevant to the application at hand without being overwhelmed by vast quantities of extraneous data.

12.2.4 Human variation and medically important mutations

Except in the cases of monozygotic twins, genetic clones, or totally inbred strains, each individual of a species has its own unique genomic sequence. Indeed, such genetic variation is known to be a critical component of individual disease susceptibility or other phenotypic variation. Moreover, even within a single individual, there may be intercellular sequence variations arising from somatic mutations, DNA replication errors, chromosomal translocations, telomeric sequence loss, and so on. Storing and displaying such population variations are increasingly important components of genome databases and browsers.

To date, the principal population variations included in the genome databases are short, common (i.e., those occurring in greater than five percent of the population) polymorphisms, as well as the linkage disequilibrium data (e.g., HapMap data) that indicate the extent of correlation or independence among these variations. These polymorphisms include SNPs and short indels (i.e., indels that are less than approximately fifty base pairs). However, data describing other types of genomic structural variations – such as large insertions and deletions, chromosomal translocations, inversions, and copy-number variations – have now become more widely available, thanks to new technologies such as array-CGH (Sebat et al., 2004) and paired-end mapping (PEM; Korbel et al., 2007).

Structural variation data promises to greatly increase our understanding of the nature of genetic variation beyond what is currently known from the more familiar SNPs and short indels. To date, structural variation data has been primarily archived in specialized databases, such as the Database of Genomic Variants (http://projects.tcag.ca/variation; Iafrate et al., 2004), and has only been available to a limited extent in the genome browsers. However, this situation is changing and in the not-too-distant future it should be possible for users of the genome browsers to visually correlate structural variations with other genomic features such as segmental duplications, retrotransposed elements, recombination hotspots, and regions of high or low gene density.[2] Being able to carry out such integrated genetic and structural sequence analyses should provide an important new source of information for medical diagnostics and treatment planning.

The second major change in population variation data is the increasing detection of rare sequence variants. As genomic resequencing costs continue to decline, one can even envision that each person may ultimately have access to their own "personal genome," consisting of a list of all the SNPs and structural variants for which one's genome differs from the reference sequence. In fact, the genome sequences of two individuals, Craig Venter and James Watson, have already been completed and are even available from public databases and several startup companies (such as Navigenics, 23AndMe, and Knome) are even beginning to offer personal genomic sequencing as a commercial service. Moreover, beyond the "personal genome," one can also envision the determination of the "personal epigenome" and "personal transcriptome" annotating an individual's epigenetic variations and patterns of transcript expression in varying cell types. If and when personal genomes (and epigenomes and transcriptomes) do become generally available, they should dramatically enhance our capabilities for medical diagnosis and treatment. With this data, it should be possible to determine whether the patient has any variations at genetic loci correlated with disease predisposition or pharmacological response.

The genome databases are potentially ideal resources for analyzing such personal genomic data. One could imagine that individual genomic variation data might be stored in a set of custom tracks that could be uploaded to a genome database. To be sure, storing and displaying all of this information in a manner that is useful to the physician and the patient – and that preserves the confidentiality of this personal information – will be challenging. However, if these challenges can be overcome, the utility of such data could be great. For example, one can envision future genome

---

[2] The UCSC Human Genome Browser already includes tracks for copy-number polymorphisms and other structural variations. However, to date, the number of annotations on these tracks is limited.

browsers containing annotation tracks with all known polymorphisms having medical implications in which one could compare these tracks with the patient's "personal genome" track to produce a list of medically relevant, patient-specific polymorphisms.

Similarly, in the next decade it should become possible to annotate the sequence, epigenetic, and transcriptional changes that occur in tumor cells. One example of such an effort is the Cancer Genome Project (http://www.sanger.ac.uk/genetics/CGP), which is developing an atlas of sequence and chromosomal changes in cancer cells. Once this data has been acquired, incorporating it into the main genome databases should be possible. With this resource, comparing a patient's genetic profile to known tumor-cell genomic-sequence variations should become possible.

12.2.5 Biological pathways and proteomic data

Two other classes of biological data that are becoming increasingly relevant to genome databases are pathway data and proteomic data. By pathway data, I include data from both metabolic and signaling pathways, whereas with proteomic data I mean not only information characterizing individual proteins but, especially, data characterizing protein-interaction networks. To date, much of this data cannot be directly accessed from the major genome databases but is instead queried via more specialized databases such as MetaCyc (Caspi et al., 2006) and BIND (Alfarano et al., 2005). However, if the genome databases are to become complete (or even partially complete) repositories of molecular biological data, which can be queried in a truly integrated manner, then these important types of data will need be more fully included.

One challenge to genome-database developers will be to determine effective interfaces for displaying biological-pathway and protein-interaction data in genome databases. Pathway and proteomics data are typically not correlated with the chromosomal locations of the pathway or protein-network components (except in prokaryotes). Consequently, conventional genome browsing along a chromosome may not be an effective way of displaying pathway or protein-network data. Instead, such data might be included via additional "views" – to use Ensembl terminology – including, perhaps, "Pathway Views" and "Protein Network Views." Such additional views would complement Ensembl's ProteinView and UCSC's Proteome Browser, which already exist for examining data from individual proteins.

An attractive, alternative approach would involve tools that facilitate exploring, in an integrated manner, all the proteins that share a pathway or interaction network. One tool that can already be used – to a limited extent – in this manner is the UCSC Gene Sorter. With the Gene Sorter, one can search for proteins that are "close" to one another in a protein network. In the future, one could well imagine extensions of this capability by which one could search for genes that encoded for proteins that are "close" to one another in a metabolic or signaling pathway, or that are homologous to genes that are "close" to a query gene in a pathway or interaction network in a different species.

12.2.6 Multiple-genome querying

Sequence alignments from multiple related species, and tools to assess the sequence conservation within these alignments, are already among the most valuable features provided by the genome browsers. However, these features represent only a fraction of the comparative-genomics capabilities that will become available as multispecies databases become more tightly integrated. In the future, it will become possible to compare annotations from multiple-genome databases. This capability will enable one to ask questions like: How do the lengths of introns vary among orthologous genes in related species? Are recombination hotspots conserved between related species? What genes are spliced in one species while their orthologs in another species are coded within a single exon? Are there systematic differences in the distances between promoter or CpG regions and transcription start sites in different species? To be sure, it is already possible to formulate some of these questions with the current genome databases using custom programs. However, in the future, performing such multi-species comparative queries should become possible to the nonprogrammer as well.

To get a sense of the kinds of analyses that are possible with multigenome querying, we can turn to the Integrated Microbial Database (IMG) of the U.S. Department of Energy Joint Genome Institute (JGI) located at http://img.jgi.doe.gov (Markowitz et al., 2006a). IMG is a relatively new genome database resource but it already includes the genome sequences of hundreds of microbial species and provides tools for performing multigenome queries of this genomic data. Figure 12.1 illustrates one of IMG's tools for multigenome querying, the Abundance Profiler. The figure depicts the variations in the number of different enzymes of various enzymatic families among several methanogenic archaeal species. Each row of the display represents a different enzyme group, and each column is a different species. Enzyme-family abundances are shown graphically by the color in the display (shown as varying shades of gray in the figure). Numerical abundances can be determined by positioning one's mouse over each enzyme class in the display. As shown in the figure, for this example, most enzyme classes have only a few members occurring in each of the species; however, two classes (EC:3.6.3.25 sulfate-transporting ATPases and EC:2.7.7.6 DNA-directed RNA polymerases) have widely varying abundances. In this purely illustrative example, the varying abundances of EC:3.6.3.25 and EC:2.7.7.6 are most likely a chance result. However, similar types of analyses, focusing on identifying classes of enzymes that seem to be missing in one species while appearing in multiple closely related species, have led to the identification of previously undetected enzymes and in the improved characterization of enzymatic pathways – see, for example, Bishop et al. (2002).

12.2.7 Phylogenomic trees and ancestral genomes

The next few years will also likely see a migration from multispecies sequence alignments to multispecies phylogenetic trees. As more sequence data from multiple species become available, phylogenetic trees provide more information on the

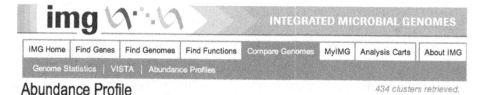

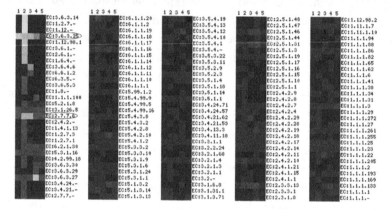

Figure 12.1 Enzymatic abundance profile of methanogenic archaeal species as displayed via the Joint Genome Institute (JGI) Integrated Microbial Genome database. Each row is a distinct enzyme class, and each column a different species. Numbers of different enzymes of each class for each species are indicated by the color. Since different colors are difficult to discern in the grayscale reproduction, the two enzyme classes with elevated abundances, EC:3.6.3.25 and EC:2.7.7.6, are highlighted by surrounding ellipses in the figure.

evolution of a genomic region than alignments can. In addition, phylogenetic trees can provide clues as to the genetic sequences of long-extinct ancestral species. Such ancestral sequence data is not only of intrinsic scientific interest but can also be of valuable practical utility. For example, at the individual gene level, phylogenetic trees have already facilitated the synthesis of enzymatically active ancestral proteins that are no longer found in nature (see Thornton et al., 2003). Genome-wide phylogenetic trees promise to make such identification of active ancestral proteins possible on a much wider scale.

Displays of phylogenetic trees, at the single gene level, already exist in the genome browsers. For example, we saw the phylogenetic tree of the CXorf34 gene in Ensembl's GeneTreeView in Figure 3.3. However, the full power of phylogenetic alignments will become evident when sufficient numbers of completely sequenced related genomes become available to enable the construction of phylogenetic trees for entire genomes. Creating such *phylogenomic trees* will not only require sequence data from multiple organisms spanning the phylogenetic tree but will probably also require novel tree-building algorithms to extend the largely single-gene oriented methods that have been used previously. Such algorithms are currently under development and initial views of the results of whole-genome phylogenetic trees and even of entire genome browsers based on inferred genomes of long-extinct ancestral species can be seen at the UCSC development site.

Along with genome browsers for inferred, ancestral-species sequences, the next decade is likely to offer the first views of experimentally acquired sequence assemblies from extinct species. In particular, DNA extraction, sequencing, and analysis techniques for genetic material isolated from fossil remains are now beginning to be sufficiently sensitive to enable the assembly of portions of the genomes of such extinct species as the cave bear (Noonan et al., 2005) or the Neanderthal (Green et al., 2006). Should such ancient DNA extraction and sequencing technologies continue to improve, integrating this data in the genome databases with the sequence and annotation data of the descendents of these extinct species should provide important insights into genomic evolution.

12.2.8 Environmental genome databases

The last emerging research area that we discuss is *metagenomics*, or *environmental genomics*. In environmental genomics, the objective is to obtain sequence data from an entire biological community of species living in a single environment. Examples of recent environmental genomic projects include the sequencing of DNA from a single environmentally important location – such as acid-mine drainage soil or oceanic seawater – or from an entire symbiotic or parasitic community that may impact the host organism's health status – for example, all of the microbes that colonize the human gut.

Some metagenomics projects are carried out primarily to sequence the genome of a single organism that cannot currently be cultured in the laboratory (it is estimated that more than ninety-nine percent of known microbes, including many with important health and environmental characteristics, cannot be cultured with current techniques). Such projects may require innovative methods for isolating the desired sequence data from the background of the remaining sequences of the sample. However, once the sequence has been obtained, annotating and displaying the data with a genome browser can be done in a relatively standard manner.

In contrast, other metagenomic projects seek to acquire sequence data from all, or at least many, species in the environment simultaneously. The motivation may be that there are so many species of comparable abundances in the environment that determining their individual genome sequences is prohibitively difficult. On the other hand, for some applications it may be more important to simply know whether a certain gene function – for example, a specific type of enzyme – is

Figure 12.2 Screenshot of the metagenome browser at the JGI Integrated Microbial Genome web site.

present somewhere in the environment and less important to identify the specific microbe that is contributing that enzyme. Such applications require not only new methods of sequence acquisition but also novel approaches for storing, analyzing, and displaying the data. For example, the "metagenomic browser" may not know from what organism a specific segment of DNA sequence has been derived or even whether two apparently related sequences on separate contigs are from the same species.

The genome browsers that we have been considering in this book do not currently provide methods for displaying "genomes" with sequences derived from multiple species. This is hardly surprising because most environmental genomic data are for microbial species and the main genome browsers are largely focused on metazoa. In contrast, the microbial IMG database does include an environmental genome component (Markowitz et al., 2008) at http://img.jgi.doe.gov/m.[3] To date, IMG's capabilities

---

[3] The Genomes OnLine (GOLD) Database at http://www.genomesonline.org is a good source of information regarding the status of metagenomics sequencing projects.

1 - <u>Mouse Gut Community lean1</u>
2 - <u>Mouse Gut Community lean2</u>
3 - <u>Mouse Gut Community ob1</u>
4 - <u>Mouse Gut Community ob2</u>

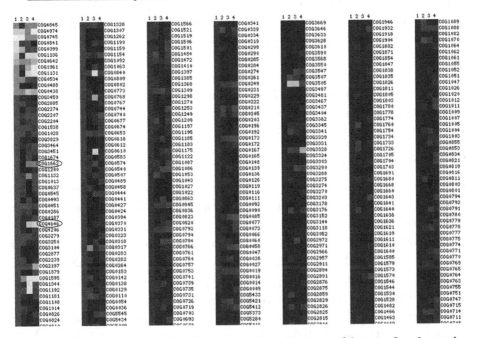

Figure 12.3 Gene-family abundance profile of environmental genomes of the guts of two lean and two obese mice. Most families show similar profiles, but a few families – for example, COG4646 (DNA methylases) and COG1662 (Transposases), which are circled – appear to differ in abundance between the lean and obese mouse gut environments.

for displaying and analyzing metagenomic data are limited, and few metagenomic annotations are available. Figure 12.2 illustrates a display from the IMG Environmental Genome Browser. At first glance, the figure looks like a conventional genome-browser display. However, the "genome" of the sequence shown in the figure is not that of a single identified species but rather the metagenome of a community of anaerobic methane oxidation (AMO) organisms extracted from a California coastal sediment.

IMG's Abundance Profiler, described in Section 12.2.6, can also be applied to environmental genome data. Figure 12.3 illustrates the application of the Abundance Profiler to the comparison of metagenomes, specifically, to the bacterial environments of the guts of lean and obese mice. In this example, the display is of the abundances of members of different gene families as indicated by membership in varying clusters of orthologous groups of genes ("COGs"). From the graph, at least two gene families (transposases and DNA methylases) appear to have different representations between the bacteria found in the guts of the obese mice and the lean ones. As in the previous example with the Abundance Profiler, one needs to be very cautious before concluding that this observation has biological significance and is not the result of random variations. Still, it should be apparent that by using careful controls to exclude chance results, this approach could lead to biologically interesting hypotheses to guide experimental testing.

Environmental genomic analyses are still in their infancy. Addressing important environmental genomics questions will most likely require additional tools for integrating data from multiple species within the metagenome. However, as these tools are developed, metagenomic databases should begin to be able to address complex biological and ecological questions involving the interactions among different species that until now have been difficult or impossible to answer.

Chapter summary

- UCSC, Ensembl, and Galaxy all provide preview or test versions of their web sites where one can get a glimpse of features and data that are likely to be included in the main web sites in the near future.

- Data that is likely to be increasingly incorporated into genome databases in the near future include regulatory region annotations, genome-wide RNA expression data, population variation data, integrated multigenome querying, ancestral genome data, and databases describing microbial communities.

- In many cases, the driving factor behind the new types of genomic data is the emergence of low-cost, high-throughput DNA-sequencing technologies.

- Storing and displaying some of this data – such as expression, epigenetic, population variation, and metagenomic data – is likely to require substantial modifications and enhancements to the data-management and data-display tools currently used by the genome browsers.

APPENDIX 1

Coordinate System Conventions

Genome databases consist largely of sequences and annotations together with genomic coordinates describing where these features are located. Consequently, specifying genomic coordinates is fundamental to using genome databases. However, interpreting coordinates in genome databases can be somewhat subtle. In this Appendix, we discuss some topics involving genomic coordinate systems that can cause confusion, especially when carrying out batch queries. These topics include absolute and relative genomic coordinates, ways for specifying the feature start and end positions, and conventions for locating negative-strand features.

A1.1 Absolute, chromosomal, contig, and clone coordinates

Because of the limited read lengths produced by current sequencing technology, genomic-sequence assembly is a gradual and iterative process. Typically, individual clones are first sequenced, and then the overlapping clone sequences are assembled into contigs, scaffolds and, ultimately, chromosomes. As a result, especially in the early stages of genome assembly, it is sometimes useful to have coordinate systems based on clones or contigs for comparing sequence assemblies or locating subfeatures (e.g., exons) relative to their parent features (e.g., transcripts). Ensembl and MapViewer offer such alternative coordinate systems in addition to a chromosomal coordinate system. In contrast, UCSC uses essentially only a single "top-level" or "absolute" coordinate system, corresponding to the entire chromosome.

The existence of multiple coordinate systems is not in itself a source of confusion. Generally, one is only interested in absolute coordinates because the chromosome is the actual biological entity on which the feature is located. Moreover, since one can visualize individual clones and contigs as annotation tracks, it is straightforward to view the sequence and annotations from the clone or contig perspective as well.

Rather, the issue is that – in contrast to clone coordinates, which are stable – a feature's chromosomal coordinates often need to be modified each time the sequence of the genome becomes more precisely determined. In fact, until sequence assembly is relatively complete, a feature's chromosomal coordinates are not even approximately known. Instead, the feature can only be located relative to a scaffold or a contig.

To deal with this issue, chromosomal coordinates are "frozen" each time a new version of the genome sequence assembly is released. Genome-sequence reassembly is performed periodically by the responsible genome-assembly organization (e.g., by the NCBI, in the case of the human genome), and at that time the new genome assembly is given a specific identifying number. In this way, absolute coordinates within any single sequence-assembly release are fixed and stable (albeit approximate).

With frozen coordinates, one can reliably compare locations of different features from a genome as long as the coordinates of all features are taken from the same assembly. However, if one does need to compare genome locations of features known only in absolute coordinates from different assemblies, then the coordinates of one of the feature sets needs to be converted so that all coordinates are from the same assembly. (Of course, if the features are in stable, clone- or contig-based coordinates, then no conversion is necessary.) Conversion of absolute coordinates of sequence features is usually not difficult using UCSC's LiftOver tool (see Section 4.4.2.2) or by directly searching for the sequence with BLAT, SSAHA, or MegaBLAST. However, if a feature occurs in multiple copies throughout the genome, coordinate conversion may not be straightforward, and locating a feature in a new sequence assembly can be quite difficult.

In this context, it is worth noting that the concept of "absolute" chromosomal coordinates is itself only an approximation, as there really is no single human-genome sequence (nor is there a unique genome sequence for any other species, for that matter). Indeed, because of insertion, deletion, duplication, and translocation polymorphisms and mutations (some of which can be quite large), the chromosomal coordinates of genes and other features vary widely among individuals of the same species. In fact, in individuals with cancer, for example, chromosomal coordinates often vary among cells of a single organism. Consequently, it is important to remember that the "absolute" chromosomal coordinates that one sees in a genome browser merely represent either those of a specific individual within the species (for example, in the mouse genome, those of a single inbred mouse strain) or the consensus positions taken from the sequencing of a relatively few individuals of a species (as is the case with the reference sequence of the human genome).

A1.2 Start- and end-numbering conventions

Conventions for specifying start and end positions of genomic features are somewhat arbitrary. In fact, the internal database coordinate representations used by the three genome database systems are not identical. Specifically, the internal representations used by the UCSC and NCBI databases are "zero-based," whereas the Ensembl coordinate system is "one-based." This can lead to confusion if one needs to directly

a) Start and End Coordinate Conventions
```
 [---------------] (Feature A)
 T   T   A   G   G   G   T   T   A   G    G    G
 1   2   3   4   5   6   7   8   9   10   11   12   Ensembl
 0   1   2   3   4   5   6   7   8   9    10   11   UCSC
|===================================================|
```
b) Ensembl Slice Coordinates
```
 [-----] (Feature B)
         [--------] (Feature C)
   |======================================| (Slice)
 A   C   T   A   A   A   T   C   T   T   G   (positive strand)
 21  22  23  24  25  26  27  28  29  30  31  32  33
 ========================================== (Chromosome or contig)
 T   G   A   T   T   T   A   G   A   A   C
                 [--------] (Feature D)
```
c) UCSC Strand Coordinates
```
 0   1   2   3   4   5   6   7   8   9    10   11
 T   T   A   G   G   G   T   T   A   G    G    G    (positive strand)
|=======================================|
 A   A   T   C   C   C   A   A   T   C   C   C
 11  10  9   8   7   6   5   4   3   2   1   0
 [---------------] (Feature E)
```

Figure A1.1 Simplified example illustrating differences between Ensembl's and UCSC's coordinate-system conventions. (a) The feature is located at the extreme end of the chromosome (i.e., the telomere), illustrating start- and end-numbering differences. (b) Three features are shown relative to a "slice," illustrating Ensembl's slice-coordinate conventions. (c) An example showing differences between UCSC's absolute and strand coordinates.

access data from the genome databases (as opposed to merely viewing the data via the genome browsers).

To better understand what is going on, let us, for simplicity, consider a feature at the extreme 5′ end of a chromosome (i.e., at a telomere) as shown schematically in Figure A1.1a. We may ask what is the coordinate of the very first nucleotide of the feature, that is, the initial "T" in the sequence. We might well imagine that the coordinate value is equal to one, and indeed this is what we would find on all three genome browsers. Similarly, we might expect that the coordinate of the final nucleotide of the feature in Figure A1.1a would be six, and that its length would be determined by

    Length = end - start + 1 (= 6 - 1 + 1 = 6)

In fact, in the Ensembl database, the start and end coordinates of our feature would be stored as "1" and "6." However the UCSC database has different coordinate conventions. In the internal representation of the UCSC database, which is used by the UCSC Table Browser and the UCSC API, our feature would begin at position zero. Also, with the UCSC coordinate conventions, the feature end-coordinate number is that of the first nucleotide beyond the actual feature. In our example, the feature would be specified internally in the UCSC database as being at positions zero through six. (Note that the UCSC Genome *Browser* adds an offset of one back to each start position so that in the browser, the feature does start at position one. Only the internal data representation starts at position zero. Additional details regarding the UCSC coordinate conventions can be found at http://genome.ucsc.edu/FAQ/FAQtracks#tracks1.)

Although the UCSC conventions for representing start and end coordinates may be somewhat unexpected, they do have some advantages. For example, accessing data from database tables is easier because tables are conventionally stored in computer memory starting at relative position zero. Moreover, feature-length arithmetic is also made simpler because one does not need to add an offset of one when computing feature lengths:

```
Length = end - start ( = 6 - 0 = 6)
```

Finally, the locations of "zero-length" features, such as sequence insertions, are arguably less awkward to annotate in the UCSC than in the Ensembl coordinate system. For example, a sequence insertion between nucleotides 1000 and 1001 on chromosome X would be described in the UCSC system with coordinates:

```
chrX 1000 1000
```

With Ensembl, the location of such a zero-length feature is described by using an end coordinate that is one *less* than its start coordinate, or

```
X 1000 999
```

The point here is not whether one coordinate system is "better" than the other. Each approach has its merits. Rather, what is important is to remember is that these differences in coordinate definitions do exist, and although when using genome browsers these details can usually be safely ignored, when creating custom tracks or using batch and programmed querying, not being aware of them can lead to puzzling inconsistencies. For example, if one has a custom track of SNPs in BED format for use with Ensembl,[1] a sample record might look like

```
X 1000 1000
```

In contrast, to describe this same SNP in a UCSC custom track, one would need to describe it as

```
chrX 1000 1001
```

[1] Making custom tracks in Ensembl using BED format is described at http://www.ensembl.org/common/helpview?se=1;ref=;kw=urlsource.

It is worth noting that several of the tools of the GMOD system for generic genome-database development described in Chapter 11 (e.g., the GBrowse Genome Browser and the Chado database schema) use another coordinate system called *interbase* coordinates. In the interbase system, position one is located between the most 5′ and the second-most 5′ nucleotide of the sequence, so that the most 5′ nucleotide would be described as extending from position zero to position one. Although defined slightly differently, in practice, interbase coordinates are essentially the same as UCSC coordinates.

A1.3 Slice coordinates

In the UCSC system, all genome features are specified in top-level (chromosomal) coordinates. In contrast, Ensembl uses both top-level and relative coordinates. Such relative coordinates, called *slice coordinates*, enable one to specify the location of one feature relative to some other feature or genomic region. Slice coordinates can be useful and are frequently used in Ensembl's software. However, slice coordinates can also be a bit confusing.

For example, consider the features shown in Figure A1.1b. In slice coordinates, a feature's coordinates are given by

```
SliceCoordinate(a) = AbsoluteCoordinate(a) - SliceOffset + 1
```

In Figure A1.1b, the slice offset is equal to 22, so Feature B's start position (in slice coordinates) is $21 - 22 + 1 = 0$. Similarly, Feature B ends, in slice coordinates, with coordinate equal to 2. Slice coordinates operate identically for features on either strand, so the start and end slice coordinates for Features C and D are (3,6) and (5,8), respectively. For more information on slice coordinates and when they are useful, see the Ensembl API Tutorial at http://www.ensembl.org/info/using/api/core/core_tutorial.html.

A1.4 Strand coordinates

As noted in Section A1.1, UCSC only uses top-level coordinates. However, UCSC does use two different top-level coordinate systems. One is what we have been calling *absolute* coordinates. The second UCSC top-level coordinate system is called *strand* coordinates. Absolute coordinates, which we have already described, are always measured with respect to the positive DNA strand. In contrast, strand coordinates are measured with respect to the strand on which the feature of interest is located. For features (e.g., genes) on the positive strand, absolute coordinates and strand coordinates are identical. However, for features on the negative strand, this is not the case.

We illustrate this first with the mini-chromosome example of Figure A1.1c. This entire "chromosome" is twelve base pairs long. If we consider Feature E (the sequence

CCCTAA) on the negative strand, its absolute start and end coordinates are (0,6). However, its strand coordinates are (6,12). Extending this concept to a more realistic chromosome of, say, six megabases, we can consider a thousand-base feature whose 5′ end aligns to the negative strand of a chromosome one third of the distance from the beginning of chromosome's positive strand. In absolute coordinates, the feature's coordinates are 1,999,000–2,000,000. However, in strand coordinates, the feature's coordinates are 4,000,000–4,001,000. More generally, to convert a negative-strand feature's location from absolute to strand coordinates, we can use the formulae:

```
featureStart(strand coordinates) =
      chromosomeSize - featureEnd(absolute coordinates)
featureEnd(strand coordinates) =
      chromosomeSize - featureStart(absolute coordinates)
```

Strand coordinates are quite useful for describing alignments of transcripts to the negative DNA strand, and are used for this purpose in the PSL sequence-alignment format. However, strand coordinates can also be confusing, especially when they are mixed together with absolute coordinates. This can be particularly true when using PSL format, described in Appendix 2, because some fields in PSL format are specified in absolute coordinates whereas other PSL fields are defined in strand coordinates.

A related source of confusion arises in the numbering of exons and in the identification of start- and stop-codon positions for genes that are located on the negative chromosomal strand. Because gene locations are stored in the UCSC database in absolute coordinates, the cdsStart position of a negative-strand gene is actually the position of the *stop* codon of the gene, whereas the cdsEnd position is that of the start codon. Similarly, for a negative-strand gene, the "first exon," which is called "exon 0" in the UCSC database, is the exon that corresponds to the 5′ end of gene location on the positive strand, and hence corresponds to the 3′ end of the mRNA transcript of the gene. Although the genome browser generally shields the user from these issues, in many batch-querying analyses, such as the identification of targets of NMD described in Chapter 5, handling these details properly is essential.

Exercise

1. The long isoform of the DRD2 gene is located on the negative strand at chr11:112,785,528–112,851,091 (according to NCBI human genome assembly 36, e.g., UCSC build hg18 or Ensembl release 42). What are the coordinates of this gene as stored internally in the UCSC database in absolute coordinates? What are the gene's coordinates in UCSC strand coordinates? If one created a genomic slice from chr11:112,800,000 to chr11:112,900,000, what would the gene's coordinates be in Ensembl slice coordinates?

Genome Data Formats

To read data from a genome database, one needs to understand the format in which the data are stored and transmitted. In the past, the use of myriads of data formats was one of the major stumbling blocks to effectively using biological databases. By reformatting the data from these databases into a few standard formats, the genome databases have facilitated the analysis of large quantities of biological data in new and powerful ways.

However, the data formats used by the genome databases can themselves be somewhat confusing, especially initially. This is because the data are complex – there is a vast amount of data (quantities are currently measured in terabytes and are rapidly growing), the data are of many different types, and the data are highly interrelated. Moreover, new types of biological information are continually being discovered, and our understanding of the biological relationships among the different types of data is continually evolving. As a result, data formats that were adequate to describe data previously must be extended in ways that may not have been planned, or else entirely new data formats must be developed.

Each time new data formats are designed, design trade-offs must be addressed, including choosing between formats that are easier for computers to read (and therefore generally faster and more compact) and ones that are easier for humans to read. Other trade-offs are between formats that are more portable between multiple computer systems and ones that run more efficiently (i.e., more rapidly) on a single system. Moreover, some formats capture detailed relationships among the data more completely, whereas other formats can be processed more quickly. Consequently, even a single genome database system may include multiple formats for storing and transmitting a single type of genomic data.

An additional source of potential confusion is that four distinct classes of data formats – table, track, file, and program – exist. *Table formats* describe how data is stored in the (relational) databases. *Track formats* describe how the data is presented on the browser. *File formats* describe how the data is stored in conventional computer files, and how the data is formatted for transmission between computer systems. Finally, computer programs reformat data into computer objects or other data structures during program execution. For example, in the UCSC system, there are "BED" and "PSL" data formats for files, database tables, browser tracks, and C structures, all of which are quite similar to one another but not quite identical. Moreover, under certain situations, data stored in a database table in one type of table format (e.g., PSL or genePred table format) may be extracted from the database in a different type of file format (e.g., BED file format.)

In this appendix, we describe the principal genomic file formats. In Appendix 3, we describe UCSC table formats (because Ensembl's software performs database-table conversion automatically, knowledge of Ensembl's table formats is rarely needed). Program formats and track formats are described as they are needed within the main text.

A2.1 File data formats

Numerous file formats have been developed for flat-file storage and the transmission of genomic data. Among the most important for genome databases are the UCSC-developed BED, PSL, MAF, and WIG data formats, and the GFF/GTF/GFF3 and DAS formats that are used by Ensembl and the single-genome databases.[1]

A2.2 BED file format

Browser Extensible Description (BED) format is the basic format used to specify a genomic location in the UCSC system. In its simplest form, the BED format consists of just three fields: the name of the entire sequence on which the feature is located (e.g., the chromosome name, or the contig or scaffold name) and the start and end positions of the feature within the sequence. In the case of a flat file in BED format, each of these fields is a string of ASCII characters with the individual fields separated by "white space" (i.e., spaces or tab marks). Several extensions of the BED format exist. The most common variant is the BED-12 format, shown in Table A2.1, which is often used for formatting gene data in flat files. For transferring gene data in BED-12, the thickStart and thickEnd fields are typically used to store the coordinates of the start and stop codons, and the comma-separated blockSizes and blockStarts fields are for storing exon locations.

BED format is used throughout the UCSC database system. A BED-like format is used in the specification of custom-track annotations. In addition, variants of the BED format are used internally by UCSC for various genomic annotations, such as conserved transcription-factor binding sites (tfbsConsSites.bed), cytogenetic band locations (cytobands.bed), and mapped bacterial artificial chromosome ends (bacEndPairs.bed).

---

[1] The distinction here between "UCSC" and "Ensembl" formats is only approximate; for compatibility purposes, the UCSC system does provide GFF and DAS output, and Ensembl software does read BED and PSL files as one of its options for displaying custom tracks.

Table A2.1 BED-12 file format

Field	Example	Description
chrom	chrX	Reference sequence chromosome or scaffold
chromStart	1541467	Start position in chromosome
chromEnd	1616000	End position in chromosome
name	NM_178129	Name of item
score	0	Score from 0 to 1000
strand	−	+ or −
thickStart	1544371	Start of where display should be thick (start codon)
thickEnd	1545451	End of where display should be thick (stop codon)
reserved	0	This should always be set to zero
blockCount	2	Number of blocks (exons) in the BED line
blockSizes	4008,187	Comma-separated list of the block sizes. Number of items should equal blockCount.
blockStarts	0,74346	Comma-separated list of block starts. BlockStart positions are calculated relative to chromStart. Number of items should equal blockCount.

Note that BED format does not contain the sequence of the feature. Consequently, retrieving the sequence of a feature is generally a two-step process – first retrieving the feature coordinates from the database and then obtaining actual sequence data. When one is using a genome browser, or a tool like Ensembl BioMart or the UCSC Table Browser, these two steps are performed automatically without the user even being aware that they are occurring. However, for performing direct database queries within the UCSC system, this two-step process must be carried out explicitly by the user (or, more typically, by the user's computer program).

A2.3 PSL file format

Pattern Space Layout (PSL) format is the principal UCSC file format for storing pairwise sequence alignment data.[2] PSL describes an alignment in terms of the two sequences being aligned, referred to as the "query sequence" (typically a transcript, e.g., mRNA or EST) and the "target sequence" (in most cases, a chromosome). The PSL file format is shown in Table A2.2. As with BEDs, PSLs generally do not contain the actual aligned sequences, just the coordinates from which one could retrieve those sequences. There is an extended version of PSL format (called PSLx) that includes the actual sequence as well; however, PSLx is rarely used within the UCSC database.

---

[2] An alternative format for storing pairwise alignments, called AXT, also exists in the UCSC database. AXT format is used with the BLASTZ program, which generates UCSC's genomic cross-species alignments. We will not be using AXT-formatted files in this text, but the interested reader is referred to http://genome.ucsc.edu/goldenPath/help/axt.html for a description of the AXT format. In addition, there is a UCSC utility program, axtToPsl, for converting files from AXT to PSL.

Table A2.2 PSL file format

Field	Example	Description
matches	557	Number of bases that match that are not repeats
misMatches	113	Number of bases that do not match
repMatches	0	Number of bases that match but are part of repeats
nCount	0	Number of "N" bases
qNumInsert	3	Number of inserts in query
qBaseInsert	194	Number of bases inserted in query
tNumInsert	7	Number of inserts in target
tBaseInsert	1984	Number of bases inserted in target
strand	+−	+ or − for strand. First character query, second target (optional)
qName	BC101888	Query sequence name
qSize	1300	Query sequence size
qStart	1	Alignment start position in query
qEnd	865	Alignment end position in query
tName	chr1	Target sequence name
tSize	245522847	Target sequence size
tStart	4558	Alignment start position in target
tEnd	7212	Alignment end position in target
blockCount	2	Number of gapless blocks in alignment
blockSizes	54,59,	Size of each block
qStarts	1,55,	Start of each block in query.
tStarts	245515635, 245515692,	Start of each block in target.

To completely describe an alignment, a PSL record must store three principal data components. First, the record needs to specify the regions of the query and target sequences that need to be retrieved from the database. The query and target sequences are indicated by the qName and tName PSL fields; the regions within those sequences that need to be retrieved are specified by the qStart and qEnd PSL fields, and the tStart and tEnd fields, respectively.

Second, the PSL annotation must include the orientations of both the query and target sequences in the alignment. These orientations are specified by the PSL strand field. Finally, the PSL record must identify the set of sequence blocks that need to be extracted to build the alignment. These sequence blocks are specified by a list of their sizes (the blockSizes field) and two lists indicating the start positions in the query and target sequences of each of the blocks (the qStarts and tStarts fields).[3]

---

[3] Note that in contrast to the BED format, the list of PSL blockSizes, qStarts, and tStarts have trailing commas. Moreover, there are other PSL fields, including ones that count the number of matches and inserts in the alignment, that we do not discuss here; we will not need these fields for the applications described in this book.

PSL format can be somewhat difficult to understand at first. One reason is that some PSL fields are in absolute (positive-strand) coordinates, whereas other PSL fields are in strand coordinates. Two coordinate systems are used because different PSL fields are used in two separate tasks. One the one hand, constructing an alignment involves retrieving sequences from the database. In the UCSC system, sequences are nearly always stored in absolute coordinates, so the PSL fields used in sequence retrieval (qStart, qEnd, tStart, tEnd) are all in absolute coordinates. However, building the actual alignment from the retrieved sequences is more easily accomplished using strand coordinates. Consequently, the fields used to build the alignment from the two sequences (i.e., the qStarts and tStarts fields) are specified in strand coordinates.

The second reason for the complexity of PSL stems from the requirement that PSL be flexible enough to represent both nucleotide and translated-nucleotide alignments.[4] In particular, PSL needs to be able to represent both same-species (nucleotide) alignments of mRNAs and ESTs, which are implemented on the UCSC system with BLAT,[5] as well as xeno-mRNA/EST translated alignments, which are implemented with translated BLAT.

One important difference between nucleotide and translated-nucleotide alignments is that with nucleotide alignments, there are only two possible orientations of the target and the query sequences. Either the query aligns to the positive strand of the target (i.e., the chromosome) or to the negative strand. By UCSC convention, nucleotide PSL alignments are always to the positive strand of the chromosome, and the PSL strand field indicates whether the query needs to be reverse-complemented or not to align to the positive chromosomal strand. Consequently, nucleotide PSLs require only a single-character strand field; a strand field equal to "-"[6] means the query is reverse-complemented, whereas a strand field equal to "+" means the query is not reverse-complemented.

In contrast, for a translated alignment it is possible that either the query or the target sequence (or both) need to be reverse-complemented to produce the (translated) alignment.[7] Consequently, for translated alignments, the PSL strand field needs to be able to take on four values, which are labeled "++," "+-," "-+," or "--." In each case, the first character in the strand field indicates whether the query sequence needs

---

[4] The UCSC system rarely uses protein alignments, and the PSL format is generally not used to represent protein alignments.

[5] In some cases, sequences from closely related species, such as human and chimpanzee, are aligned with nucleotide BLAT, as well.

[6] Note: that the "minus" character for PSL is the hyphen ("-") rather than the dash ("–") since ASCI does not support the dash/minus symbol.

[7] The fact that the query sequence may need to be reverse-complemented in a translated alignment may not be immediately obvious. The point is that query sequences may include ESTs, which are often fragmentary and taken from the 3′ ends of transcripts. Consequently, ESTs are often found reverse-complemented in the archival EST databases (e.g., dbEST) and, hence, also in the genome databases. In contrast, mRNAs only need to be reverse-complemented in the rare cases in which they were originally incorrectly deposited in the archival databases in reverse orientation.

to be reverse-complemented in the translated alignment, and the second character indicates whether the target sequence needs to be reverse complemented. Note that because the reverse complements of two synonymous codons are generally not synonymous themselves, a "+ -" PSL alignment is different from a "- +" alignment, and a "++" alignment is not the same as a "- -" one.

The "toy" examples in Figure A2.1 should help explain the use of PSL format for different types of alignments. Figure A2.1a shows the thirty base pairs at the extreme 5′ end of a thousand base-pair "chromosome." The figure also shows two transcripts, one aligning to each strand of the same, discontinuous, six-nucleotide region of the chromosome. At the top of the schematic in Figure A2.1a are the coordinates of the region in the absolute coordinate system. At the bottom of the figure, the coordinates of the negative strand of the region are shown in strand coordinates.

In Figure A2.1b, we consider a nucleotide alignment of a 9-nt "query sequence" (GCCCCTTGG) to the chromosome. The transcript aligns to the positive strand, as shown in the figure. The principal PSL field values required to represent this alignment are shown below the alignment schematic. In addition, the tStarts (equal to 12 and 25) and qStarts (equal to 1 and 5) field values are illustrated at the top and bottom of the figure, respectively.

Figure A2.1c is a nucleotide alignment of a different query sequence (CCAAGGGGC) to the chromosome. This query aligns to the negative chromosomal strand; hence, in this case, the PSL strand field is equal to "-." The region of the query that is included in the alignment extends (in absolute coordinates) from 2 to 8. Hence, qStart is equal to 2 and qEnd is equal to 8. The other PSL fields as well as the schematic representation of the alignment are the same as in part (b).

Figure A2.1d shows a "++" translated alignment of the sequence GCCGCTGGG to the chromosome. In Figure A2.1d, we include the amino acid translations of the two codons in the 6-nt alignment. The alignment shows that both amino acids match in the translated alignment, even though a nucleotide alignment would have two mismatches. Figure A2.1e represents a translated alignment of the query sequence CCAGCGGC for which the query sequence must be reverse-complemented for the sequences to align. The PSL strand field is "- +" in this case.

Figure A2.1f represents a translated alignment of the sequence CCAAAGGAC, in which the *chromosome* sequence must be reverse-complemented. In this case, the PSL tStarts field will be in negative-strand coordinates (tStarts = 974, 985), as shown at the top of the alignment schematic. Note that the order of the block sizes in the blockSizes field is reversed compared to the previous examples. The chromosomal region that needs to be retrieved from the database, however, is still the same as in each of the previous cases, so the tStart = 12 and tEnd = 26 values will be the same (in absolute coordinates) as they were before. Finally, Figure A2.1g shows the PSL field values for a transcript (CTCCTTTCC), which aligns only if both the query and the chromosome are reverse-complemented in the translated alignment. Hence, the PSL strand field is equal to "- -" in this case.

```
Positions shown are at 5' end of 1000 bp "chromosome".
a)


0         1         2         3 ten's position in target (absolute coordinates)
01234567890123456789012345678901234567890 one's position in target
            >>>>            >>      plus strand alignment
AAAAAAAAAAAAACCCCAAAAAAAAAATTAAAA
TTTTTTTTTTTTTGGGGTTTTTTTTTTAATTTT
            <<<<            <<      minus strand alignment
09876543210987654321098765432109876543210 one's position (negative strand coordinates)
0         9         8         7 ten's position
0         9         9         9 hundred's position
1


b) nucleotide positive orientation alignment
Sequence to align: gCCCCTTgg
            1         2       ten's position in target
            234567890123456   one's position in target
AAAAAAAAAAAAACCCCAAAAAAAAAATTAAAA
            ****            **
            gCCCC---------TTgg
            1234---------56   one's position in query

qSize=9, qStart=1, qEnd=7
tSize=1000, tStart=12, tEnd=26
blockSizes=4,2
qStarts=1,5
tStarts=12,25
strand='+'


c) nucleotide negative query orientation alignment
Sequence to align: ccAAGGGGc
reverse complemented query:  gCCCCTTgg

qSize=9, qStart=2, qEnd=8
strand='-'
Other fields same as in (b)
```

Figure A2.1 PSL representations of simple transcript alignments to the 5' end of a thousand base-pair "chromosome." (a) Schematic of the alignment of two ESTs to a genome, one aligning to the positive genomic strand (shown by ">" symbols), the other to the negative strand (shown by "<" symbols). Both strands of the chromosomal sequence are shown. The chromosome's coordinates are shown in absolute coordinates above the sequence and in (negative) strand coordinates below. (b) PSL fields and alignment schematic for "+" nucleotide alignment. Aligning (nonaligning) nucleotides are shown in uppercase (lowercase) letters, respectively. (c) PSL fields for "-" nucleotide alignment, that is, for a transcript that aligns to the negative chromosomal strand. Both the transcript (query) sequence and the reverse-complemented query sequence are shown. (d) PSL fields and schematic of a "++" translated alignment. Schematic also includes the amino acids coded for by the target and query sequences. (e) PSL fields of "- +" translated alignment. (f) PSL fields and schematic of a "+ -" translated alignment. The reverse-complemented chromosomal region is shown. Target start positions are shown in the schematic in negative-strand coordinates (tStarts = 974, 985). (g) PSL fields for a "--" translated alignment.

```
d) translated ++ orientation alignment (EST/mRNA)
Sequence to align: gCCGCTGgg
              1              2          ten's position in target
           234567890123456             one's position in target
              P              L          target amino acid
AAAAAAAAAAAAACCCCAAAAAAAAAATTAAAA
              *              *
           gCCGC---------TGgg
              P              L          query amino acid
           1234---------56             one's position in query
strand='++'
Other fields same as in (b)


e) translated -+ orientation alignment (EST)
Sequence to align: ccCAGCGGc
reverse complemented query: gCCGCTGgg

qSize=9, qStart=2, qEnd=8
strand='-+'
Other fields same as in (b); same alignment as in (d)


f) translated +- orientation alignment (EST/mRNA)
Sequence to align: ccAAAGGAc
reverse complemented target: TTTTAATTTTTTTTTGGGGTTTTTTTTTTTTT

      9              9          hundred's position in target
      7              8          ten's position in target
    456789012345678            one's position in target
      K              G          target amino acid
TTTTAATTTTTTTTTGGGGTTTTTTTTTTTTT
      *              *
    ccAA---------AGGAc
      K              G          query amino acid
      23---------4567          one's position in query

qSize=9, qStart=2, qEnd=8
tSize=1000, tStart=12, tEnd=26
blockSizes=2,4
qStarts=2,4
tStarts=974,985
strand='+-'


g) translated -- orientation alignment (EST)
Sequence to align: cTCCTTTcc
reverse complemented query: ggAAAGGAg
reverse complemented target: TTTTAATTTTTTTTTGGGGTTTTTTTTTTTTT

qSize=9, qStart=1, qEnd=7
strand='--'
Other fields and alignment same as in (f)
```

Figure A2.1 (continued)

```
##maf version=1
a score=9584.000000
s hg18.chr14                75000019 8 + 106368585 TCTAGGCT
s panTro1.chr15             75034079 8 + 106954593 TCTAGGCT
s rheMac2.chr7             138580786 8 + 169801366 TCTAGGCT
s mm8.chr12                 86522660 1 + 120463159 T-------
s oryCun1.scaffold_174157       7566 8 +     54562 TCGAGGCT
s bosTau2.chr10             57039273 8 +  70001009 TCGAGACT
s canFam2.chr8              51520568 8 +  77315194 TCTAGGCT
s loxAfr1.scaffold_140471       3502 8 +      5079 TCTAGGCT

##maf version=1
a score=6657.000000
s hg18.chr9               129250759 11 + 140273252 ATCTGACATGG
s panTro1.chr11          112107791 11 + 123086034 ATCTGACATGG
s rheMac2.chr15           10987254 11 - 110119387 ATCTGACATGG
s mm8.chr2               149191489  8 - 181976762 ---CAGCAGCC
s bosTau2.chr11           12621147 11 -  87172399 ACCTTACATGG
s canFam2.chr9             5176735 11 -  64418924 ATCTTACACAG
s dasNov1.scaffold_349      130322 11 -    188029 ATCTGACATGT
s loxAfr1.scaffold_8295       8936 11 +    100154 ACCTGACAtat
s echTel1.scaffold_174690     5038 11 +      6760 TACTCACATAA
s galGal2.chr17            9322292 11 +  10632206 CTTTGAGGATG
s danRer3.chr5           29470428 11 +  73302350 ATATTTTAGCA
```

Figure A2.2 File with two multiple-sequence alignments in MAF format. The file consists of two MAF records separated by a blank line.

A2.4 MAF file format

Numerous formats have been developed for the representation of multiple-sequence alignments. These include ClustalW, Pfam, PHYLIP, and MAF. The UCSC system uses the *Multiple Alignment Format (MAF) format* because it is easily read (i.e., parsed) by computer code while being human-readable as well.

An example of the basic MAF file format is shown in Figure A2.2. In contrast to BEDs and PSLs, for which each record consists of a single line and does not include sequence data, a MAF record (describing a single multiple-sequence alignment) consists of multiple lines and does include the actual sequence alignments. Consequently, a blank line is used as part of the format specification to indicate the separation of MAF records. For example, Figure A2.2 shows a single MAF file with two MAF records.

Each line within a MAF record starts with a single-letter code indicating the type of information within the line, for example, "a" indicates an alignment score and "s" a sequence from the alignment itself. Each "s" line specifies the database from which the sequence was extracted, as well as the sequence ID (e.g., chromosome name), the location of the aligned segment within the sequence, and the sequence itself. Note that in contrast to the PSL representation, MAFs may contain gaps (indicated by dashes) within an alignment block. We also note that there are additional line types besides "a" and "s" included in the MAF specification. However, these additional line types are rarely needed and will not be used or described further (see http://genome. ucsc.edu/FAQ/FAQformat#format5 for a detailed description of MAF syntax).

A2.5 WIG file format

The final UCSC file format we will describe is the *WIG format*, which is used for numerical annotations that vary along the genome. Examples include local GC% or multiple-species conservation scores. (Such annotations typically appear in the web browser as wiggly lines, from which the format name originates.)

In its simplest form, a WIG file is similar to a BED file, with each line including the name of the underlying sequence (e.g., the chromosome), the start and end coordinates of a region on that sequence, and a numerical value (e.g., the local GC%) that is associated with that region. In addition to the basic WIG format, variations of the format exist that enable regions of constant or periodic numerical values to be stored in a more compact form. These format modifications are straightforward and are described at http://genome.ucsc.edu/google/goldenPath/help/wiggle.html.

A2.6 GFF and DAS formats

Outside of the UCSC database system, the BED, PSL, and MAF formats are less frequently used. Instead, genomic annotation files typically use either the GFF/GTF/GFF3 or DAS data formats. Tools for converting data files between the GFF and BED formats exist, for example, on the Galaxy web site.

Both GFF and DAS formats are capable of describing multiple types of genomic features including genes, transcripts, and pairwise alignments, though they do so in a manner that is different from the BED and PSL approaches. In particular, the UCSC formats all use tab-delimited fields in which each file contains a single type of data (e.g., PSL files contain pairwise alignments, MAF files contain multiple-sequence alignments). In addition, most of the UCSC annotation formats, such as BED and PSL (MAF is an exception), are line-oriented, meaning that all the data required to describe a complex feature is contained within a single (sometimes rather long) line.

In contrast, a GFF or DAS file may contain multiple types of annotations (e.g., gene structures and alignments may be stored in the same file), with the type of annotation stored as one of the fields in each record. Data records are not necessarily all contained on a single line, nor need they be implemented using tab-delimited fields. In particular, field boundaries may be specified using an XML Document Type Definition (DTD) file. As a result of using these more flexible data formats, GFF and DAS records can store more complex data relationships than BED or PSL records.

However, the price is that GFF and DAS data parsers are generally more difficult to write than BED or PSL parsers.

A2.6.1 GFF/GTF/GFF3

The *GFF format*, and its more recent variants GFF2.5 (also known as GTF) and GFF3, consists of records with nine fields, the first eight of which are simple fields, including name, reference sequence (e.g., chromosome name), start, end, strand, and score fields. GFF also includes fields indicating the type of feature (e.g., gene, transcript, or exon), the source of the annotation and, for protein-coding features, the codon reading frame, or "phase," of the feature. However, in contrast to the BED-12 format – in which hierarchical data information is stored in the blockStarts and blockLengths fields, the GFF format does not include any fields for directly describing data subfeatures.

Consequently, the GFF record requires some other mechanism for indicating hierarchical relationships among features, for example, the set of exons that are associated with a single transcript. In GFF, this hierarchical information is stored in the ninth field of the record. Precisely how this hierarchical data is represented in the ninth data field is one of the principal differences between the original version of GFF and its GFF2.5 (GTF) and GFF3 variants. In the original GFF format, the ninth field was called the "group" field, and all GFF records with the same group field value were components of the same larger feature (e.g., all exons of a single transcript would have the same group field value).

However, there were limitations with this form of data representation. First, only a single level of annotation grouping could be represented. In addition, the group field was also used for storing other types of feature annotation, such as feature descriptions. The GTF and GFF3 formats overcame these limitations. First, GTF extended GFF to allow multiple, semicolon-separated values to be included in the group field. GFF3 then specified how the multiple values in the group field could be used to describe data hierarchies with more than a single level. In addition, the GFF3 specification restricted the values that could be used in the various data fields to be those from a controlled vocabulary of genomic terms established by the Sequence Ontology Consortium (Eilbeck et al., 2005), thereby eliminating ambiguous annotations and making it easier to develop data indexes for fast data retrieval. An example of the GFF3 representation, using tab-delimited fields, of a gene structure is shown in Figure A2.3. Note that the two lines starting with a single "#" symbol are comments, which indicate how the same gene information would be annotated in BED-12 format.

The main limitation of the GFF3 format is that it is still somewhat new and, consequently, not that widely used. In particular, many programs that generate GFF output or parse GFF files do not yet take advantage of the newer features provided by the GFF3 format. For a more detailed description of the GFF/GTF/GFF3 data formats, the reader is referred to http://www.sequenceontology.org/gff3.shtml.

```
##gff-version  3
##sequence-region  chrX 1 154824264
#NM_138636 chrX  +  12684414  12700943  12700943  12684414,12696819, 12684485,12700943,
#NM_016610 chrX  +  12684414  12700079  12700079  12684414,12688081,12696819, 12684485,12688218,12700079,
chrX . gene  12684414  12700943  .  +  .  ID=gene00001;Name=TLR8
chrX . mRNA  12684414  12700943  .  +  .  ID=mRNA00001;Parent=gene00001;Name=NM_138636
chrX . mRNA  12684414  12700079  .  +  .  ID=mRNA00002;Parent=gene00001;Name=NM_016610
chrX . exon  12684414  12684485  .  +  .  ID=exon00001;Parent=mRNA00001,mRNA00002
chrX . exon  12696819  12700943  .  +  .  ID=exon00002;Parent=mRNA00001
chrX . exon  12688081  12688218  .  +  .  ID=exon00003;Parent=mRNA00002
chrX . exon  12696819  12700079  .  +  .  ID=exon00004;Parent=mRNA00002
chrX . CDS   12684414  12684485  .  +  .  ID=cds00001;Parent=mRNA00001;Name=TLR8.protein.1
chrX . CDS   12696819  12699942  .  +  .  ID=cds00001;Parent=mRNA00001;Name=TLR8.protein.1
chrX . CDS   12688161  12684485  .  +  .  ID=cds00002;Parent=mRNA00002;Name=TLR8.protein.2
chrX . CDS   12688081  12688218  .  +  .  ID=cds00002;Parent=mRNA00002;Name=TLR8.protein.2
chrX . CDS   12696819  12699942  .  +  .  ID=cds00002;Parent=mRNA00002;Name=TLR8.protein.2
```

Figure A2.3 Gene with two transcripts and three exons represented as tab-delimited GFF3 record. Lines starting with a single "#" are comments and show how these gene transcripts would be annotated in BED-12 format.

Gene	NM_138636	transcript	curated	chrX	12684414	12700943	+	.	.
Gene	NM_138636	exon	curated	chrX	12684414	12684485	+	.	.
Gene	NM_138636	exon	curated	chrX	12696819	12700943	+	.	.
Gene	NM_138636	CDS	curated	chrX	12684414	12684485	+	.	.
Gene	NM_138636	CDS	curated	chrX	12696819	12699942	+	.	.
Gene	NM_016610	transcript	curated	chrX	12684414	12700079	+	.	.
Gene	NM_016610	exon	curated	chrX	12684414	12684485	+	.	.
Gene	NM_016610	exon	curated	chrX	12688081	12688218	+	.	.
Gene	NM_016610	exon	curated	chrX	12696819	12700079	+	.	.
Gene	NM_016610	CDS	curated	chrX	12688161	12684485	+	.	.
Gene	NM_016610	CDS	curated	chrX	12688081	12688218	+	.	.
Gene	NM_016610	CDS	curated	chrX	12696819	12699942	+	.	.

Figure A2.4 Representation of same gene from Figure A2.3, in lightweight DAS format.

A2.6.2 Lightweight DAS

The *DAS format* specification (Dowell et al., 2001) is one component of the larger DAS protocol describing the process of annotating a genomic sequence via a remote annotation server. A complete description of the DAS protocol and format can be found at http://www.biodas.org. A streamlined version of the DAS annotation format, known as the "lightweight" DAS format, is supported by several genome-database systems including UCSC, Ensembl, and the Chado database schema. Lightweight DAS format is similar to tab-delimited GFF. Data fields include annotation type, reference sequence, start, end, strand, phase, score, and subtype (analogous to the GFF source field). DAS also includes explicit "name" and "class" fields, and optional alignment start and end fields for alignment annotations. Figure A2.4 shows the lightweight DAS representation of the same gene structure as shown in Figure A2.3 in GFF format. Similarly to GFF, lightweight DAS has limitations for describing multilevel data hierarchies. More complex versions of DAS, such as DAS2.0, have been proposed and are described in detail at the DAS web site. We will not need these newer DAS specifications in this book.

APPENDIX 3

UCSC Table Formats

In this appendix, we describe the most frequently used UCSC table formats (BED, PSL, genePred, and MAF) that are needed to perform batch querying with the UCSC system. Note that as Ensembl's software performs database-table conversion automatically, knowledge of Ensembl's table formats is rarely needed.

A3.1 BED and PSL table formats

UCSC's BED and PSL table formats are very similar to their corresponding file formats, with almost identical fields generally occurring in the same order. For example, BED tables, like BED files, come in several varieties, all of which contain the basic BED fields – reference-sequence name, feature start, and feature end – whereas some of them include additional fields such as strand, feature name, score, block count, block starts, or block lengths.

However, BED and PSL table formats are not identical to their corresponding file formats. One difference is that in BED or PSL table format, numerical data such as the start and end positions are stored as integers rather than as character strings. Another difference is that some BED and PSL tables have an additional (initial) field called the bin field. The bin field is used internally by the UCSC system to speed up table lookup and indexing. In general, the bin field is not used in batch querying. However, it is important to know whether a BED or PSL table has a bin field because it will be extracted along with all the other table fields in a "SELECT * FROM table" SQL query or a "Select all fields from table" Table Browser request. In such cases, the bin field will typically need to be removed from the obtained output file before the file can be processed by any UCSC BED file or PSL file programs because these programs expect data in BED or PSL file format, which do not have initial bin fields.

It is also worth noting that data stored in the database in PSL *table* format, such as mRNA- or EST-alignment data, can be accessed in BED-12 *file* format. This can happen, for example, when retrieving PSL alignments using the "BED output" option of the Table Browser. An example of the result of accessing a PSL table with BED-file output is illustrated in Figure A3.1a. The figure shows records for several mRNAs from the

a) PSL table records read out in BED-12 format

```
chrX 1625294 1627388 AK092956 0 - 1625294 1627388 0 2 403,1689,     0,405,
chrX 1625319 1627325 BC012792 0 - 1625319 1627325 0 2 378,1627,     0,379,
chrX 1754312 1758200 BCC16935 0 + 1754312 1758200 0 3 177,781,97,   0,1851,3791,
chrX 1764000 1765205 CRE20922 0 + 1764000 1765205 0 1 1205,         0,
```

b) GenePred table records read out in BED-12 format

```
chrX 1556494 1558274 AK123701  0 + 1556749 1557466 0 1 1780,         0,
chrX 2397815 2412369 NM_004729 0 - 2400036 2402121 0 2 4356,151,     0,14403,
chrX 6004134 6008068 AK_25309  0 + 6004490 6004877 0 1 3934,         0,
chrX 6311395 6312895 NM_016379 0 - 6311521 6312274 0 3 585,248,161,  0,777,1339,
```

Figure A3.1 Data in PSL or genePred table formats may be written to files in BED file format, for example, using the Table Browser. (a) Part of the all_rna table (which is in PSL table format) written out in BED-12 format. Note that some PSL data has been lost in the conversion to BED file format (e.g., match and mismatch counts), as can be seen by comparing the BED output in the figure with PSL data fields shown in Table A2.2. (b) Portion of the Known Genes table, which is in genePred table format, as read out in BED file format. Again, some annotations, such as ProteinID, are lost in the conversion, as can be seen by comparing with the fields in Table A3.1.

all_mrna (PSL) table after they have been converted to BED-12 file format. Although some information is lost in the change from PSL to BED, such format conversion is often useful because other data-processing tools, such as Galaxy, are designed to use BED input.

A3.2 genePred table format

The genePred table format is used to store gene or gene-prediction features in the UCSC database. In addition to the familiar reference-sequence name, feature name, feature start and end, and strand fields, there are additional fields for storing the start and end of the coding sequence, the number of exons, and comma-separated strings of exon-start positions and exon lengths (see Table A3.1). In contrast to the BED-12 format, exon starts are relative to the entire reference sequence (e.g., the chromosome), whereas in BED-12 the block starts are relative to the feature start coordinates. We also note that genePred fields are in absolute (i.e., positive strand) coordinates. This means, for example, that if the gene is on the negative strand, then cdsStart is actually the location of the stop codon of the gene. Also, as with BED format, the actual sequence of the feature (i.e., the gene) is not stored directly in the database but instead must be retrieved indirectly via the chromosome name and sequence coordinates. Last, as illustrated in Figure A3.1b, we note that data in genePred format tables can also be output as a BED file, although again some annotation data is lost in the process.

A3.3 MAF table format

Whereas BED and PSL table formats are very similar to their corresponding file formats, MAF table format is quite different from MAF file format. For example, the MAF record for the MULTIZ alignment data in UCSC database hg18 at location chr14: 75000019–75000027 is shown, in MAF file format, as the first record in Figure A2.2. In contrast, if we use the Table Browser or SQL to extract the MAF table record from the hg18 multiz17Way table corresponding to this position, we obtain

# bin	chrom	chromStart	chromEnd	extFile	offset	score
1157	chr14	74999993	75000106	2389771	764900017	0.505048

This is quite different from Figure A2.2. The first difference is that in a relational database, each record must fit on a single line of a single table. In contrast, as we have already noted, each alignment in a MAF file consists of multiple lines: some lines containing overall properties of the alignment (e.g., the alignment score), whereas other lines contain the actual aligned sequence data. Consequently, to store an entire MAF alignment in a single table record with the UCSC architecture, the actual MAF

Table A3.1 genePred table format

Field	Example	Description
name	BC073913	Name of gene
chrom	chr1	Reference sequence chromosome or scaffold
strand	−	+ or − for strand
txStart	4268	Transcription start position
txEnd	7438	Transcription end position
cdsStart	6607	Coding region start
cdsEnd	7173	Coding region end
exonCount	6	Number of exons
exonStarts	4268,4832,5658,6469,6720,7095,	Exon start positions
exonEnds	4692,4901,5810,6628,6918,7438,	Exon end positions
proteinID	Q6GMS0_HUMAN	SWISS-PROT ID
alignID	G220323	Unique identifier for each (known gene, alignment position) pair

sequences are stored in separate data files outside of the relational database. For example, the table record shows extFile is equal to 2389771 and an offset equal to 764900017. The record does not show the actual aligned sequences. Rather, the extFile and offset parameters indicate the ID of the external file and an offset within that file from which one could obtain the aligned sequences.

The second difference between MAF file and table output is that the MAF file corresponds precisely to the query region (in our case, chr14:75000019–75000027), whereas the MAF table in the UCSC database corresponds to the *entire* alignment that overlaps the specified region (in the present example, chr14: 74999993–75000106). To find the MAF alignment, or even the alignment score, for the specified subregion from the database table record requires additional work. If one accesses the database via the UCSC Genome Browser or Table Browser, the browser will retrieve the actual MAF sequence alignments from the external files and extract the proper subregion without one needing to understand any of the "behind the scenes" file access and data offset protocols. However, if one needs to directly access MAF records (e.g., as in example 5 in Chapter 10), one does need to be concerned about these details.

APPENDIX 4

Genomic Sequence Alignments

Alignment annotations – including both alignments of transcripts to the genome and alignments of genomic DNA from other species – are among the most useful annotations found in the genome browsers. Ensembl, MapViewer, and UCSC each use somewhat different algorithms for generating these alignments. As a result, their alignment displays sometimes differ and, consequently, it is important to have some understanding of the various approaches used so as to better interpret the varying (and sometimes conflicting) alignments you may find in the different browser displays.

In this appendix, we present an overview of the various strategies used to build the alignments found in the browsers and a description of how the different assumptions used in the underlying alignment algorithms can affect the resulting browser displays. This appendix assumes that the reader is familiar with the basics of biological sequence alignment. Readers unfamiliar with biological sequence alignment are referred to the extensive literature describing this subject, such as the texts by Mount (2004) and Durbin et al. (1998).

A4.1 Aligning transcripts to the genome

Browser transcript alignments include both "cis" alignments – that is, alignments of protein sequences or mRNA- or EST-transcripts to the region from which they were generated – and "trans" alignments, alignments from other regions of the genome or from the genomes of other species. cis alignments provide information on the diversity of splicing and transcriptional isoforms that may be produced from a single genomic locus. In contrast, trans alignments can provide clues as to the function of an unannotated gene. In addition, identification of domains within a trans alignment that are highly conserved may indicate which regions are important for gene function.

To generate transcript alignments, Ensembl uses the program Exonerate (Slater and Birney, 2005) and NCBI uses Splign (Kapustin et al., 2004). Both of these programs include algorithms to aid in the identification of canonical splice sites as

276

part of their alignment. In contrast, UCSC currently uses the BLAT (Kent, 2002) and translated BLAT programs, which are not sensitive to splice-site signals and, consequently, BLAT-derived genomic mRNA alignments may have inaccurate intron-exon boundaries. Improved alignment tools that include splice constraints are currently under development at UCSC that will make intron-exon boundaries displayed in the browser more reliable.

No matter what alignment algorithm is used, intron-exon boundaries in browser alignments can be misleading if there are polymorphisms between the transcript data and the genome, or if there are sequencing errors in the transcript or genome sequence data. Although sequencing errors are uncommon, they do occur (Furey et al., 2004) and, along with splice-site polymorphisms, they can produce unexpected results (see, for example, Figures 5.4 and 5.5). In addition, approximately 0.1 to 0.3% eukaryotic splice sites are created by the U11/U12 "minor" spliceosome. The minor spliceosome is known to use a variety of noncanonical 3'- and 5'-splice sites (Will and Luhrmann, 2005). Identifying minor spliceosome splice sites computationally is challenging, and current gene-annotation programs may miss some of them. Consequently, additional care should be exercised in accepting intron-exon boundaries if one has reason to believe that the splicing might be produced by the minor spliceosome.

A4.2 Genome alignments

Genomic (i.e., DNA) alignments also provide important information in a genome browser. Such genomic alignments include both "localized" alignments,[1] which extend over a limited region (typically less than one megabase), or "extended" alignments, which may extend over many tens of megabases or even the entire length of a chromosome. Genomic alignments typically also include some set of alignment scores that indicate the level of similarity among the sequences for each subregion of the alignment. Such similarity measures, in turn, indicate the level of evolutionary constraint on sequence variations within the region and, hence, the functional importance of the region.

For localized, pairwise alignments, UCSC and Ensembl use the BLASTZ tool (Schwartz et al., 2003). BLASTZ is similar to BLAST, however, with two significant differences. First, the parameters in the similarity scoring matrix are selected to reflect biological similarity in non-protein-coding regions. This choice reflects the fact that many metazoan genomes consist largely of non-protein-coding sequences. Second, BLASTZ uses the technique of "discontinuous seeds," which has been shown

---

[1] I use the somewhat awkward-sounding term "localized" alignment because a "local" alignment has a specific (and different) meaning in biological sequence-alignment terminology. In fact, both what I call localized alignments and extended alignments are examples of local alignments in the usual terminology of biological sequence analysis – see, for example, Durbin et al. (1998) for the standard definitions of local and global alignments.

Figure A4.1 Example of UCSC "nets" and "chains" tracks, illustrating an apparent chromosomal translocation subsequent to the evolutionary divergence of human and mouse. On the UCSC Browser, the best BLASTZ homolog to human snoRNA ACA66 is shown in red indicating that it is on mouse chromosome 5, whereas the best homologs to the exons of its host gene are displayed in yellow-green showing that they are located on mouse chromosome 11. The colors corresponding to these different mouse chromosomes are difficult to distinguish in the grayscale figure reproduction. However the different chromosomal locations of the two homologs can also be seen from the chr11 and chr5 annotations at the left side of the Mouse Chained Alignments track.

to be somewhat more sensitive than the contiguous seeds used by conventional BLAST (Ma et al., 2002).

BLASTZ identifies regions of high local similarity. To link together high-scoring BLASTZ alignments into extended alignments, UCSC uses the chain and net algorithms (Kent et al., 2003), whereas Ensembl uses the Mercator program (Dewey and Pachter, 2006). The UCSC chain algorithm accepts gaps between localized alignments that are longer than those allowed in BLAST or BLASTZ. The net algorithm then ranks the chains and allows for gaps in one chained alignment to be filled with shorter chains that may be from other genomic regions, such as those that may have been created by chromosomal inversions or translocations. The result of this process is a global view of the syntenic history of a genome, and can be viewed on the "nets" and "chains" tracks of the UCSC Browser. An example of nets and chains can be seen in Figure A4.1, which shows the genomic region surrounding the human snoRNA gene ACA66, including the intron of the USP32 gene in which ACA66 is located. The figure shows that the best mouse homolog of ACA66 is on a different chromosome from the mouse homolog of USP32, as shown by the different colors – or shadings in the figure – of the snoRNA and host-gene exons on the mouse-net track. The figure suggests that the region surrounding ACA66 has been translocated to another chromosome in the human or the mouse genome or both subsequent to the divergence of the two species.

The Mercator program used by Ensembl builds extended alignments by a different approach from UCSC nets and chains. Rather than build chains from localized BLASTZ nucleotide alignments, Mercator builds a "synteny" map by locating strings of similar protein-coding exons between the two genomes. These exon strings serve as anchors for the extended nucleotide alignments. Because Mercator and the nets-and-chains method use different algorithms, they may produce rather different genomic alignments. Consequently, it is wise to examine both UCSC and Ensembl's alignments for one's region of interest. For more details on the nets-and-chains approach, see the paper by Jim Kent (Kent et al., 2003). For additional information on Mercator, see Dewey and Pachter (2006).

A4.2.1 Multiple-species genome alignments

For many applications, it is useful to obtain genome alignments of homologous regions from more than two species. For example, when looking for functionally important regions, a region that is highly conserved among multiple species provides stronger evidence of the presence of constraining evolutionary pressure than a region that is only known to be conserved between two species.

Genomic multiple-sequence alignment (MSA) is a challenging and active field of research, and several genomic MSA programs have been developed. These programs include MULTIZ (Blanchette et al., 2004), PECAN (see http://www.ebi.ac.uk/~bjp/pecan), TBA (Blanchette et al., 2004), MAVID (Bray and Pachter, 2004), and MLAGAN (Brudno et al., 2003a). Each employs a different approach for adding sequences to the alignment, a different algorithm for selecting the anchors for the underlying localized pairwise alignments, and a different approach for linking the anchors together to create extended alignments.

At the time of this writing, Ensembl performs MSA using the PECAN algorithm, whereas UCSC uses MULTIZ. PECAN's approach is to initially use the Exonerate program to search for orthologous exons in each of the genomes and then to use the Mercator program to link those exons into syntenic maps. PECAN then uses a progressive-alignment algorithm with consistency constraints to build the actual MSA from Mercator's syntenic map. The UCSC approach, in contrast, begins with pairwise extended alignments generated by BLASTZ and the nets-and-chains approach. Once the pairwise extended alignments have been generated, the MULTIZ program then uses a progressive-alignment approach to build the final MSA.

To fully exploit the evolutionary information contained in a MSA also requires tools that measure local sequence conservation. As with MSA itself, multiple programs have been developed to score MSA similarity and thereby identify regions of evolutionary conservation. These algorithms are generally based on probabilistic models to identify regions of high conservation in MSAs. Two widely used probabilistic MSA scoring algorithms are the phastCons program (Siepel et al., 2005), which is used by UCSC, and GERP (Cooper et al., 2005), which is used by Ensembl.

As Ensembl and UCSC use different genomic MSA and MSA-scoring algorithms, their resulting alignments and predictions of highly conserved regions may be different. Moreover, few benchmarks are available to guide one as to the relative accuracy of the different multiple-alignment and conservation programs – see Prakash and Tompa (2007) and Margulies et al. (2007) for two interesting attempts to assess MSA algorithms. Consequently, it may be worthwhile to check the results of more than one MSA program, especially if the answers returned by one algorithm differ from one's biological intuition. For comparing MULTIZ and MLAGAN alignments, the VISTA Genome Browser at the VISTA web site (http://pipeline.lbl.gov/cgi-bin/gateway2) can

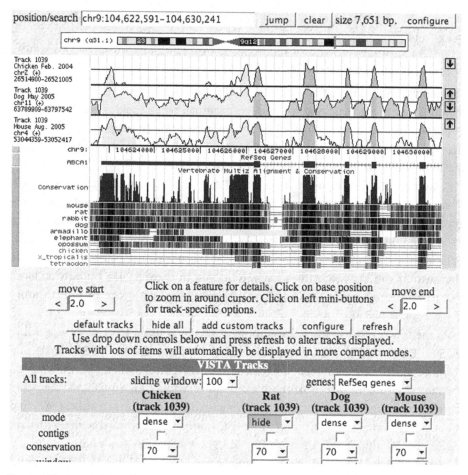

Figure A4.2 Comparison of levels of conservation of MULTIZ and MLAGAN alignments as viewed on the VISTA web site. Tracks labeled "Track 1039" indicate the level of sequence conservation as determined by MLAGAN. The track labeled "conservation" shows sequence conservation when the phastCons conservation program is applied to the MULTIZ alignment.

be informative. With this tool, one can simultaneously view the results of MULTIZ and MLAGAN alignments on the UCSC Browser. An example of such a display is shown in Figure A4.2. Another useful tool for getting a sense of the variety of alignments and conservation predictions is the set of conservation tracks for the ENCODE regions, which is available on the hg17 build of the UCSC Browser. Using these tracks along with the standard UCSC conservation track, one can compare the output of the MULTIZ, MAVID, MLAGAN, and TBA alignment algorithms, as well as of the phastCons, GERP, and SCONE[2] MSA-scoring algorithms. For example, Figure A4.3 shows the level of sequence conservation at a region within the cystic fibrosis gene generated

---

[2] SCONE (Sequence CONservation Evaluation) is yet another program for evaluating the conservation of a multiple-sequence alignment (Asthana et al. 2007).

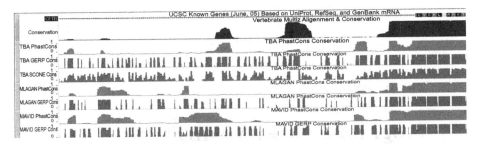

Figure A4.3 Varying patterns of multiple-sequence conservation in the region of the cystic fibrosis gene as determined by the MULTIZ, MLAGAN, MAVID, and TBA alignment programs and the phastCons, GERP, and SCONE conservation-scoring programs.

by several of these tools. From the figure, one notices differences in the regions of constrained sequence identified by the phastCons, GERP, and SCONE algorithms.

Despite the remarks here concerning the different results produced by various alignment procedures and the lack of rigorous techniques for comparing and assessing these results, genomic MSA can be very useful in identifying homologous and conserved regions in multiple genomes, especially if the level of conservation is relatively high. As just one anecdotal example using a MULTIZ alignment of three mammalian genomes, it was possible to use the location of snoRNA genes in one species to identify homologous snoRNA genes in the other two mammalian species (Schattner et al., 2006). This result indicated not only that snoRNA genes are conserved but also that the MULTIZ alignment procedure is able to properly group these homologs.

Program Code README File

This archive contains the code and data files for the examples used in this book. Before accessing the files in the archive, you need to execute: tar -xvzf gbd1.0.tar.gz from the Unix command line in the directory where you want to store the included programs and data files.

Chapters 4, 5, and 6

Chapters 4 and 5 include examples using BioMart, the UCSC Table Browser and Gene Sorter, and Galaxy. In the subdirectory data/martViewTbGalaxyTests, you will find some track and bed files that may be useful as input data for the examples in these chapters. In Chapter 6, there is a program, ucscPerlDbiExample.pl, that illustrates direct SQL access of the UCSC databases with Perl's DBI module. This program is located in the top-level directory. You will need to install the Perl DBI module to run this program.

Chapters 7 and 8

These examples are run using the Ensembl API. You will need to have installed MySQL, BioPerl, and the Ensembl API as described in the text. In addition, you will need to modify the "use lib" statements in each of the scripts to point to the locations of Bioperl and the Ensembl API as indicated in the text or use the Perl -I directive. The three scripts described in the text are in the ensemblScripts subdirectory:

ensemblTest1.pl
ensemblIntronLengths.pl
ensemblComparaExample.pl
Input data is located in the data subdirectory:
hacaWgRna.hg18.bed (for ensemblIntronLengths.pl)
ensemblCompara.test.bed (for ensemblComparaExample.pl)

Chapters 9 and 10

These examples are run using the UCSC API. You will need to have installed MySQL and the UCSC API as described in the text. The six programs described in the text are each in a separate subdirectory of the ucscExamples subdirectory:

helloWorld
ucscDbConnTest
ucscIntronLengths1
ucscIntronLengths2
mafWriteRegions
pslDisplaySeqs

Input data is located in the data subdirectory:

hacaWgRna.hg18.bed (for ucscIntronLengths1 and ucscIntronLengths2)
ensGene.hg18.txt (also needed to run ucscIntronLengths2 in "file" mode)
ensemblCompara.test.bed (for mafWriteRegions)
pslDisplay.hg18.bed (for pslDisplaySeqs)

The programs all need to be compiled and linked before they can be executed. Instructions for using "make" to compile and link these programs can be found in the text. The mafWriteRegions example in the text will only work if you have the UCSC auxiliary files (see the text for details) located in /gbdb/hg18 (and preferably also the UCSC hg18 database table multiz17way) installed on your system. The pslDisplaySeqs example in the text will only work if you have the UCSC auxiliary files (see the text for details) located in /gbdb/hg18 and /gbdb/genbank and the UCSC hg18 database tables chromInfo, gbSeq, and gbExtFile installed on your system.

Chapter 11

Input and expected output data for the autoSql example in the text can be found in the subdirectory data/autoSql.

Selected General References for Genome Databases and Browsers

A6.1 Genome database overview

Stein, L. D. (2003). "Integrating biological databases." *Nat Rev Genet* 4(5): 337–45.

Furey, T. S. (2006). "Comparison of human (and other) genome browsers." *Hum Genomics* 2(4): 266–70.

Baxevanis, A. D. (2003). "Using genomic databases for sequence-based biological discovery." *Mol Med* 9(9–12): 185–92.

A6.2 UCSC Genome Database and Browser

Karolchik, D., R. M. Kuhn, et al. (2008). "The UCSC Genome Browser Database: 2008 Update." *Nucleic Acids Res* 36(Database issue): D773–9.

Thomas, D. J., K. R. Rosenbloom, et al. (2007). "The ENCODE Project at UC Santa Cruz." *Nucleic Acids Res* 35(Database issue): D663–7.

Karolchik, D., A. S. Hinrichs, et al. (2004). "The UCSC Table Browser data retrieval tool." *Nucleic Acids Res* 32(Database issue): D493–6.

Kent, W. J., F. Hsu, et al. (2005). "Exploring relationships and mining data with the UCSC Gene Sorter." *Genome Res* 15(5): 737–41.

A6.3 Ensembl Genome Database and Browser

Stabenau, A., G. McVicker, et al. (2004). "The Ensembl core software libraries." *Genome Res* 14(5): 929–33.

Flicek. P., B. L. Aken, et al. (2008). "Ensembl 2008." *Nucleic Acids Res* 36(Database issue): D707–14.

Curwen, V., E. Eyras, et al. (2004). "The Ensembl automatic gene annotation system." *Genome Res* 14(5): 942–50.

Birney, E. (2003). "Ensembl: a genome infrastructure." *Cold Spring Harb Symp Quant Biol* 68: 213–15.

Birney, E., T. D. Andrews, et al. (2004). "An overview of Ensembl." *Genome Res* 14(5): 925–8.

A6.4 MapViewer Genome Database and Browser

Wheeler, D. L., T. Barrett, et al. (2005). "Database resources of the National Center for Biotechnology Information." *Nucleic Acids Res* 33(Database issue): D39–45.

Wheeler, D. L., T. Barrett, et al. (2007). "Database resources of the National Center for Biotechnology Information." *Nucleic Acids Res* 35(Database issue): D5–12.

Wheeler, D. L., T. Barrett, et al. (2008). "Database resources of the National Center for Biotechnology Information." *Nucleic Acids Res* 36(Database issue): D13–21.

A6.5 Other genome databases

Yeast genome database: Christie, K. R., S. Weng, et al. (2004). "Saccharomyces Genome Database (SGD) provides tools to identify and analyze sequences from *Saccharomyces cerevisiae* and related sequences from other organisms." *Nucleic Acids Res* 32(Database issue): D311–14.

Worm genome database: Schwarz et al. (2006). "WormBase: better software, richer content." *Nucleic Acids Res* 34 (Database issue): D475–8.

FlyBase genome database: Gilbert, D. G. (2007). "DroSpeGe: rapid access database for new Drosophila species genomes." *Nucleic Acids Res* 35(Database issue): D480–5.

Gramene genome database: Jaiswal, P., J. Ni, et al. (2006). "Gramene: a bird's eye view of cereal genomes." *Nucleic Acids Res* 34(Database issue): D717–23.

Mouse genome database: Eppig, J. T., J. A. Blake, et al. (2007). "The mouse genome database (MGD). new features facilitating a model system." *Nucleic Acids Res* 35(Database issue): D630–7; and Blake, J. A., J. E. Richardson, et al. (2003). "MGD: the Mouse Genome Database." *Nucleic Acids Res* 31(1): 193–5.

UCSC prokaryote genome browser: Schneider, K. L., K. S. Pollard, et al. (2006). "The UCSC Archaeal Genome Browser." *Nucleic Acids Res* 34(Database issue): D407–10.

Microbial metagenomic databases: Markowitz, V. M., F. Korzeniewski, et al. (2006a). "The integrated microbial genomes (IMG) system." *Nucleic Acids Res* 34(Database issue): D344–8; and Markowitz, V. M., N. Ivanova, et al. (2006b). "An experimental metagenome data management and analysis system." *Bioinformatics* 22(14): e359–67.

A6.6 Genome browser and related tools

Overview: Stajich, J. E. and H. Lapp (2006). "Open source tools and toolkits for bioinformatics: significance, and where are we?" *Brief Bioinform* 7(3): 287–96.

Galaxy: Blankenberg, D., J. Taylor, et al. (2007). "A framework for collaborative analysis of ENCODE data: making large-scale analyses biologist-friendly." *Genome Res* 17(6): 960–4.

VISTA: Frazer, K. A., L. Pachter, et al. (2004). "VISTA: computational tools for comparative genomics." *Nucleic Acids Res* 32(Web Server issue): W273–9.

BioPerl: Stajich, J. E., D. Block, et al. (2002). "The Bioperl toolkit: Perl modules for the life sciences." *Genome Res* 12(10): 1611–18.

GBrowse: Stein, L. D., C. Mungall, et al. (2002). "The generic genome browser: a building block for a model organism system database." *Genome Res* 12(10): 1599–610.

BioMart: Kasprzyk, A., D. Keefe, et al. (2004). "EnsMart: a generic system for fast and flexible access to biological data." *Genome Res* 14(1): 160–9.

DAS: Dowell, R. D., R. M. Jokerst, et al. (2001). "The distributed annotation system." *BMC Bioinformatics* 2: 7.

BioPipe: Hoon, S., K. K. Ratnapu, et al. (2003). "Biopipe: a flexible framework for protocol-based bioinformatics analysis." *Genome Res* 13(8): 1904–15.

A6.7 Molecular biology and genomics

Primrose, S. B. and R. M. Twyman (2006). *Principles of Gene Manipulation and Genomics,* 7th Edition, Blackwell Publishing.

A6.8 Sequence alignment and conservation

Durbin, R., S. Eddy, et al. (1998). *Biological Sequence Analysis,* Cambridge University Press.

Mount, D. W. (2004). *Bioinformatics: Sequence and Genome Analysis*, 2nd Edition, Cold Spring Harbor Laboratory Press.

Siepel, A., G. Bejerano, et al. (2005). "Evolutionarily conserved elements in vertebrate, insect, worm, and yeast genomes." *Genome Res* 15(8): 1034–50.

Ma, B., J. Tromp, et al. (2002). "PatternHunter: faster and more sensitive homology search." *Bioinformatics* 18(3): 440–5.

Kent, W. J., R. Baertsch, et al. (2003). "Evolution's cauldron: duplication, deletion, and rearrangement in the mouse and human genomes." *Proc Natl Acad Sci U S A* 100(20): 11484–9.

Kent, W. J. (2002). "BLAT – the BLAST-like alignment tool." *Genome Res* 12(4): 656–64.

Blanchette, M., W. J. Kent, et al. (2004). "Aligning multiple genomic sequences with the threaded blockset aligner." *Genome Res* 14(4): 708–15.

Brudno, M., C. B. Do, et al. (2003a). "LAGAN and Multi-LAGAN: efficient tools for large-scale multiple alignment of genomic DNA." *Genome Res* 13(4): 721–31.

Brudno, M., S. Malde, et al. (2003b). "Glocal alignment: finding rearrangements during alignment." *Bioinformatics* 19 Suppl 1: i54–62.

Margulies, E. H., G. M. Cooper, et al. (2007). "Analyses of deep mammalian sequence alignments and constraint predictions for 1% of the human genome." *Genome Res* 17(6): 760–74.

A6.9 Programming references

Perl:

Holzner, S. (1999). *Perl Core Language*, Coriolis.

Tisdall, J. D. (2001). *Beginning Perl for Bioinformatics*, O'Reilly.

Tisdall, J. D. (2003). *Mastering Perl for Bioinformatics*, O'Reilly.

C:

Reek, K. A. (1998). *Pointers on C*, Addison-Wesley.

MySQL:

DuBois, P. (2005). *MySQL*, Sams Developer's Library.

Online Documentation and Useful Web Sites for Genome Databases and Browsers

A7.1 General web tutorials

Open Helix online tutorials: http://www.openhelix.com/tutorials.shtml

A7.2 UCSC Genome Database and Browser

A7.2.1 UCSC Genome Browser

Genome Browser home: http://genome.ucsc.edu
Genome Browser Gateway: http://genome.ucsc.edu/cgi-bin/hgGateway
Genome Browser User's Guide:
    http://genome.ucsc.edu/goldenPath/help/hgTracksHelp.html
Genome Browser FAQ: http://genome.ucsc.edu/FAQ
Genome Browser custom tracks:
    http://genome.ucsc.edu/goldenPath/help/hgTracksHelp.html#CustomTracks
Wiki site: http://genomewiki.ucsc.edu
UCSC development site: http://genome-test.cse.ucsc.edu
Browser mailing lists: http://genome.ucsc.edu/contacts.html

A7.2.2 Table Browser

Table Browser: http://genome.ucsc.edu/cgi-bin/hgTables
Table Browser User's Guide:
    http://genome.ucsc.edu/google/goldenPath/help/hgTablesHelp.html
Database table descriptions (this resource is no longer actively maintained):
    http://genome.ucsc.edu/goldenPath/gbdDescriptionsOld.html

A7.2.3 UCSC software and API

UCSC software: http://hgdownload.cse.ucsc.edu/admin/jksrc.zip
CVS software site: http://genome.ucsc.edu/admin/cvs.html
Web software installation instructions: http://genome.ucsc.edu/admin/jk-install.html

Software installation instructions (included in UCSC download files):

    README.building.source in the kent/src/product/ subdirectory

Program list: http://genome-test.cse.ucsc.edu/eng/useMessageIndex.html

A7.2.4 UCSC mirror databases and installation

Public mirror database:

    host = genome-mysql.cse.ucsc.edu, user = genome (no password)

Public mirror database for use with UCSC API:

    host = genome-mysql.cse.ucsc.edu, user = genomep, password = password

Public mirror information:

    http://genome.ucsc.edu/FAQ/FAQdownloads/download29#download29

Mirror installation: http://genome.ucsc.edu/admin/mirror.html

Installation documentation:

    README.install and ex.InstallExample in the kent/src/product subdirectory

Minimal browser installation:

    http://genomewiki.ucsc.edu/index.php/Browser_installation

Database files: http://hgdownload.cse.ucsc.edu/downloads.html

Database table files (hg18): http://hgdownload.cse.ucsc.edu/goldenPath/hg18/database

Updating partial mirrors: http://genomewiki.ucsc.edu/index.php/Browser_Mirrors

Documentation describing the building of the various databases (e.g., for hg18):

    kent/src/hg/makeDb/trackDB/README and kent/src/hg/makeDb/doc/hg18.txt

A7.3 Ensembl Genome Database and Browser

A7.3.1 Genome Browser

Genome Browser home: http://www.ensembl.org

Human Genome Browser: http://www.ensembl.org/Homo sapiens

Site map: http://www.ensembl.org/sitemap.html

Ensembl General Help page: http://www.ensembl.org/info

Ensembl HelpView page (with links to documentation describing different views):

    http://www.ensembl.org/common/helpview

Ensembl ContigView documentation:

    http://www.ensembl.org/common/helpview?se=1; ref=;kw=contigview

Animated tutorials: http://www.ensembl.org/common/Workshops_Online

Browser custom tracks: http://www.ensembl.org/info/using/external_data

Using Ensembl with a DAS server:

    http://www.ensembl.org/info/using/external_data/das/das_server.html

DAS registry: http://www.dasregistry.org

Ensembl Pre! site: http://pre.ensembl.org

Ensembl mailing lists and mailing list archives:

    http://www.ensembl.org/info/about/contact.html

A7.3.2 Ensembl software and API

Ensembl MartView: http://www.biomart.org/biomart/martview

Ensembl API: http://www.ensembl.org/info/using/api

Ensembl Perl API tutorial: http://www.ensembl.org/info/using/api/core/core_tutorial. html

API pdoc pages: http://www.ensembl.org/info/using/api/Pdoc

Archival Java API (no longer maintained):
   http://oct2006.archive.ensembl.org/info/software/java

A7.3.3 Ensembl mirror databases and installation

Ensembl mirror: host = ensembldb.ensembl.org, user = anonymous (no password)

Ensembl MartDb mirror: host = martdb.ensembl.org, user = anonymous, port = 3316 (no password)

Ensembl installation: http://www.ensembl.org/info/webcode

Ensembl registry configuration: http://www.ensembl.org/info/using/api/registry

A7.4 MapViewer Genome Browser

Genome Browser home: http://www.ncbi.nlm.nih.gov/mapview

MapViewer Help: http://www.ncbi.nlm.nih.gov/mapview/static/MapViewerHelp.html

MapViewer Human Maps Help:
   http://www.ncbi.nlm.nih.gov/mapview/static/humansearch.html

MapViewer tutorial:
   http://www.ncbi.nlm.nih.gov/About/outreach/gettingstarted/mapviewer

NCBI Handbook Introduction to MapViewer:
   http://www.ncbi.nlm.nih.gov/books/bv.fcgi?rid=handbook.chapter.ch20

NCBI Handbook MapViewer exercises:
   http://www.ncbi.nlm.nih.gov/books/bv.fcgi?rid=handbook.chapter.ch24

A7.5 Other genome databases

Yeast genome database: http://www.yeastgenome.org

Worm genome database: http://www.wormbase.org

FlyBase genome database: http://www.flybase.org

Gramene genome database: http://www.gramene.org/genome_browser

Mouse Genome Database:
   http://gbrowse.informatics.jax.org/cgi-bin/gbrowse/mouse_current

Cat genome browser: http://lgd.abcc.ncifcrf.gov/cgi-bin/gbrowse/cat

Farm animal genome browser: http://public-contigbrowser.sigenae.org:9090

UCSC prokaryote genome browser: http://archaea.ucsc.edu

Ensembl genome reviews: http://www.ebi.ac.uk/GenomeReviews

Tribolium genome database: http://www.bioinformatics.ksu.edu/BeetleBase
Dictyostelium genome database:
  http://dictybase.org/db/cgi-bin/ggb/gbrowse/dictyBase
Joint Genome Institute microbial genomes: http://img.jgi.doe.gov
Joint Genome Institute environmental genomes: http://img.jgi.doe.gov/m
NCBI microbial browser:
  http://www.ncbi.nlm.nih.gov/genomes/MICROBES/microbial_taxtree.html
Genomes OnLine Database: http://www.genomesonline.org

A7.6 Genome browser and related tools

Galaxy home page: http://main.g2.bx.psu.edu
Galaxy Documentation: http://g2.trac.bx.psu.edu
Galaxy Screencast tutorials: http://g2.trac.bx.psu.edu/wiki/ScreenCasts
Galaxy genomic interval operations: http://g2.trac.bx.psu.edu/wiki/GopsDesc
Galaxy test server: http://test.g2.bx.psu.edu
Galaxy mailing list: http://mail.bx.psu.edu/cgi-bin/mailman/listinfo/galaxy-user
Taverna: http://taverna.sourceforge.net/
VISTA multiple-sequence alignment tools: http://genome.lbl.gov/vista
VISTA Browser: http://pipeline.lbl.gov/cgi-bin/gateway2
Perl module repository: http://www.cpan.org
BioPerl home: http://www.bioperl.org
BioPerl pdoc pages: http://doc.bioperl.org
BioPerl tutorial: http://www.bioperl.org/Core/Latest/bptutorial.html
BioPerl Deobfuscator: http://www.bioperl.org/cgi-bin/deob_interface.cgi
BioJava home: http://www.biojava.org
NCBI (genomic) BLAST: http://www.ncbi.nlm.nih.gov/sutils/genom_table.cgi
NCBI genomes: http://www.ncbi.nlm.nih.gov/Genomes
NCBI (same-species) trace archives BLAST:
   http://www.ncbi.nlm.nih.gov/blast/mmtrace.shtml
NCBI (xeno) trace archives BLAST: http://www.ncbi.nlm.nih.gov/blast/tracemb.shtml
Ensembl trace server: http://trace.ensembl.org/cgi-bin/tracesearch
DAS: http://www.biodas.org
Bio::DAS API: http://search.cpan.org/~lds/Bio-Das-1.06/Das.pm

A7.7 Tools for creating custom genome databases

autoSql and autoXml: http://www.linuxjournal.com/article/5949
sqlToXml, autoDtd, xmlToSql, and autoXml:
   http://www.ddj.com/web-development/193402895
GMOD project: http://www.gmod.org
Chado database schema: http://www.gmod.org/chado

Chado manual: http://www.gmod.org/wiki/index.php/Chado_Manual
GBrowse: http://www.gmod.org/ggb
GBrowse tutorial: http://www.gmod.org/nondrupal/tutorial/tutorial.html
GBrowse installation guide:
    http://www.gmod.org/wiki/index.php/GBrowse_Install_HOWTO
BioSQL: http://www.biosql.org
BioMart: http://www.biomart.org
BioMart manual: http://www.biomart.org/user-docs.pdf

A7.8 File and track format references

UCSC file formats (BED, PSL, MAF, WIG): http://genome.ucsc.edu/FAQ/FAQformat
GFF3: http://www.sequenceontology.org/gff3.shtml
Lightweight DAS: http://biodas.org/servers/LDAS.html
Embedded UCSC documentation describing UCSC track format:
    README in kent/src/hg/makeDb/track DB subdirectory
Documentation describing search-field component of UCSC track format:
    http://hgwdev.cse.ucsc.edu/admin/hgFindSpecHowTo.html

A7.9 Useful software programs

MySQL: http://www.mysql.com
CVS: http://www.nongnu.org/cvs
rsync: http://samba.anu.edu.au/rsync
NcFTP: http://www.ncftp.com
grep: http://www.gnu.org/software/grep
Apache: http://www.apache.org
Cywin: http://www.cygwin.com
Active State Perl (Perl for Windows users): http://www.activestate.com

Glossary of Biological and Computer Terms Used in the Text

This glossary contains definitions of terms used in the text. More information on most of these terms can be obtained from Wikipedia, the online encyclopedia at http://www.wikipedia.org. For additional information on biological terms, see Primrose and Twyman (2006). For information on terms from computational biology, see Mount (2004) or Durbin et al. (1998).

ab initio gene prediction: Computational gene prediction not based on transcript data.

absolute coordinates: See "top-level" coordinates.

aCGH: "array Comparative Genomic Hybridization." Micro-array technology that uses genomic DNA probes for applications such as determining DNA copy number variations between individuals and DNA copy number alterations in cancer.

ADAR: "Adenosine deaminase acting on RNA." RNA-editing enzyme that converts adenosines to inosines.

alignment: The association of residues between the two or more sequences in a way that maximizes the "similarity" between the sequences in a biologically meaningful way.

allele: A variant of a gene observed between individuals of the same species; a gene polymorphism.

alternative splicing: Process in which a single gene can be spliced in multiple ways to produce varying transcripts, each yielding a protein with a potentially different function.

Alu: Family of SINEs that are widespread in primate genomes.

Apache: A widely used, free, open-source web server program.

API: "Application Programming Interface." A standard interface to a programming library that enables a programmer to easily use the functions in the library without needing to understand their implementations.

Archaea: Third major kingdom of life (along with bacteria and eukaryotes). Includes unicellular prokaryotes, which, on the basis of the similarity of their ribosomal RNA sequences and other features to those of eukaryotes, are believed to be more closely related to eukaryotes than to bacteria.

array-CGH: See "aCGH."

ArrayExpress: Micro-array database maintained by EBI and available at http://www. ebi.ac.uk/microarray-as/aer.

ASCII: "American Standard Code for Information Interchange." Widely used data format for storing text data on computer systems.

assembly (genome): The process of linking together individual overlapping sequence "reads" into a sequence of an entire chromosome or genome; also refers to the resulting chromosomal sequence produced by the assembly process.

AXT: Pairwise-alignment format used by the UCSC genome database. Most pairwise alignments are stored at UCSC in PSL format; however, those alignments derived directly from the BLASTZ program are stored in AXT format.

base: Adenine, cytosine, guanine, or thymine (uracil) subunit of DNA (RNA) nucleotide.

batch querying: Process of executing multiple requests (queries) of a database in a single command or a single invocation of a computer program.

BED: "Browser Extensible Description." Flexible data format for describing genomic locations and gene structure used by the UCSC genome database.

BioMart: Tool for interactive batch querying of genome databases used by Ensembl, GMOD, and other genome database projects.

BioPerl: Suite of Perl software objects and subroutines designed to facilitate the writing of Perl software for bioinformatics.

BLAST: Universally used, heuristic algorithm for rapidly aligning sequences by using short exact sequence matches to "seed" the alignment process.

blastp: Version of the BLAST program optimized for aligning protein sequences.

BLASTZ: Version of the BLAST program optimized for aligning long nucleotide sequences that do not necessarily code for proteins.

BLAT: "BLAST-like Alignment Tool." Very fast sequence-alignment tool designed for aligning extremely similar sequences, such as sequence fragments within a genome.

build (database): A complete reconstruction of a genome database, typically in response to the release of a revised genome assembly.

cDNA: "Complementary DNA." DNA that is synthesized from mRNA by reverse transcriptase.

C/D snoRNA (or C/D Box snoRNA): Class of snoRNAs characterized by two sequence motifs known as the "C Box" and the "D Box." Most C/D snoRNAs are involved in the methylation of ribosomal or spliceosomal RNAs.

Celera assembly: Refers to the assembly of the human genome using the data and tools from Celera's Human Genome Project.

centromere: Region near the center of the chromosome to which the spindles attach during cell division.

CG%: Percentage of nucleotide bases in a genomic region that are either cytosine or guanine.

CGI: "Common Gateway Interface." Software protocol for designing programs that can be executed over a web-page interface.

ChIP/chip: Chromatin immumoprecipitation on a (micro-array) chip. A high-throughput technique for determining DNA sites that are bound by chromosomal proteins.

chromatin: Complex of DNA and associated proteins making up the chromosome.

chromosomal coordinates: Coordinate system derived from the actual chrosomomal feature locations on a reference genome; for a completely assembled genome, the "top-level" coordinates will be chromosomal coordinates.

clade: Collection of related species, for example, vertebrates, nematodes, insects.

clone (genomic): A section of DNA that is stored and amplified within a bacterial or eukaryotic chromosome, often as part of the DNA sequencing process.

clone mapping: Strategy for genome sequence assembly in which one builds genomic maps onto which the sequences of individual clones are placed.

Clustalw: Widely used program for aligning multiple protein or nucleotide sequences.

coding sequence: Portion of an mRNA transcript that is translated by the ribosome into an amino acid sequence.

codon: Three consecutive nucleotides in the coding sequence of an mRNA that are translated by the ribosome into a single amino acid.

compiler: Computer program used to parse a set of instructions to a computer in a "high-level language" (which is human-readable, and referred to as "source code") into machine-readable code.

Consensus Coding Sequence (CCDS) Project: Project that seeks to identify human gene annotations for which there is 100% agreement between RefSeq and VEGA annotations.

consensus sequence: Description of a multiple-sequence alignment by means of a single sequence consisting of the most commonly occurring nucleotide (or amino acid) at each position in the alignment.

conserved site: Position in a sequence alignment at which all, or most, of the nucleotides (or amino acids) are identical, or represent amino acids with similar biochemical properties, or nucleotides that code for similar or identical amino acids.

contig: Group of overlapping clones that represent a single continuous region of a chromosome.

copy number polymorphism: Polymorphism involving variation in the number of copies of a section of chromosomal DNA in the genomes of different individuals.

coverage (genome): In a sequencing project, the ratio of the total number of bases sequenced to the size of the target genome.

CpG island: Region of vertebrate genome with unusually high occurrence of CG dinucleotides; often found near gene promoters.

custom track: Track on a genome browser containing a user's private data.

CVS: "Concurrent Versioning System." A computer utility program designed to assist multiple software developers in accessing and modifying a set of computer programs without interfering with one another.

DAS: "Distributed Annotation System." A set of software specifications enabling genome browsers to display sequence annotations stored on remote computer systems.

data warehouse: Data storage and management system where data from multiple, remote, "primary" databases are copied and stored locally in a single integrated data structure.

data focus: One of the central concepts or subjects in relation to which data is stored and retrieved in a data mart.

data mart: Data storage and management system where data are stored as multiple data foci, each focus corresponding to a particular subject (such as "genes" or "SNPs") central to the data being stored.

dbEST: Database of EST sequences, maintained at NCBI.

dbSNP: Database of known SNPs, maintained at NCBI.

DDBJ: DNA Databank of Japan.

deoxyribonucleotide: Monomer subunit of DNA molecule, generally identified with its associated base of adenine (A), cytosine (C), guanine (G), or thymine (T).

discontiguous MegaBLAST: Version of NCBI's MegaBLAST program using discontiguous "seeds."

DNase hypersensitive site: Region of the genome that is particularly susceptible to digestion by the DNase enzyme; often found near gene promoters.

draft sequence assembly: Intermediate-level genomic assembly; usually defined as having approximately 4x to 5x coverage.

DTD: "Document Type Definition." In XML, specification of the set of "tags" and their properties used in an XML document.

dynamic programming: Class of rigorous optimization algorithms useful in multiple bioinformatics applications, including sequence alignment.

EBI: European Bioinformatics Institute.

EMBL: European Molecular Biology Laboratory.

EMBOSS: Open-source software sequence-manipulation package developed by EMBL.

ENCODE: "Encyclopedia of DNA Elements." Name of a project to classify all features found in genomic (initially human) sequence. The pilot phase of the project, recently completed, characterized approximately one percent of the human genome.

Ensembl: Genome browser and database hosted by the European Bioinformatics Institute (EBI).

epigenetics: Study of DNA and chromosomal changes not involving changes to the DNA sequence (histone modifications, DNA methylation, and so on).

EST: "Expressed Sequence Tag." Single read of a portion of an mRNA or cDNA sequence.

eukaryote: Organism with cells that contain a cell nucleus.

exon: Parts of a gene transcript that are spliced together during the processing of many eukaryote genes.

Exonerate: Program used by Ensembl for aligning transcripts to genomic sequence.

extended alignment: Alignment of two or more sequences that may include long (greater than 100 kilobase) gaps and that may extend over multiple megabases of the genomes.

FASTA: Both a computer program for sequence alignment and database searching as well as a widely used data format for representing nucleotide and protein sequences.

federated database: Data management system where data from multiple, remote, "primary" databases are accessed in a unified manner by an integrated querying system.

finished sequence assembly: Genomic sequence assembly considered to be as complete as possible within the limits of current technology; typically having 8x to 9x coverage.

flat-file data structure: Conventional computer data structure in which data is organized in files that are located with a hierarchical structure of directories with subdirectories.

Galaxy: Web-based tool set for manipulating genome-size datasets, such as those produced by batch queries of the genome databases.

GBrowse: Browser program provided by the GMOD project for visualizing data from a GMOD-compatible genome database.

GC%: See "CG%."

GenBank: One of three main sequence data repositories (along with EMBL and DDBJ); maintained by NCBI.

GENCODE genes: Genes identified and experimentally validated in one percent of the human genome by the ENCODE Project.

gene structure: Annotation of a gene (or, more precisely, of a transcript) indicating the locations of all of the exon-intron boundaries.

genetic association study: Genetic study of a large population of unrelated individuals that attempts to find statistically significant associations between an observed phenotype (e.g., a disease) and a specific genomic polymorphism.

genetic map: Map of a chromosome as inferred by the shared inheritance patterns of specific genetic loci (contrast with "physical map").

genome browser: Graphical tool for visualizing data in a genome database as annotation tracks along the chromosomes of the genome of a species.

genome database: Database of biological data in which data is indexed on the basis of genomic (chromosomal) location as well as by keyword.

genotype: Specific genetic sequence of an individual member of a population.

GENSCAN: Widely used computer program for identifying the locations of genes within the chromosomes of eukaryotic genomes.

GEO: "Gene Expression Omnibus." Micro-array database maintained by NCBI and available at http://www.ncbi.nlm.nih.gov/geo.

GERP: "Genomic Evolutionary Rate Profiling." Program used by Ensembl to identify regions of constrained evolution.

GFF: "Generic Feature Format." Widely used genomic data format capable of representing multiple types of annotations, with the type of annotation stored as one of the fields in the record.

GFF3: Extension of GFF format that enables the description of data hierarchies and requires that data fields be described using terms from the controlled vocabulary of the Sequence Ontology Project.

global alignment: Form of sequence alignment that seeks to find the high-scoring alignments between the entire input sequences (in contrast to "local alignment").

GMOD: "Generic Model Organism Database." Project to develop software tools to facilitate the development of genome databases for newly sequenced genomes.

GNF database: Database of mammalian gene expression developed by the Genomics Institute of the Novartis Research Foundation.

GO: "Gene Ontology." Project to develop a standardized "controlled vocabulary" of terms for the description of genes.

grep: Computer utility program for locating specific items within a text file.

GTF: "Gene Transfer Format." Extension of GFF format allowing multiple, semicolon-separated values to be included in the GFF group field.

H/ACA snoRNA (or H/ACA Box snoRNA): Class of snoRNAs characterized by two sequence motifs known as the "Hinge" or H Box" and the "ACA Box." Most H/ACA snoRNAs are involved in the pseudouridylation of ribosomal or spliceosomal RNAs.

haplotype: A set of genetic polymorphisms, typically inherited together, that are associated with a single chromosomal region.

HapMap: International project to determine the common genetic variants (in particular, haplotypes) existing among humans.

hash: Computer data structure in which data items are indexed and can be retrieved via unique keys.

heuristic algorithm: Algorithm that has been empirically shown to be useful and is generally fast but which cannot be rigorously proven to always yield the optimal result.

histone: Any of a class of proteins that are physically associated with eukaryotic chromosomes.

homologous genes or proteins: Refers to genes (or proteins) that are similar because the genes share a common ancestor; includes both orthologous and paralogous genes.

homopolymer regions: Regions consisting of a single repeated nucleotide, for example, AAAAA.

host gene: Gene that has another genomic feature (such as a snoRNA) embedded in one of its introns.

host intron: Intron that has another genomic feature (such as a snoRNA) embedded within it.

HyPhy: Open-source phylogenetics program suite.

immunoprecipitation: Biochemical technique by which molecules are extracted or purified on the basis of their physical association with specific antibodies or other binding proteins.

indel: A genetic polymorphism consisting of either a sequence insertion or deletion.

in silico: Performed via computer, as in an "in silico analysis."

in situ: "In place." Refers to any experimental technique where the phenomenon is observed in the location where it naturally occurs, as in "fluorescent in situ hybridization."

inheritance (software): In an object-oriented language, inheritance refers to the fact that objects automatically possess all the functionality (e.g., methods) associated with any of their parent objects.

interactive batch querying: Batch querying that does not require the writing of a computer program.

integrated database: Single database that includes data from multiple other (primary) databases in a unified manner.

interpreter: Computer program used to sequentially parse and immediately execute each line of a set of instructions to a computer.

intron: Segment of a gene transcript that is removed during the processing of many eukaryote genes.

inversion (chromosomal): Section of a chromosome whose sequence has become inverted.

isoform: Specific alternative splice form of a gene.

join (database): SQL command for retrieving data from multiple tables of a relational database.

key (database): Field or column from a database table that can be used to link the database record to the data for the same record in another database table.

link integration: Form of database integration in which the databases are connected solely by hypertext links.

linkage disequilibrium: Term used to describe the correlated inheritance of alleles.

local alignment: Form of sequence alignment that seeks to find the high-scoring alignments between subsequences of the input sequences (in contrast to "global alignment").

localized alignment: Local sequence alignment in which only relatively short gaps are allowed between matching subsequences.

low-coverage assembly: Genomic sequence assembly with 1x to 2x coverage.

MAF: "Mutiple Alignment Format." Principal data format for storing multiple alignment data in the UCSC database.

makefile: Data file used by the "make" utility program that specifies the tasks to be performed.

map: Term used by NCBI MapViewer to describe a genome annotation; similar to a track in Ensembl and the UCSC Genome Browser.

MapViewer: NCBI's genome browser program.

MAVID: Multiple-sequence alignment algorithm developed by Bray and Pachter (2004).

MegaBLAST: Fast variant of the BLAST program for aligning nearly identical nucleotide sequences.

Mercator: Computer program used by Ensembl to extend localized alignments by building a map of homologous exons.

metagenomics: Simultaneous study of the genomes of all species from a single environmental habitat; also referred to as environmental genomics.

metazoa: Multicelluar organisms.

method: Subroutine associated with a computer object (see "object").

micro-array: Collection of biological-polymer sequences, which may be synthetic or natural sequences, that have been arrayed on a solid surface; may include RNA, DNA, or peptide sequences.

minor spliceosome: Spliceosome found in many eukaryotic cells that removes introns with atypical splice-junction sequences.

miRNA/microRNA: Short (typically twenty-one to twenty-three bases in length) naturally occurring RNA sequence involved in gene regulation in eukaryotic cells.

mirror: Fully functional copy of a computer database or web site.

MLAGAN: Multiple-sequence alignment algorithm developed by Brudno et al. (2003a).

mRNA: "Messenger RNA." A transcribed section of RNA that is capable of being translated into an amino acid sequence.

multiple-sequence alignment (MSA): Alignment of three or more nucleotide or amino-acid sequences.

MULTIZ: Computer program that uses multiple pairwise sequence alignments produced by the BLASTZ program to create a single multiple-sequence alignment.

MySQL: Open-source relational database management system widely used by genome databases, including UCSC and Ensembl.

NCBI: "National Center for Biotechnology Information." U.S. government agency with the responsibility of maintaining multiple archival and curated molecular biology databases and developing tools for facilitating the analysis of the data within them.

ncRNA: See "noncoding RNA."

negative selection: Removal over evolutionary time of variations in the genomic sequence that diminish the fitness of an organism; also referred to as "purifying selection."

nets-and-chains: Program used by UCSC to link together high-scoring, localized BLASTZ alignments into extended alignments.

nonsense mediated decay (NMD): Cellular process by which mRNA transcripts that have premature stop codons are selectively degraded.

non-coding RNA (ncRNA): Transcribed functional RNA that is not translated into protein.

N-SCAN: Gene prediction program developed at Washington University that uses the characteristics of multiple-sequence alignments as part of its gene-prediction algorithm.

nucleosome: Histone-protein structures that are wrapped around DNA.

nucleotide (nt): Ribonucleotide or deoxyribonucleotide.

object: In computer science, a data structure and a set of associated subroutines that are available for manipulating the data within the data structure.

object-oriented language: Computer language in which all or most data are stored and manipulated via objects.

oligonucleotide (oligo): Short (e.g., less than 20 nt) sequence of nucleotides.

OMIM: "Online Mendelian Inheritance in Man." Database of human genes and diseases maintained by NCBI.

open reading frame (ORF): Section of DNA that begins with an initiation codon (generally ATG), terminates with a stop codon (TAA, TAG, and TGA), and has a minimum length (typically around twenty codons).

orthologous genes or proteins: Homologous genes or proteins resulting from a speciation event.

paired end mapping: High-resolution technology for detecting insertion and deletion polymorphisms using sequencing-by-synthesis.

paired sequence alignment: Alignment of two nucleotide or amino-acid sequences.

PAML: "Phylogenetic Analysis by Maximum Likelihood." Software package for phylogenetic analyses.

paralogous genes or proteins: Homologous genes or proteins resulting from a gene or chromosomal duplication event.

parsing: Process (typically implemented by specialized computer programs) of converting instructions or data written in human-readable format into a format that can be processed by a computer.

PCR: "Polymerase Chain Reaction." A technique for the amplification of small quantities of DNA.

PECAN: Program used by Ensembl for multiple-sequence alignment.

Pfam: Database of protein structural domains.

phastCons: Computer program for estimating the extent of evolutionary conservation among a set of genome sequences from a sequence alignment.

phenotype: Observable properties of an organism or cell.

phylogenetic tree: Tree diagram showing the relationships between species, genomes, or parts of genomes (e.g., genes) showing their descent from a common ancestor.

physical map: Map of a chromosome in terms of physically identifiable biochemical landmarks, such as specific subsequences or restriction-enzyme cutting sites (contrast with "genetic map").

pipeline: Suite of computer programs that are executed sequentially on a set of data.

polymorphism (sequence): Variation between two sequences taken of the same region of the genomes of two individuals of the same species.

porting: Modifying computer programs that have been designed to operate on one computer platform or operating system to function properly on another computer system.

positive selection: Increase in the frequency of a genetic variant over evolutionary time as a result of its enhancing the fitness of the organism.

post-transcriptional processing: Refers to all of the RNA modifications – such as splicing, polyadenylation, RNA-editing, and so on – that are performed by the cell on an RNA after it is transcribed.

pragma: Single-line directive to a computer compiler or interpreter.

primary database: Database that stores a single or a small number of types of data.

primer (sequence): Short section of single-stranded DNA that can hybridize with a nucleic acid substrate and is required for the initiation of certain enzymatic reactions, such as PCR and reverse transcription.

programmed batch querying (or programmed querying): Batch database querying and data post-processing carried out by means of a computer program.

prokaryote: Unicellular organisms lacking a cell nucleus; includes bacteria and archaea.

promoter: Region of a gene that binds RNA polymerase and signals the initiation of transcription.

pseudogene: Section of the genome that resembles a gene but which is either not transcribed or whose RNA or protein product is nonfunctional. Often the remnant of a previously functional gene.

PSIBLAST: Variant of BLAST that uses position-specific scoring matrices and that enables one to search for sequences that are similar to a set of related sequences rather than to only search for sequences that are similar to a single sequence.

PSL: "Pattern Space Layout." Principal data format used by the UCSC genome database for storing pairwise sequence alignment data.

QTL: "Quantitative trait locus." Region of a genome for which the occurrence of any one of a set of alleles correlates with the value of some quantitatively measurable phenotype.

radiation hybrid map: Ordering of genetic markers along a chromosome made by fusing irradiated donor cells with host cells from another species.

read: Raw sequence data acquired from a single data acquisition of a DNA-sequencing instrument; currently ranges from about 30 to 1,000 nt, depending on the technology.

reading frame: One of the six possible ways of translating a DNA sequence into a peptide sequence.

recombination hotspot: Region of a genome in which high rates of genetic recombination during meiosis have been detected.

record: Single entry in a data structure, such as a single row in a relational database table or a single line in data file.

reference assembly: Genome assembly of a single individual (or a single well-defined composite) that is used to specify a reference sequence and coordinate system for the genome.

RefSeq: Database of mRNA sequences maintained by the NCBI.

registry: Data structure with associated querying software that allows a user to determine all the services that are provided by a computer or network system.

relational database: Computer data-storage system in which the data is stored in multiple tables and in which an efficient querying system for data retrieval is implemented.

relational database management system (RDBMS): Set of computer programs controlling the management of, and access to, a relational database.

repeat-sequence: Genome subsequence that occurs in multiple locations throughout the genome.

resequencing: Sequencing of a previously sequenced section of a genome in additional individuals for the purpose of identifying sequence variations occurring within a population.

residue: Individual unit of a biological polymer; a single amino acid in the case of proteins, and a single nucleotide in the case of nucleic acids.

retrotransposon: Mobile genetic element whose transposition includes the reverse transcription of an intermediate RNA molecule.

Rfam: Database of families of ncRNAs.

ribonucleotide: Monomer subunit of RNA molecule, generally identified by its associated base of adenine (A), cytosine (C), guanine (G), or uracil (U).

ribosome: Ribonuclear-protein complex that mediates the translation of an mRNA into a polypeptide.

RNA editing: Originally, a post-transcriptional modification of one or more nucleotides of an mRNA leading to the production of a modified peptide during translation. Currently, also used to describe other post-transcriptional nucleotide modifications, including those not resulting in modified peptides.

rsID: SNP ID numbering system used by the dbSNP database.

rsync: Computer utility program widely used to update, compare, and synchronize related computer files and directories.

same-species sequence-alignment: Alignment of two (or more) sequences from the same species (e.g., alignment of a mouse EST to the mouse genome).

Sanger sequencing: Widely used method of sequencing of DNA involving chain termination of the DNA sequence followed by electrophoresis.

scaffold: Set of contigs of DNA sequences from a chromosome with known relative order, orientation, and spacing.

schema (database): Specification of the set of tables

SCONE: "Sequence CONservation Evaluation." Computer program for scoring evolutionary conservation; used with TBA alignment program.

scripting language: High-level computer language that is executed directly by another program (the "interpreter") without needing to be initially processed by a compiler; examples include Perl, Python, and Bash.

seed: In sequence-similarity search programs, a short region which must exactly match between the query and target sequences.

segmental duplication: An extended region of a chromosome (generally 1,000 nt or longer) that has an almost exact duplicate somewhere else in the genome.

sequence tagged site (STS): A unique short (typically 200 to 500 base pairs) region of DNA within a genome with known location and sequence.

sequence trace: Output from a single "read" of a conventional DNA sequencing instrument.

sequencing-by-hybridization: Class of technologies in which the sequence of a "target" DNA is determined by its pattern of hybridization to a large number of short DNA elements of known sequence.

sequencing-by-synthesis: Class of DNA sequencing technologies in which the DNA sequence is determined by detecting light signals generated as nucleotides are incorporated during DNA replication.

SINE: "Short Interspersed Nuclear Element." Short repetitive, transposable element found throughout eukaryotic genomes.

singly linked list: Software data structure in which a set of related data are organized by having each piece of data include a pointer to the location of the next piece of data in the list.

slice coordinates: Genomic coordinate system in which a feature's coordinates are specified in relation to the location of another, generally larger, feature (the "slice").

snoRNA: Small RNAs typically found in the nucleolus of eukaryotes. Also found in Cajal bodies of eukaryotes, and in archaea. See also "C/D snoRNAs" and "H/ACA snoRNAs."

SNP: "Single Nucleotide Polymorphism." A variation between a DNA sequence and the sequence of a reference sequence in which the variation consists of a single-nucleotide difference.

splice junction: Location in the genomic sequence containing to an intron-exon boundary.

spliced EST: EST sequence derived from a spliced RNA transcript.

spliceosome: Ribonuclear-protein complex that mediates the removal of introns from an RNA.

splicing enhancer: Sequence or structural motif in an RNA, the presence of which increases the likelihood of a splicing event during post-transcriptional RNA processing.

splicing silencer: Sequence or structural motif in an RNA, the presence of which decreases the likelihood of a splicing event during post-transcriptional RNA processing.

Splign: Program used by NCBI for aligning transcripts to genomic sequence.

SQL: "Sequence Query Language." Widely used, standard programming language for querying relational databases.

SSAHA: "Sequence Search and Alignment by Hashing Algorithm." Fast, heuristic sequence-alignment algorithm used by Ensembl.

start codon: Initial codon of an mRNA coding sequence that is translated by the ribosome; nearly always AUG.

stop codon: Trinucleotide mRNA subsequence that signals the ribosome to terminate mRNA translation; generally either UGA, UAA, or UAG.

strand: Either of the two nucleotide biopolymers that make up the DNA molecule; the two strands are conventionally referred to as the "Watson" or "positive" strand and the "Crick" or "negative" strand.

strand coordinates: Coordinates measured with respect to the strand on which a feature of interest is located.

struct: General data structure used in the C programming language.

structural polymorphism: Genetic polymorphism in which the variation involves not only a change of one or more nucleotides but also the structure of the DNA molecule, generally via a large scale indel, inversion, or translocation.

synonymous codons: Two (or more) distinct codons that code for the same amino acid.

syntenic: Having the same gene order along the chromosome; often used to describe structural relatedness between chromosomes of different species.

table (database): Two-dimensional array of data; the fundamental unit of data storage in a relational database.

Taverna: Computer scripting language and tool set designed for creating bioinformatics data-processing pipelines.

TBA: See "threaded blockset aligner."

tagged SNP: SNP that is generally inherited together as a set of other polymorphisms and is used as a marker to indicate whether the entire set of polymorphisms has been inherited.

telomere: End of a linear chromosome; often characterized by a specific repetitive DNA sequence.

threaded blockset aligner (TBA): Genomic mutiple-sequence alignment program used by UCSC.

tiling array: DNA micro-array in which the micro-array probes are designed to hybridize with each region of a genome or chromosome in a well-specified and, generally, high-resolution manner.

top-level coordinates: Coordinates derived from (the positive strand of) the longest contiguous pieces in the assembly.

track (browser): Means of depicting a genomic property in a genome browser in which the annotation consists of a series of line segments along a chromosome, which indicate the sections of the chromosome that possess the property.

transition (mutation): Mutation involving a change between A ↔ G or T ↔ C.

transcript: RNA product of transcription by means of RNA polymerase.

transcription: Copying of a section of chromosomal DNA into RNA by the enzyme, RNA polymerase.

transcriptional enhancer: DNA sequence or structural motif that increases the likelihood of a genomic region being transcribed.

translation: Synthesis of a sequence of amino acids by the ribosome from an mRNA coding sequence.

translocation (chromosomal): Chromosomal alteration in which a piece of a chromosome is moved to a new location either on the same chromosome or on another chromosome.

transposase: Enzyme (or enzyme complex) involved in the movement of DNA sequence elements between different parts of the genome.

transversion (mutation): Any mutation other than a transition.

tRNA: "transfer RNA." One of a family of RNA molecules that are used by the cell in conjunction with the ribosome in the translation of mRNA to protein.

UTR: "Untranslated Region." Section of spliced mRNA that is not translated by the ribosome into protein.

vector (in molecular biology): Section of DNA – typically of bacterial, viral, plasmid, or yeast origin – used for the maintenance, amplification, or transfer of genetic material of another species.

Vertebrate Genome Annotation Project (VEGA): Project for the identification of vertebrate genes by the manual curation of alignments of experimentally confirmed transcripts to the genome.

VISTA: Web site providing access to tools for multiple genomic-sequence alignment.

web service: Software system designed to facilitate communication between machines over the Internet; in particular, a system by which a "client" program on one machine can execute a "service" (i.e., another program) on a remote machine.

whole genome shotgun assembly (WGSA): Strategy for genome sequence assembly in which individual sequence "reads" are directly linked together without creating a genomic map prior to building the sequence scaffold.

WIG: Data format used by UCSC for numerical annotations that vary along the genome, such as local GC% or multiple-species sequence-conservation scores.

xeno sequence-alignment: Alignment of two (or more) sequences from different species, for example, alignment of a mouse EST to the human genome; cf. same-species alignment.

XML: "Extensible Markup Language." Computer language developed to facilitate the transfer of machine-readable data between computers with differing operating systems and hardware over the Internet.

yeast two-hybrid system: Experimental system for identifying proteins that interact with one another.

References

Alfarano, C., C. E. Andrade, et al. (2005). "The Biomolecular Interaction Network Database and related tools 2005 update." *Nucleic Acids Res* 33(Database issue): D418–24.

Ashurst, J. L., C. K. Chen, et al. (2005). "The Vertebrate Genome Annotation (Vega) database." *Nucleic Acids Res* 33(Database issue): D459–65.

Asthana, S., M. Roytberg, et al. (2007). "Analysis of Sequence Conservation at Nucleotide Resolution." *PLoS Comp Bio* 3(12): e254.

Barrett, J. C., B. Fry, et al. (2005). "Haploview: analysis and visualization of LD and haplotype maps." *Bioinformatics* 21(2): 263–5.

Baxevanis, A. D. (2003). "Using genomic databases for sequence-based biological discovery." *Mol Med* 9(9–12): 185–92.

Birkland, A. and G. Yona (2006). "BIOZON: a hub of heterogeneous biological data." *Nucleic Acids Res* 34(Database issue): D235–42.

Birney, E. (2003). "Ensembl: a genome infrastructure." *Cold Spring Harb Symp Quant Biol* 68: 213–15.

Birney, E., T. D. Andrews, et al. (2004). "An overview of Ensembl." *Genome Res* 14(5): 925–8.

Birney, E., J. A. Stamatoyannopoulos, et al. (2007). "Identification and analysis of functional elements in 1% of the human genome by the ENCODE pilot project." *Nature* 447(7146): 799–816.

Bishop, A. C., J. Xu, et al. (2002). "Identification of the tRNA-dihydrouridine synthase family." *J Biol Chem* 277(28): 25090–5.

Blake, J. A., J. E. Richardson, et al. (2003). "MGD: the Mouse Genome Database." *Nucleic Acids Res* 31(1): 193–5.

Blanchette, M., W. J. Kent, et al. (2004). "Aligning multiple genomic sequences with the threaded blockset aligner." *Genome Res* 14(4): 708–15.

Blankenberg, D., J. Taylor, et al. (2007). "A framework for collaborative analysis of ENCODE data: making large-scale analyses biologist-friendly." *Genome Res* 17(6): 960–4.

Bray, N. and L. Pachter (2004). "MAVID: constrained ancestral alignment of multiple sequences." *Genome Res* 14(4): 693–9.

Brent, M. R. (2007). "How does eukaryotic gene prediction work?" *Nat Biotechnol* 25(8): 883–5.

Brudno, M., C. B. Do, et al. (2003a). "LAGAN and Multi-LAGAN: efficient tools for large-scale multiple alignment of genomic DNA." *Genome Res* 13(4): 721–31.

Brudno, M., S. Malde, et al. (2003b). "Glocal alignment: finding rearrangements during alignment." *Bioinformatics* 19 Suppl 1: i54–62.

Burge, C. and S. Karlin (1997). "Prediction of complete gene structures in human genomic DNA." *J Mol Biol* 268(1): 78–94.

Caspi, R., H. Foerster, et al. (2006). "MetaCyc: a multiorganism database of metabolic pathways and enzymes." *Nucleic Acids Res* 34(Database issue): D511–16.

Choi, K., Y. Ma, et al. (2005). "PLATCOM: a Platform for Computational Comparative Genomics." *Bioinformatics* 21(10): 2514–16.

Christie, K. R., S. Weng, et al. (2004). "Saccharomyces Genome Database (SGD) provides tools to identify and analyze sequences from *Saccharomyces cerevisiae* and related sequences from other organisms." *Nucleic Acids Res* 32(Database issue): D311–14.

Cooper, G. M., E. A. Stone, et al. (2005). "Distribution and intensity of constraint in mammalian genomic sequence." *Genome Res* 15(7): 901–13.

Curwen, V., E. Eyras, et al. (2004). "The Ensembl automatic gene annotation system." *Genome Res* 14(5): 942–50.

Dewey, C. N. and L. Pachter (2006). "Evolution at the nucleotide level: the problem of multiple whole-genome alignment." *Hum Mol Genet* 15 (Spec No 1): R51–6.

Dowell, R. D., R. M. Jokerst, et al. (2001). "The distributed annotation system." *BMC Bioinformatics* 2: 7.

Drmanac, R., I. Labat, et al. (1989). "Sequencing of megabase plus DNA by hybridization: theory of the method." *Genomics* 4(2): 114–28.

DuBois, P. (2005). *MySQL*, Sams Developer's Library.

Durbin, R., S. Eddy, et al. (1998). *Biological Sequence Analysis*, Cambridge University Press.

Eilbeck, K., S. E. Lewis, et al. (2005). "The Sequence Ontology: a tool for the unification of genome annotations." *Genome Biol* 6(5): R44.

Eppig, J. T., J. A. Blake, et al. (2007). "The mouse genome database (MGD): new features facilitating a model system." *Nucleic Acids Res* 35(Database issue): D630–7.

Flicek. P., B. L. Aken, et al. (2008). "Ensembl 2008." *Nucleic Acids Res* 36(Database issue): D707–14.

Frazer, K. A., D. G. Ballinger, et al. (2007). "A second generation human haplotype map of over 3.1 million SNPs." *Nature* 449(7164): 851–61.

Frazer, K. A., L. Pachter, et al. (2004). "VISTA: computational tools for comparative genomics." *Nucleic Acids Res* 32(Web Server issue): W273–9.

Furey, T. S. (2006). "Comparison of human (and other) genome browsers." *Hum Genomics* 2(4): 266–70.

Furey, T. S., M. Diekhans, et al. (2004). "Analysis of human mRNAs with the reference genome sequence reveals potential errors, polymorphisms, and RNA editing." *Genome Res* 14(10B): 2034–40.

Gerstein, M. B., C. Bruce, et al. (2007). "What is a gene, post-ENCODE? History and updated definition." *Genome Res* 17(6): 669–81.

Giardine, B., C. Riemer, et al. (2005). "Galaxy: a platform for interactive large-scale genome analysis." *Genome Res* 15(10): 1451–5.

Gilbert, D. G. (2007). "DroSpeGe: rapid access database for new Drosophila species genomes." *Nucleic Acids Res* 35(Database issue): D480–5.

Green, P. (2007). "2x genomes: does depth matter?" *Genome Res* 17(11): 1547–9.

Green, R. E., J. Krause, et al. (2006). "Analysis of one million base pairs of Neanderthal DNA." *Nature* 444(7117): 330–6.

Green, R. E., B. P. Lewis, et al. (2003). "Widespread predicted nonsense-mediated mRNA decay of alternatively-spliced transcripts of human normal and disease genes." *Bioinformatics* 19(Suppl 1): i118–21.

Griffiths-Jones, S., S. Moxon, et al. (2005). "Rfam: annotating non-coding RNAs in complete genomes." *Nucleic Acids Res* 33(Database issue): D121–4.

Gross, S. S. and M. R. Brent (2006). "Using multiple alignments to improve gene prediction." *J Comput Biol* 13(2): 379–93.

Harrow, J., F. Denoeud, et al. (2006). "GENCODE: producing a reference annotation for ENCODE." *Genome Biol* 7(Suppl 1): S4 1–9.

Holzner, S. (1999). *Perl Core Language*, Coriolis.

Hoon, S., K. K. Ratnapu, et al. (2003). "Biopipe: a flexible framework for protocol-based bioinformatics analysis." *Genome Res* 13(8): 1904–15.

Hsu, F., T. H. Pringle, et al. (2005). "The UCSC Proteome Browser." *Nucleic Acids Res* 33(Database issue): D454–8.

Hubbard, T. J., B. L. Aken, et al. (2007). "Ensembl 2007." *Nucleic Acids Res* 35(Database issue): D610–17.

Hull, D., K. Wolstencroft, et al. (2006). "Taverna: a tool for building and running workflows of services." *Nucleic Acids Res* 34(Web Server issue): W729–32.

Hüttenhofer, A., P. Schattner, et al. (2005). "Non-coding RNAs: hope or hype?" *Trends Genet* 21(5): 289–97.

Iafrate, A. J., L. Feuk, et al. (2004). "Detection of large-scale variation in the human genome." *Nat Genet* 36(9): 949–51.

Jaiswal, P., J. Ni, et al. (2006). "Gramene: a bird's eye view of cereal genomes." *Nucleic Acids Res* 34(Database issue): D717–23.

Kapustin Yu., A. Souvorov, et al. (2004). "Splign – a hybrid approach to spliced alignments." *RECOMB 2004 – Currents in Comp Mol Bio*: 741.

Karolchik, D., A. S. Hinrichs, et al. (2004). "The UCSC Table Browser data retrieval tool." *Nucleic Acids Res* 32(Database issue): D493–6.

Karolchik, D., R. M. Kuhn, et al. (2008). "The UCSC Genome Browser Database: 2008 Update." *Nucleic Acids Res* 36(Database issue): D773–9.

Kartalov, E. P. and S. R. Quake (2004). "Microfluidic device reads up to four consecutive base pairs in DNA sequencing-by-synthesis." *Nucleic Acids Res* 32(9): 2873–9.

Kasprzyk, A., D. Keefe, et al. (2004). "EnsMart: a generic system for fast and flexible access to biological data." *Genome Res* 14(1): 160–9.

Kent, W. J. (2002). "BLAT – the BLAST-like alignment tool." *Genome Res* 12(4): 656–64.

Kent, W. J., R. Baertsch, et al. (2003). "Evolution's cauldron: duplication, deletion, and rearrangement in the mouse and human genomes." *Proc Natl Acad Sci U S A* 100(20): 11484–9.

Kent, W. J., F. Hsu, et al. (2005). "Exploring relationships and mining data with the UCSC Gene Sorter." *Genome Res* 15(5): 737–41.

Korbel, J. O., A. E. Urban, et al. (2007). "Paired-end mapping reveals extensive structural variation in the human genome." *Science* 318(5849): 420–6.

Kuhn, R. M., D. Karolchik, et al. (2007). "The UCSC Genome Browser database: update 2007." *Nucleic Acids Res* 35(Database issue): D668–73.

Leamon, J. H. and J. M. Rothberg (2007). "Cramming more sequencing reactions onto microreactor chips." *Chem Rev* 107(8): 3367–76.

Lee, T. J., Y. Pouliot, et al. (2006). "BioWarehouse: a bioinformatics database warehouse toolkit." *BMC Bioinformatics* 7: 170.

Lestrade, L. and M. J. Weber (2006). "snoRNA-LBME-db, a comprehensive database of human H/ACA and C/D box snoRNAs." *Nucleic Acids Res* 34(Database issue): D158–62.

Lev-Maor, G., R. Sorek, et al. (2003). "The birth of an alternatively spliced exon: 3′ splice-site selection in Alu exons." *Science* 300(5623): 1288–91.

Levanon, E. Y., E. Eisenberg, et al. (2004). "Systematic identification of abundant A-to-I editing sites in the human transcriptome." *Nat Biotechnol* 22(8): 1001–5.

Lewis, S. E., S. M. Searle, et al. (2002). "Apollo: a sequence annotation editor." *Genome Biol* 3(12): RESEARCH0082.

Ma, B., J. Tromp, et al. (2002). "PatternHunter: faster and more sensitive homology search." *Bioinformatics* 18(3): 440–5.

Margulies, E. H., G. M. Cooper, et al. (2007). "Analyses of deep mammalian sequence alignments and constraint predictions for 1% of the human genome." *Genome Res* 17(6): 760–74.

Markowitz, V. M., F. Korzeniewski, et al. (2006a). "The integrated microbial genomes (IMG) system." *Nucleic Acids Res* 34(Database issue): D344–8.

Markowitz, V. M., N. Ivanova, et al. (2006b). "An experimental metagenome data management and analysis system." *Bioinformatics* 22(14): e359–67.

Markowitz, V. M., N. N. Ivanova, et al. (2008). "IMG/M: a data management and analysis system for metagenomes." *Nucleic Acids Res* 36(Database issue): D534–8.

Mount, D. W. (2004). *Bioinformatics: Sequence and Genome Analysis*, 2nd Edition, Cold Spring Harbor Laboratory Press.

Mungall, C. J. and D. B. Emmert (2007). "A Chado case study: an ontology-based modular schema for representing genome-associated biological information." *Bioinformatics* 23(13): i337–46.

Noonan, J. P., M. Hofreiter, et al. (2005). "Genomic sequencing of Pleistocene cave bears." *Science* 309(5734): 597–9.

Oinn, T., M. Addis, et al. (2004). "Taverna: a tool for the composition and enactment of bioinformatics workflows." *Bioinformatics* 20(17): 3045–54.

Olson, M. (2007). "Enrichment of super-sized resequencing targets from the human genome." *Nat Methods* 4(11): 891–2.

Pedersen, J. S., G. Bejerano, et al. (2006). "Identification and classification of conserved RNA secondary structures in the human genome." *PLoS Comput Biol* 2(4): e33.

Pond, S. L., S. D. Frost, et al. (2005). "HyPhy: hypothesis testing using phylogenies." *Bioinformatics* 21(5): 676–9.

Pontius, J. U., J. C. Mullikin, et al. (2007). "Initial sequence and comparative analysis of the cat genome." *Genome Res* 17(11): 1675–89.

Potter, S. C., L. Clarke, et al. (2004). "The Ensembl analysis pipeline." *Genome Res* 14(5): 934–41.

Prakash A. and M. Tompa (2007). "Measuring the accuracy of genome-size multiple alignments." *Genome Biol* 8(6): R124.

Primrose, S. B. and R. M. Twyman (2006). *Principles of Gene Manipulation and Genomics*, 7th Edition, Blackwell Publishing.

Pruitt, K. D., T. Tatusova, et al. (2007). "NCBI reference sequences (RefSeq): a curated non-redundant sequence database of genomes, transcripts and proteins." *Nucleic Acids Res* 35(Database issue): D61–5.

Rampp, M., T. Soddemann, et al. (2006). "The MIGenAS integrated bioinformatics toolkit for web-based sequence analysis." *Nucleic Acids Res* 34(Web Server issue): W15–19.

Reek, K. A. (1998). *Pointers on C*, Addison-Wesley.

Reich, M., T. Liefeld, et al. (2006). "GenePattern 2.0." *Nat Genet* 38(5): 500–1.

Rice, P., I. Longden, et al. (2000). "EMBOSS: the European Molecular Biology Open Software Suite." *Trends Genet* 16(6): 276–7.

Rogers, Y. H. and J. C. Venter (2005). "Genomics: massively parallel sequencing." *Nature* 437(7057): 326–7.

Schattner, P., S. Barberan-Soler, et al. (2006). "A computational screen for mammalian pseudouridylation guide H/ACA RNAs." *RNA* 12(1): 15–25.

Schneider, K. L., K. S. Pollard, et al. (2006). "The UCSC Archaeal Genome Browser." *Nucleic Acids Res* 34(Database issue): D407–10.

Schwartz, S., W. J. Kent, et al. (2003). "Human-mouse alignments with BLASTZ." *Genome Res* 13(1): 103–7.

Schwarz, E. M., I. Antoshechkin, et al. (2006). "WormBase: better software, richer content." *Nucleic Acids Res* 34(Database issue): D475–8.

Sebat, J., B. Lakshmi, et al. (2004). "Large-scale copy number polymorphism in the human genome." *Science* 305(5683): 525–8.

Shah, S. P., Y. Huang, et al. (2005). "Atlas – a data warehouse for integrative bioinformatics." *BMC Bioinformatics* 6: 34.

Siepel, A., G. Bejerano, et al. (2005). "Evolutionarily conserved elements in vertebrate, insect, worm, and yeast genomes." *Genome Res* 15(8): 1034–50.

Slater, G. S. and E. Birney (2005). "Automated generation of heuristics for biological sequence comparison." *BMC Bioinformatics* 6: 31.

Southern, E. M., U. Maskos, et al. (1992). "Analyzing and comparing nucleic acid sequences by hybridization to arrays of oligonucleotides: evaluation using experimental models." *Genomics* 13(4): 1008–17.

Stabenau, A., G. McVicker, et al. (2004). "The Ensembl core software libraries." *Genome Res* 14(5): 929–33.

Stajich, J. E., D. Block, et al. (2002). "The Bioperl toolkit: Perl modules for the life sciences." *Genome Res* 12(10): 1611–18.

Stajich, J. E. and H. Lapp (2006). "Open source tools and toolkits for bioinformatics: significance, and where are we?" *Brief Bioinform* 7(3): 287–96.

Stein, L. D. (2003). "Integrating biological databases." *Nat Rev Genet* 4(5): 337–45.

Stein, L. D., C. Mungall, et al. (2002). "The generic genome browser: a building block for a model organism system database." *Genome Res* 12(10): 1599–610.

Stevens, R. D., H. J. Tipney, et al. (2004). "Exploring Williams-Beuren syndrome using myGrid." *Bioinformatics* 20(Suppl 1): i303–10.

Subramaniam, S. (1998). "The Biology Workbench – a seamless database and analysis environment for the biologist." *Proteins* 32(1): 1–2.

Sundquist, A., M. Ronaghi, et al. (2007). "Whole-genome sequencing and assembly with high-throughput, short-read technologies." *PLoS ONE* 2(5): e484.

Thomas, D. J., K. R. Rosenbloom, et al. (2007). "The ENCODE Project at UC Santa Cruz." *Nucleic Acids Res* 35(Database issue): D663–7.

Thornton, J. W., E. Need, et al. (2003). "Resurrecting the ancestral steroid receptor: ancient origin of estrogen signaling." *Science* 301(5640): 1714–17.

Tisdall, J. D. (2001). *Beginning Perl for Bioinformatics*, O'Reilly.

Tisdall, J. D. (2003). *Mastering Perl for Bioinformatics*, O'Reilly.

Wheeler, D. L., T. Barrett, et al. (2005). "Database resources of the National Center for Biotechnology Information." *Nucleic Acids Res* 33(Database issue): D39–45.

Wheeler, D. L., T. Barrett, et al. (2006). "Database resources of the National Center for Biotechnology Information." *Nucleic Acids Res* 34(Database issue): D173–80.

Wheeler, D. L., T. Barrett, et al. (2007). "Database resources of the National Center for Biotechnology Information." *Nucleic Acids Res* 35(Database issue): D5–12.

Wheeler, D. L., T. Barrett, et al. (2008). "Database resources of the National Center for Biotechnology Information." *Nucleic Acids Res* 36(Database issue): D13–21.

Will, C. L. and R. Luhrmann (2005). "Splicing of a rare class of introns by the U12-dependent spliceosome." *Biol Chem* 386(8): 713–24.

Zdobnov, E. M., R. Lopez, et al. (2002). "The EBI SRS server – recent developments." *Bioinformatics* 18(2): 368–73.

Zimmerman, O., M. Tomlinson, et al. (2005). *Perspectives on Web Services*, Springer.

Index

Printed in the United States
by Baker & Taylor Publisher Services